Hao Hua.
hhua@broadcom.com
23332

Power Distribution Networks with On-Chip Decoupling Capacitors

Mikhail Popovich · Andrey V. Mezhiba ·
Eby G. Friedman

Power Distribution Networks with On-Chip Decoupling Capacitors

 Springer

Mikhail Popovich
University of Rochester
Rochester, NY
USA

Andrey V. Mezhiba
Intel Corporation
Hillsboro, OR
USA

Eby G. Friedman
University of Rochester
Rochester, NY
USA

Library of Congress Control Number: 2007931772

ISBN 978-0-387-71600-8   e-ISBN 978-0-387-71601-5

9 8 7 6 5 4 3 2 1

springer.com

To Oksana and Elizabeth

To Elizabeth

To Laurie, Joseph, and Samuel

Preface

The purpose of this book is to provide insight and intuition into the behavior and design of power distribution systems with decoupling capacitors for application to high speed integrated circuits. The primary objectives are threefold. First, to describe the impedance characteristics of the overall power distribution system, from the voltage regulator through the printed circuit board and package onto the integrated circuit to the power terminals of the on-chip circuitry. The second objective of this book is to discuss the inductive characteristics of on-chip power distribution grids and the related circuit behavior of these structures. Finally, the third primary objective is to present design methodologies for efficiently placing on-chip decoupling capacitors in nanoscale integrated circuits.

Technology scaling has been the primary driver behind the amazing performance improvement of integrated circuits over the past several decades. The speed and integration density of integrated circuits have dramatically improved. These performance gains, however, have made distributing power to the on-chip circuitry a difficult task. Highly dense circuitry operating at high clock speeds have increased the distributed current to many tens of amperes, while the noise margin of the power supply has shrunk consistent with decreasing power supply levels. These trends have elevated the problems of power distribution and allocation of the on-chip decoupling capacitors to the forefront of several challenges in developing high performance integrated circuits.

This book is based on the body of research carried out by Mikhail Popovich from 2001 to 2007 and Andrey V. Mezhiba from 1998 to 2003 at the University of Rochester during their doctoral studies under the supervision of Professor Eby G. Friedman. It is apparent to the authors that although various aspects of the power distribution problem have been addressed in numerous research publications, no text exists that provides a unified focus on power distribution systems and related design problems. Furthermore, the placement of on-chip decoupling capacitors has traditionally been treated as an algorithmic oriented problem. A more electrical perspective, both circuit models and design techniques, has been used in this book for presenting how to efficiently allocate on-chip decoupling capacitors. The fundamental objective of this book is to provide a broad and cohesive treatment of these subjects.

Another consequence of higher speed and greater integration density has been the emergence of inductance as a significant factor in the behavior of on-chip global interconnect structures. Once clock frequencies exceeded several hundred megahertz, incorporating on-chip inductance into the circuit analysis process became necessary to accurately describe signal delays and waveform characteristics. Although on-chip decoupling capacitors attenuate high frequency signals in power distribution networks, the inductance of the on-chip power interconnect is expected to become a significant factor in multi-gigahertz digital circuits. An important objective of this book, therefore, is to clarify the effects of inductance on the impedance characteristics of on-chip power distribution grids and to provide an understanding of related circuit behavior.

The organization of the book is consistent with these primary goals. The first eight chapters provide a general description of distributing power in integrated circuits with decoupling capacitors. The challenges of power distribution are introduced and the principles of designing power distribution systems are described. A general background to decoupling capacitors is presented followed by a discussion of the use of a hierarchy of capacitors to improve the impedance characteristics of the power network. An overview of related phenomena, such as inductance and electromigration, is also presented in a tutorial style. The following seven chapters are dedicated to the impedance characteristics of on-chip power distribution networks. The effect of the interconnect inductance on the impedance characteristics of on-chip power distribution networks is investigated. The implications of these impedance characteristics on circuit behavior are also discussed. On-chip power distribution grids are described, exploiting multiple power supply voltages and multiple grounds. Techniques and algorithms for the computer-aided design and analysis of power distribution networks are also described; however, the emphasis of the book is on developing circuit intuition and understanding the electrical principles that govern the design and operation of power distribution systems. The remaining five chapters focus on the design of a system of on-chip decoupling capacitors. Methodologies for designing power distribution grids with on-chip decoupling capacitors are also presented. These techniques provide a solution for determining the location and magnitude of the on-chip decoupling capacitance to mitigate on-chip voltage fluctuations.

Acknowledgments

The authors would like to thank Alex Greene and Katelyn Stanne from Springer for their support and assistance. We are particularly thankful to Bill Joyner and Dale Edwards from the Semiconductor Research Corporation, and Marie Burnham, Olin Hartin, and Radu Secareanu from Freescale Semiconductor Corporation for their continued support of the research project that culminated in this book. The authors would also like to thank Emre Salman for his corrections and suggestions on improving the quality of the book. Finally, we are grateful to Michael Sotman and Avinoam Kolodny from Technion — Israel Institute of Technology for their collaboration and support.

The original research work presented in this book was made possible in part by the Semiconductor Research Corporation under Contract Nos. 99–TJ–687 and 2004–TJ–1207, the DARPA/ITO under AFRL Contract F29601–00–K–0182, the National Science Foundation under Contract Nos. CCR–0304574 and CCF–0541206, grants from the New York State Office of Science, Technology & Academic Research to the Center for Advanced Technology in Electronic Imaging Systems, and by grants from Xerox Corporation, IBM Corporation, Lucent Technologies Corporation, Intel Corporation, Eastman Kodak Company, Intrinsix Corporation, Manhattan Routing, and Freescale Semiconductor Corporation.

Rochester, New York   *Mikhail Popovich and Eby G. Friedman*
Hillsboro, Oregon   *Andrey V. Mezhiba*

June 2007

Contents

List of Figures

List of Tables

1

Introduction

In July 1958, Jack Kilby of Texas Instruments suggested building all of the components of a circuit completely in silicon [1]. By September 12, 1958, Kilby had built a working model of the first "solid circuit," the size of a pencil point. A couple of months later in January 1959, Robert Noyce of Fairchild Semiconductor developed a better way to connect the different components of a circuit [2], [3]. Later, in the spring of 1959, Fairchild Semiconductor demonstrated the first planar circuit — a "unitary circuit." The first monolithic integrated circuit (IC) was born, where multiple transistors coexisted with passive components on the same physical substrate [4]. Microphotographs of the first IC (Texas Instruments, 1958), the first monolithic IC (Fairchild Semiconductor, 1959), and the recent high performance dual core Montecito microprocessor (Intel Corporation, 2005) are depicted in Fig. 1.1.

In 1960, Jean Hoerni invented the planar process [5]. Later, in 1960, Dawon Kahng and Martin Atalla demonstrated the first silicon based MOSFET [6], followed in 1967 by the first silicon gate MOSFET [7]. These seminal inventions resulted in the explosive growth of today's multi-billion dollar microelectronics industry. The fundamental cause of this growth in the microelectronics industry has been made possible by technology scaling, particularly in CMOS technology.

The goal of this chapter is to provide a brief perspective on the development of integrated circuits (ICs), introduce the problem of power distribution in the context of this development, motivate the use of on-chip decoupling capacitors, and provide guidance and perspective to the rest of this book. The evolution of integrated circuit technology from the first ICs to highly scaled CMOS technology is described in Section 1.1. As manufacturing technologies supported higher integration

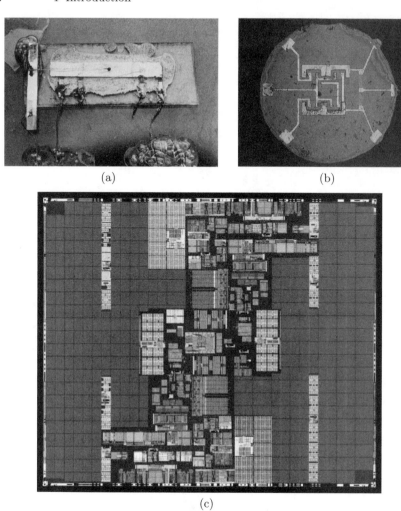

(a)   (b)

(c)

Fig. 1.1. Microphotographs of the first integrated circuit (IC) and first monolithic IC along with a high performance, high complexity IC (the die size is not to scale). (a) The first IC (Texas Instruments, 1958), (b) the first monolithic IC (Fairchild Semiconductor, 1959), and (c) the high performance dual core Montecito microprocessor (Intel Corporation, 2005).

densities and switching speeds, the primary constraints and challenges in the design of integrated circuits have also shifted, as discussed in Section 1.2. The basic nature of the problem of distributing power and ground in integrated circuits is described in Section 1.3. The adverse effects of variations in the power supply voltage on the operation of a digital integrated circuit are discussed in Section 1.4. Finally, the overall structure of the book and the content of each chapter are outlined in Section 1.5.

1.1 Evolution of integrated circuit technology

The pace of IC technology over the past three decades has been well characterized by Moore's law. As noted in 1965 by Gordon Moore, the integration density of the first commercial integrated circuits has doubled approximately every year [8]. A prediction was made that the economically effective integration density, *i.e.*, the number of transistors on an integrated circuit leading to the minimum cost per integrated component, will continue to double every year for another decade. This prediction has held true through the early 1970's. In 1975, the prediction was revised to suggest a new, slower rate of growth–doubling of the IC transistor count every two years [9]. This trend of exponential growth of IC complexity is commonly referred to as "Moore's Law." As a result, since the start of commercial production of integrated circuits in the early 1960's, circuit complexity has risen from a few transistors to hundreds of millions of transistors functioning together on a single monolithic substrate. Furthermore, Moore's law is expected to continue at a comparable pace for at least another decade [10].

The evolution of the integration density of microprocessor and memory ICs is shown in Fig. 1.2 along with the original prediction described in [8]. As seen from the data illustrated in Fig. 1.2, DRAM IC complexity has grown at an even higher rate, quadrupling roughly every three years. The progress of microprocessor clock frequencies is shown in Fig. 1.3. Associated with increasing IC complexity and clock speed is an exponential increase in overall microprocessor performance (doubling every 18 to 24 month). This performance trend is also referred to as Moore's law.

The principal driving force behind this spectacular improvement in circuit complexity and performance has been the steady decrease in the feature size of semiconductor devices. Advances in optical lithography

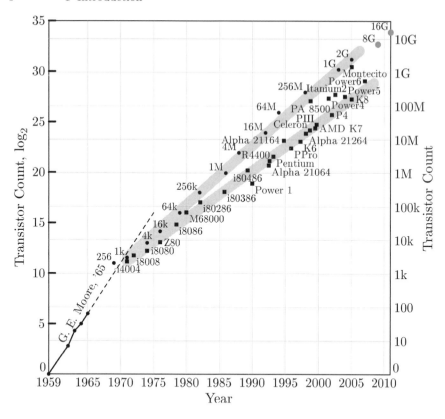

Fig. 1.2. Evolution of transistor count of CPU/microprocessor and memory ICs. In the lower left corner the original Moore's data [8] is displayed followed by the extrapolated prediction (dashed line).

have allowed manufacturing of on-chip structures with increasingly higher resolution. The area, power, and speed characteristics of transistors with a planar structure, such as MOS devices, improve with the decrease (*i.e.*, scaling) of the lateral dimensions of the devices. These technologies are therefore referred to as *scalable*. The maturing of scalable planar circuit technologies, first PMOS and later NMOS technology, has allowed the potential of technology scaling to be fully exploited as lithography has gradually improved. The development of planar MOS technology culminated in CMOS circuits. The low power characteristics of CMOS technology deferred the thermal limitations on integration complexity and permitted technology scaling to continue unabated through the 1980's and 1990's, making CMOS the digital circuit technology of choice.

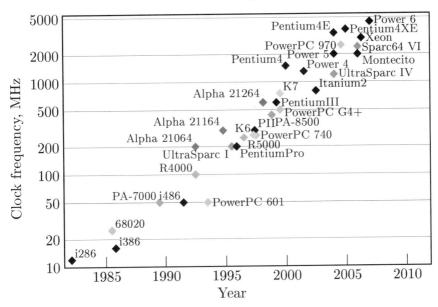

Fig. 1.3. Evolution of microprocessor clock frequency. Several lines of microprocessors are shown in shades of gray.

From a historical perspective, the development of scalable ICs was simultaneously circuitous and serendipitous, as described by Murphy, Haggan, and Troutman [11]. Although the ideas and motivation behind scalable ICs seem straightforward from today's vantage point, the emergence of scalable commercial ICs was neither inevitable nor a result of a well guided and planned pursuit. Rather, the original motivation for the development of integrated circuits was circuit miniaturization for military and space applications. Although the active devices of the time, discrete transistors, offered smaller size (and also lower power dissipation with higher reliability) as compared to vacuum tubes, much of this advantage was lost at the circuit level, as the size and weight of electronic circuits were dominated by passive electronic components. Thus, the original objective was to reduce the size of the passive elements through integration of these elements onto the same die as the transistors. The cost effectiveness and commercial success of high complexity ICs were highly controversial for several years after the IC was invented. Successful integration of a large number of transistors on the same die seemed infeasible, considering the yield of discrete devices at the time [11].

Many obstacles precluded early ICs from scaling. The bulk collector bipolar transistors used in early ICs suffered from performance degradation due to high collector resistance and, more importantly, the collectors of all of the on-chip transistors were connected through the bulk substrate. The speed of a bipolar transistor does not, in general, scale with the lateral dimensions (*i.e.*, vertical NPN and PNP doping structures most often determine the performance). In addition, early device isolation approaches were not amenable to scaling and consumed significant die area. On-chip resistors and diodes also made inefficient use of die area. Scalable schemes for device isolation and interconnection were therefore essential for truly scalable ICs. It was not until these problems were solved and the structure of the bipolar transistor was improved that device miniaturization led to dramatic improvements in IC complexity.

The concept of scalable ICs received further development with the maturation of the MOS technology. Although the MOS transistor is a contemporary of the first ICs, the rapid progress in bipolar devices delayed the development of MOS ICs at the beginning of the IC era. The MOS transistor lagged in performance as compared with existing bipolar devices and suffered from reproducibility and stability problems. The low current drive capability of MOS transistors was also an important disadvantage at low integration densities. Early use of the MOS transistor was limited to those applications that exploited the excellent switch-like characteristics of the MOS devices. Nevertheless, the circuit advantages and scaling potential of MOS technology were soon realized, permitting MOS circuits to gain increasingly wider acceptance. Gate insulation and the enhancement mode of operation made MOS technology ideal for direct-coupled logic [12]. Furthermore, MOS did not suffer from punch-through effects and could be fabricated with higher yield. The compactness of MOS circuits and the higher yield eventually resulted in a fourfold density advantage in devices per IC, as compared to bipolar ICs. Ironically, it was the refinement of bipolar technology that paved the path to these larger scales of integration, permitting the efficient exploitation of MOS technology. With advances in lithographic resolution, the MOS disadvantage in switching speed as compared to bipolar devices gradually diminished. The complexity of bipolar ICs had become constrained by circuit power dissipation. As a result, MOS emerged as the dominant digital integrated circuit technology.

1.2 Evolution of design objectives

Advances in fabrication technology and the emergence of new applications have induced several shifts in the principal objectives in the design of integrated circuits over the past thirty years. The evolution of the IC design paradigm is illustrated in Fig. 1.4.

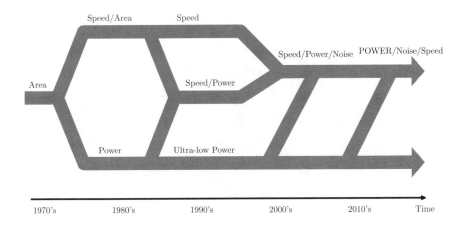

Fig. 1.4. Evolution of design criteria in CMOS integrated circuits.

In the 1960's and 1970's, yield concerns served as the primary limitation to IC integration density and, as a consequence, circuit compactness and die area were the primary criteria in the IC design process. Due to limited integration densities, a typical system at the time would contain dozens to thousands of small ICs. As a result, chip-to-chip communications traversing board-level interconnect limited the overall system performance. As compared to intra-chip interconnect, board level interconnect traces have high latency and dissipate large amounts of power, limiting the speed and power performance of the system. Placing as much functionality as possible into a yield-limited silicon die supported the realization of electronic systems with fewer ICs. A lower number of board level contacts and interconnections in systems comprised of fewer ICs improved system reliability and lowered system cost, increased system speed (due to lower communication latencies), reduced system power consumption, and decreased the size and weight of the overall system. Implementing higher functionality per silicon area with the ensuing reduction in the number of individual ICs typically achieved an improved cost/performance tradeoff at the system level. A landmark example of that era is the first Intel microprocessor, the 4004, commercialized at the end of 1971 [13]. Despite the limitation to 4-bit word processing and initially operating at a mere 108 kHz, the 4004 microprocessor was a complete processor core implemented on a monolithic die containing approximately 2300 transistors. A microphotograph of the 4004 microprocessor is shown in Fig. 1.5.

Fig. 1.5. Microphotograph of the 4004 — the first microprocessor implemented on a monolithic die.

The impact of off-chip communications on overall system speed decreased as the integration density increased with advances in fabrication technology, lowering the number of ICs comprising a system. System speed became increasingly dependent on the speed of the component ICs (and less dependent on the speed of the board-level communications). By the 1980's, circuit speed had become the design criterion of highest priority. Concurrently, a new class of applications emerged, principally restricted by the amount of power consumed. These applications included digital wrist watches, handheld calculators, pacemakers, and satellite electronics. These applications established a new design concept — design for ultra-low power, *i.e.*, power dissipation being the primary design criterion, as illustrated by the lowest path shown in Fig. 1.4.

As device scaling progressed and a greater number of components were integrated onto a single die, on-chip power dissipation began to produce significant economic and technical difficulties. While the market for high performance circuits could support the additional cost, the design process in the 1990's had focused on optimizing both speed and power, borrowing a number of design approaches previously developed for ultra-low power products. The proliferation of portable electronic devices further increased the demand for power efficient and ultra-low power ICs, as shown in Fig. 1.4.

A continuing increase in power dissipation exacerbated system price and reliability concerns, making circuit power a primary design metric across an entire range of applications. The evolution of power consumed by several lines of commercial microprocessors is shown in Fig. 1.6. Furthermore, aggressive device scaling and increasing circuit complexity have caused severe noise (or signal integrity) problems in high complexity, high speed integrated circuits. Although digital circuits have traditionally been considered immune to noise due to the inherently high noise margins, circuit coupling through the parasitic impedances of the on-chip interconnect has significantly increased with technology scaling. Ignoring the effects of on-chip noise is no longer possible in the design of high speed digital ICs. These changes are reflected in the convergence of "speed" and "speed/power" design criteria to "speed/power/noise," as depicted in Fig. 1.4.

As device scaling continued in the twenty first century, more than a billion transistors were successfully integrated onto a single die [14], keeping up with Moore's law. As a result, the overall power dissipation increased accordingly, exceeding the maximum capability of conventional cooling technologies. Any further increase in on-chip power dissipation would require expensive and challenging solutions, such as liquid cooling, significantly increasing the overall cost of a system. Moreover, an explosive growth of portable and handheld devices, such as cell phones and PDAs, resulted in a shift of design focus towards low power. As a technical solution for low power in high performance ICs, multicore systems emerged [15], [16], [17], [18], trading off silicon area with on-chip power dissipation. Since the emphasis on ultra-low power design continues into the first decade of the twenty first century, major design effort is focused on reducing power dissipation.

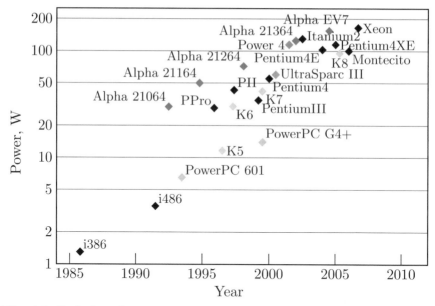

Fig. 1.6. Evolution of microprocessor power consumption. Several lines of micro-processors are shown in shades of gray.

1.3 The problem of power distribution

The problem of power delivery is illustrated in Fig. 1.7, where a basic power delivery system is shown. The system consists of a power supply, a power load, and interconnect lines connecting the supply to the load. The power supply is assumed to behave as an ideal voltage source providing nominal power and ground voltage levels, V_{dd} and V_{gnd}. The power load is modeled as a variable current source $I(t)$. The interconnect lines connecting the supply and the load are not ideal; the power and ground lines have a finite parasitic resistance R_p and R_g, respectively, and inductance L_p and L_g, respectively. Resistive voltage drops $\Delta V_R = IR$ and inductive voltage drops $\Delta V_L = L\frac{dI}{dt}$ develop across the parasitic interconnect impedances, as the load draws current $I(t)$ from the power distribution system. The voltage levels across the load terminals, therefore, change from the nominal level provided by the supply, dropping to $V_{\mathrm{dd}} - IR_p - L_p\frac{dI}{dt}$ at the power terminal and rising to $V_{\mathrm{gnd}} + IR_g + L_g\frac{dI}{dt}$ at the ground terminal, as shown in Fig. 1.7.

This change in the supply voltages is referred to as power supply noise. Power supply noise adversely affects circuit operation through several mechanisms, as described in Section 1.4. Proper design of the

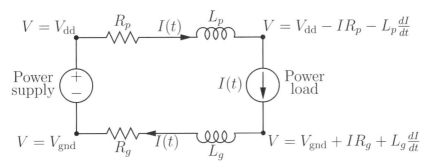

Fig. 1.7. Power delivery system consisting of the power supply, power load, and non-ideal interconnect lines.

load circuit ensures correct operation under the assumption that the supply levels are maintained within a certain range near the nominal voltage levels. This range is called the power noise margin. The primary objective in the design of the power distribution system is to supply sufficient current to each transistor on an integrated circuit while ensuring that the power noise does not exceed target noise margins.

The on-going miniaturization of integrated circuit feature size has placed significant requirements on the on-chip power and ground distribution networks. Circuit integration densities rise with each very deep submicrometer (VDSM) technology generation due to smaller devices and larger dies; the current density and the total current increase accordingly. At the same time, the higher speed switching of smaller transistors produces faster current transients in the power distribution network. Both the average current and the transient current are rising exponentially with technology scaling. The evolution of the average current of high performance microprocessors is illustrated in Fig. 1.8.

The current in contemporary microprocessors has exceeded 130 amperes and will further increase with technology scaling. Forecasted demands in the power current of high performance microprocessors are illustrated in Fig. 1.9. The rate of increase in the transient current is expected to more than double the rate of increase in the average current as indicated by the slope of the trend lines depicted in Fig. 1.9.

The faster rate of increase in the transient current as compared to the average current is due to increasing on-chip clock frequencies. The transient current in modern high performance microprocessors is approximately one teraampere per second (10^{12} A/s) and is expected to rise, exceeding a hundred teraamperes per second by 2016. A transient

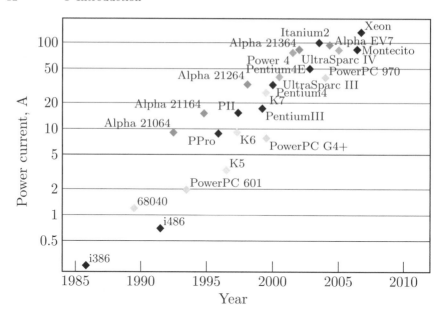

Fig. 1.8. Evolution of the average current in high performance microprocessors. Several lines of microprocessors are shown in shades of gray.

current of this high magnitude is due to switching hundreds of amperes within a fraction of a nanosecond. Fortunately, the rate of increase in the transient current has slowed with the recent introduction of lower speed multi-core microprocessors. In a multi-core microprocessor, similar performance is achieved at a lower frequency at the expense of increased circuit area.

Insuring adequate signal integrity of the power supply under these high current requirements has become a primary design issue in high performance, high complexity integrated circuits. The high average currents produce large ohmic IR voltage drops [19] and the fast current transients cause large inductive $L\frac{dI}{dt}$ voltage drops [20] (ΔI noise) in power distribution networks [21]. The power distribution networks must be designed to minimize these voltage drops, maintaining the local supply voltage within specified noise margins. If the power supply voltage sags too low, the performance and functionality of the circuit will be severely compromised. Alternatively, excessive overshoot of the supply voltage can affect circuit reliability. Further exacerbating these problems is the reduced noise margins of the power supply as the supply

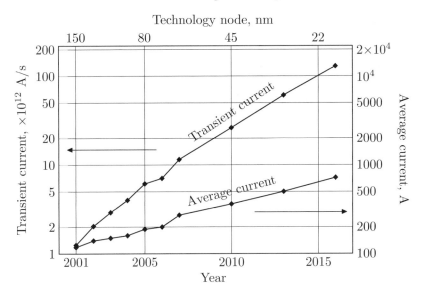

Fig. 1.9. Increase in power current requirements of high performance microprocessors with technology scaling, according to the ITRS roadmap [10]. The average current is the ratio of the circuit power to the supply voltage. The transient current is the product of the average current and the on-chip clock rate, $2\pi f_{\mathrm{clk}}$.

voltage is reduced with each new generation of VDSM process technology, as shown in Fig. 1.10.

To maintain the local supply voltage within specified design margins, the output impedance of a power distribution network should be low as seen at the power terminals of the circuit elements. IC technologies are expected to scale for at least another decade [22]. As a result, the average and transient currents drawn from the power delivery network will continue to rise. Simultaneously, the power supply voltage will be scaled to manage the on-chip power consumption. The target output impedance of a power distribution network in high speed, high complexity ICs such as microprocessors will therefore continue to drop, reaching an inconceivable level of $250\,\mu\Omega$ by the year 2017 [22], as depicted in Fig. 1.11.

With transistor switching times as short as a few picoseconds, on-chip signals typically contain significant harmonics at frequencies as high as $\sim 100\,\mathrm{GHz}$. For on-chip wires, the inductive reactance ωL dominates the overall wire impedance beyond $\sim 10\,\mathrm{GHz}$. The on-chip inductance affects the integrity of the power supply through two phenomena. First, the magnitude of the ΔI noise is directly proportional

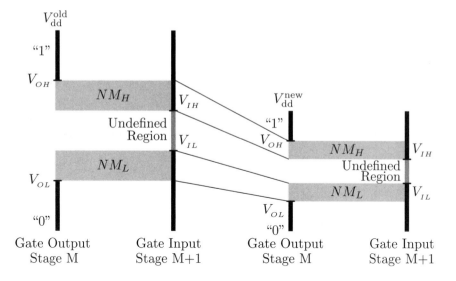

Fig. 1.10. Reduction in noise margins of CMOS circuits with technology scaling. NM_H and NM_L are the noise margins in the high and low logic state, respectively.

to the power network inductance as seen at the current sink. Second, the network resistance, inductance, and decoupling capacitance form an RLC tank circuit with multiple resonances. The peak impedance of this RLC circuit must be lowered to guarantee that target power supply noise margins are satisfied. Thus, information characterizing the inductance is needed to perform accurate analyses of power distribution networks.

Power distribution networks in high performance digital ICs are commonly structured as a multilayer grid. In such a grid, straight power/ground (P/G) lines in each metalization layer span the entire die (or a large functional unit) and are orthogonal to the lines in the adjacent layers. The power and ground lines typically alternate in each layer. Vias connect a power (ground) line to another power (ground) line at the overlap sites. The power grid concept is illustrated in Fig. 1.12, where three layers of interconnect are depicted with the power lines shown in dark grey and the ground lines shown in light grey. The power/ground lines are surrounded by signal lines.

A significant fraction of the on-chip resources is committed to insure the integrity of the power supply levels. The global on-chip power distribution networks are typically determined at the early stages of the design process, when little is known about the power demands at

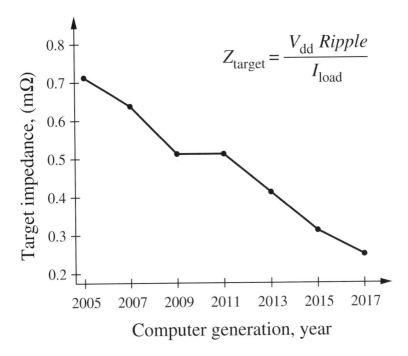

$$Z_{target} = \frac{V_{dd}\,Ripple}{I_{load}}$$

Fig. 1.11. Projections of the target impedance of a power distribution system. The target impedance will continue to drop for future technology generations at an aggressive rate of 1.25 X per computer generation [22].

specific locations on an IC. Allocating additional metal resources for the global power distribution network at the later stages of the design process in order to improve the local electrical characteristics of the power distribution network is likely to necessitate a complete redesign of the surrounding global signals and, therefore, can be prohibitively expensive. For these reasons, power distribution networks tend to be conservatively designed [23], sometimes using more than a third of the on-chip metal resources [24], [25]. Overengineering the power supply system is, therefore, costly in modern interconnect limited, high complexity digital integrated circuits.

Performance goals in power distribution networks, such as low impedance (low inductance and resistance) to satisfy noise specifications under high current loads, small area, and low current densities (for improved reliability) are typically in conflict. Widening the lines to increase the conductance and improve the electromigration reliability

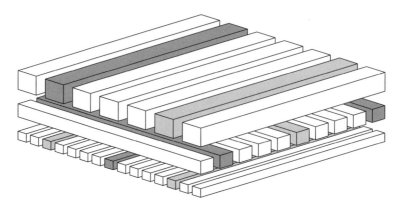

Fig. 1.12. A grid structured power distribution network. The ground lines are light grey, the power lines are dark grey, and the signal lines are white.

also increases the grid area. Replacing wide metal lines with narrow interdigitated P/G lines increases the line resistance if the grid area is maintained constant or increases the area if the net cross section of the lines is maintained constant. It is therefore important to make a balanced choice under these conditions. A quantitative model of the inductance/area/resistance tradeoff in high performance power distribution networks is therefore needed to achieve an efficient power distribution network. Another important goal is to provide quantitative tradeoff guidelines and intuition in the design of high performance power distribution networks.

Decoupling capacitors are often used to reduce the impedance of a power distribution system and provide the required charge to the switching circuits, lowering the power supply noise [26]. At high frequencies, however, only on-chip decoupling capacitors can be effective due to the high parasitic impedance of the power network connecting a decoupling capacitor to the current load [27]. On-chip decoupling capacitors, however, reduce the self-resonant frequency of a power distribution system, resulting in high amplitude power supply voltage fluctuations at the resonant frequencies. A hierarchical system of on-chip decoupling capacitors should therefore be carefully designed to provide a low impedance, resonant-free power distribution network over the entire range of operating frequencies, while delivering sufficient charge to the switching circuits to maintain the local power supply voltages within target noise margins [28].

1.4 Deleterious effects of power distribution noise

Power distribution noise adversely affects the operation of an integrated circuit through several mechanisms. These mechanisms are discussed in this section. Power supply noise produces variations in the delay of the clock and data signals, as described in Section 1.4.1. Power supply noise increases the uncertainty of the timing reference signals generated on-chip, lowering the clock frequency of the circuit, as discussed in Section 1.4.2. The reduction of noise margins is discussed in Section 1.4.3. Finally, power supply variations diminish the maximum supply voltage of the circuit, degrading the circuit frequency of operation, as described in Section 1.4.4.

1.4.1 Signal delay uncertainty

The propagation delay of on-chip signals depends on the power supply level during the signal transition. The source of the PMOS transistors in pull-up networks within logic gates is connected to the highest supply voltage directly or through other transistors. Similarly, the source of the NMOS transistors within a pull-down networks is connected to the ground distribution network. The drain current of a MOS transistor increases with the voltage difference between the transistor gate and source. When the rail-to-rail power voltage is reduced due to power supply variations, the gate-to-source voltage of the NMOS and PMOS transistors is decreased, thereby lowering the output current of the transistors. The signal delay increases accordingly as compared to the delay under a nominal power supply voltage. Conversely, a higher power voltage and a lower ground voltage will shorten the propagation delay. The net effect of the power noise on the propagation of the clock and data signals is, therefore, an increase in both delay uncertainty and the delay of the data paths [29], [30]. Consequently, power supply noise limits the maximum operating frequency of an integrated circuit [31], [32], [33].

1.4.2 On-chip clock jitter

A phase-locked loop (PLL) is often used to generate the on-chip clock signal. An on-chip PLL generates an on-chip clock signal by multiplying the system clock signal. Various changes in the electrical environment of a PLL, power supply level variations in particular, affect the phase of the on-chip clock signal. A feedback loop within the PLL controls the phase of the PLL output and aligns the output signal phase with the phase of the system clock. Ideally, the edges of the on-chip clock signal are at precisely equidistant time intervals determined by the system clock signal. The closed loop response time of the PLL is typically hundreds of nanoseconds. Disturbances of shorter duration than the PLL response time result in deviations of the on-chip clock phase from the ideal timing. These deviations are referred to as clock jitter [34], [35]. The clock jitter is classified into two types: cycle-to-cycle jitter and peak-to-peak jitter.

Cycle-to-cycle jitter refers to *random* deviations of the clock phase from the ideal timing, as illustrated in Fig. 1.13 [36]. Deviation from the ideal phase at one edge of a clock signal is independent of the deviations at other edges. That is, the cycle-to-cycle jitter characterizes the variation of the time interval between two adjacent clock edges. The average cycle-to-cycle jitter asymptotically approaches zero with an increasing number of samples. This type of jitter is therefore characterized by a mean square deviation. This type of phase variation is produced by disturbances of duration shorter or comparable to the clock period. Active device noise and high frequency power supply noise (*i.e.*, of a frequency higher or comparable to the circuit clock frequency) contribute to the cycle-to-cycle jitter. Due to the stochastic nature of phase variations, the cycle-to-cycle jitter directly contributes to the uncertainty of the time reference signals across an integrated circuit. Increased uncertainty of an on-chip timing reference results in a reduced operating frequency [36].

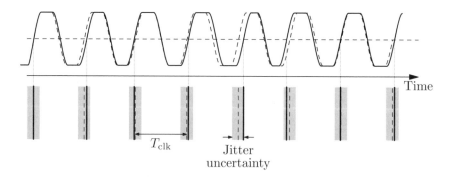

Fig. 1.13. Cycle-to-cycle jitter of a clock signal. The phase of the clock signal (solid line) randomly deviates from the phase of an ideal clock signal (dashed line).

The second type of jitter, peak-to-peak jitter, refers to *systematic variations of on-chip clock phase as compared to the system clock.* Consider a situation where several consecutive edges of an on-chip clock signal have a positive cycle-to-cycle variation, *i.e.*, several consecutive clock cycles are longer than the ideal clock period, as illustrated in Fig. 1.14 (due to, for example, a prolonged power supply variation from the nominal voltage). The timing requirements of the on-chip circuits are not violated provided that the cycle-to-cycle jitter is sufficiently small. The phase difference between the system clock and the on-chip clock, however, accumulates with time. Provided the disturbance persists, the phase difference between the system and the on-chip clocks can accumulate for tens or hundreds of clock periods, until the PLL feedback adjustment becomes effective. This phase difference degrades the synchronization among different clock domains (*i.e.*, between an integrated circuit and other system components controlled by different clock signals). Synchronizing the clock domains is critical for reliable communication across these domains. The maximum phase difference between two clock domains is characterized by the peak-to-peak jitter.

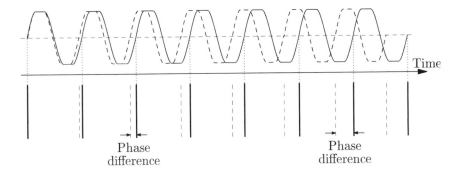

Fig. 1.14. Peak-to-peak jitter of a clock signal. The period of the clock signal (the solid line) systematically deviates from the period of the reference clock (the dashed line), leading to accumulation of the phase difference.

The feedback response time is highly sensitive to the power supply voltage [37]. For example, the PLL designed for the 400 MHz IBM S/390 microprocessor exhibits a response time of approximately 50 clock cycles when operating at 2.5 V power supply and disturbed by a 100 mV drop in supply voltage. The recovery time from the same disturbance increases manyfold when the supply voltage is reduced to 2.3 V and below [37].

1.4.3 Noise margin degradation

In digital logic styles with single-ended signaling, the power and ground supply networks also serve as a voltage reference for the on-chip signals. If a transmitter communicates a low voltage state, the output of the transmitter is connected to the ground distribution network. Alternatively, the output is connected to the power distribution network to communicate the high voltage state. At the receiver end of the communication line, the output voltage of the transmitter is compared to the power or ground voltage *local to the receiver*. Spatial variations in the supply voltage create a discrepancy between the power and ground voltage levels at the transmitter and receiver ends of the communication line. The power noise induced uncertainty in these reference voltages degrades the noise margins of the on-chip signals. As the operating speed of integrated circuits rise, crosstalk noise among on-chip signals has increased. Sufficient noise margins of the on-chip signals has therefore become a design issue of primary importance.

1.4.4 Degradation of gate oxide reliability

The performance characteristics of a MOS transistor depend on the thickness of the gate oxide. The current drive of the transistor increases as the gate oxide thickness is reduced, improving the speed and power characteristics. Reduction of the gate oxide thickness in process scaling has therefore been instrumental in improving transistor performance. A thin oxide layer, however, poses the problems of electron tunneling and oxide layer reliability [38]. As the thickness of the gate silicon oxide has reached several molecular layers (tens of angstroms) in contemporary digital CMOS processes, the power supply voltage has become limited by the maximum electric field within the gate oxide layer [33]. Variations in the power supply levels can increase the voltage applied across the gate oxide layer above the nominal power supply, degrading the long term reliability of the devices [39]. Overshoots of the power and ground voltages should be limited to avoid significant degradation in the transistor reliability characteristics.

1.5 Book outline

The book consists of two parts. In the first part, an overview of power distribution networks in integrated circuits and relevant background information are provided. The electrical properties of on-chip power distribution networks are also described. In the second part, design methodologies are presented for efficient placement of on-chip decoupling capacitors in nanoscale ICs.

The inductive properties of interconnect are described in Chapter 2. Different methods of characterizing the inductance of complex interconnect systems as well as the limitations of these methods are discussed. The concept of a partial inductance is reviewed. This concept is helpful in describing the inductive properties of complex structures. The distinction between the absolute inductance and the inductive behavior is emphasized and the relationship between these concepts is discussed.

The inductive properties of interconnect structures where current flows in long loops are investigated in Chapter 3. The variation of the partial inductance with line length is compared to that of the loop inductance. The inductance of a long current loop is shown to increase linearly with loop length. Similarly, the effective inductance of several long loops connected in parallel is shown to decrease inversely linearly with the number of loops. Exploiting these properties to enhance the efficiency of the circuit analysis process is discussed.

The phenomenon of electromigration and related circuit reliability implications are the subject of Chapter 4. As the current density in on-chip interconnect lines increases, the transport of metal atoms under the electric driving force, known as electromigration, becomes increasingly important. Metal depletion and accumulation occurs at the sites of electromigration atomic flux divergence. Voids and protrusions are formed at the sites of metal depletion and accumulation, respectively, causing open circuit and short-circuit faults in interconnect structures. The mechanical characteristics of the interconnect structures are critical in determining the electromigration reliability. Power and ground lines are particularly susceptible to electromigration damage as these lines carry unidirectional current.

An overview of power distribution systems for high performance integrated circuits is presented in Chapter 5. The inductive nature of the board and package interconnect is identified as the primary obstacle toward achieving low impedance characteristics at high frequencies. The effect of decoupling capacitances on the impedance characteristics of power distribution systems is also examined. The use of a hierarchy of decoupling capacitors to reduce the output impedance of power distribution systems is described. Finally, design guidelines for improving the impedance characteristics of a power distribution system are discussed.

Decoupling capacitance is introduced in Chapter 6. A historical perspective of capacitance is provided. The decoupling capacitor is shown to be analogous to a reservoir of charge. A hydraulic analogy of a hierarchical placement of decoupling capacitors is introduced. It is demonstrated that the impedance of a power distribution system can be maintained below target specifications over an entire range of operating frequencies by utilizing a hierarchy of decoupling capacitors. Antiresonance in the impedance of a power distribution system with decoupling capacitors is also intuitively explained in this chapter. Different types of on-chip decoupling capacitors are compared. Several allocation strategies for placing on-chip decoupling capacitors are reviewed.

The focus of Chapter 7 is on-chip power and ground distribution networks. The topological variations of the structure of on-chip power distribution networks are described, highlighting the benefits and disadvantages of each network topology. Techniques are reviewed to reduce the impedance of the on-chip power interconnect. A discussion of strategies for allocating on-chip decoupling capacitances concludes the chapter.

The process of analyzing power distribution networks is the topic of the next chapter — Chapter 8. The flow of the computer-aided design process for on-chip power distribution networks is described. The objectives and challenges of power network analysis at each stage of the design process are identified. A description of efficient numerical techniques for analyzing complex power distribution networks closes the chapter.

An analysis of the inductive properties of power distribution grids is presented in Chapter 9. Three types of power grids are considered. The dependence of the grid inductance on grid type, grid line width, and grid dimensions is discussed. The concept of a sheet inductance to characterize the inductive properties of a power grid is introduced.

The variation of the inductance of power distribution grids with frequency is discussed in Chapter 10. The physical mechanisms underlying the dependence of inductance on frequency are discussed. The variation of the inductance of the power grids is interpreted in terms of these mechanisms. The variation of inductance with frequency in paired, interdigitated, and non-interdigitated grid types is compared. The dominant mechanisms for the variation of inductance with frequency are identified for each grid type. The dependence of the frequency variation characteristics on grid type and line width is discussed.

Inductance/area/resistance tradeoffs in high performance power distribution grids are analyzed in Chapter 11. Two tradeoff scenarios are considered. In the first scenario, the area overhead of a power grid is maintained constant and the resistance versus inductance tradeoff is explored as the width of the grid lines is varied. In the second scenario, the total metal area occupied by the grid lines (and, consequently, the grid resistance) is maintained constant and the area versus inductance tradeoff is explored as the width of the grid lines is varied.

Scaling trends of on-chip power distribution noise are discussed in Chapter 12. A model for analyzing the scaling of power distribution noise is described. Two scenarios of interconnect scaling are analyzed. The effects of scaling trends on the design of next generation CMOS circuits are discussed.

The impedance characteristics of on-chip multi-layer power distribution grids are described in Chapter 13. A circuit model of a multi-layer power distribution grid is described. Analytic expressions describing the variation in the resistance and inductance of multilayer grids with frequency are developed. An intuitive explanation of the electrical behavior of power grids is presented. The results are supported with a case study.

In Chapter 14, systems with multiple power supply voltages are described. Several multi-voltage structures are reviewed. Primary challenges in integrated circuits with multiple power supplies are discussed. The power savings is shown to depend upon the number and magnitude of the available power supply voltages. Rules of thumb are presented to determine the appropriate number and magnitude of the multiple power supplies, increasing the savings in power.

On-chip power distribution grids with multiple power supply voltages are discussed in Chapter 15. A power distribution grid with multiple power supplies and multiple grounds is presented. It is shown that this power distribution grid structure results in reduced voltage fluctuations as seen at the terminals of the current load, as compared to traditional power distribution grids with multiple supply voltages and a single ground. It is noted that a multi-power and multi-ground power distribution grid can be an alternative to a single supply voltage and single ground power distribution system.

Decoupling capacitors for power distribution systems with multiple power supply voltages is the topic of the following chapter — Chapter 16. With the introduction of a second power supply, the noise at one power supply can propagate to the other power supply, producing power and signal integrity problems in the overall system. The interaction between the two power distribution networks should therefore be considered. The dependence of the impedance and magnitude of the voltage transfer function on the parameters of the power distribution system is investigated. Design techniques to cancel and shift the antiresonant spikes out of the range of the operational frequencies are also presented.

On-chip power noise reduction techniques in high performance ICs are the primary subject of Chapter 17. A design technique to lower ground bounce in noise sensitive circuits is described. An on-chip noise-free ground is added to divert ground noise from the sensitive nodes. An on-chip decoupling capacitor tuned in resonance with the parasitic inductance of the interconnects is shown to provide an additional low impedance ground path, reducing the power noise. The dependence of ground noise reduction mechanisms on various system parameters is also discussed.

On-chip decoupling capacitors have traditionally been allocated into the available white space on a die. The efficacy of the on-chip decoupling capacitors depends upon the impedance of the power/ground lines connecting the capacitors to the current loads and power supplies. A design methodology for placing on-chip decoupling capacitors is presented in Chapter 18. The maximum effective radii of an on-chip decoupling capacitor as determined by the target impedance (during discharge) and the charge time are described. Two criteria to estimate the minimum required on-chip decoupling capacitance are also presented.

As the minimum feature size continues to scale, additional on-chip decoupling capacitance will be required to support increasing current demands. A larger on-chip decoupling capacitance requires a greater area which cannot conveniently be placed in the proximity of the switching load circuits. Moreover, a large decoupling capacitor exhibits a distributed behavior. A lumped model of an on-chip decoupling capacitor, therefore, results in underestimating the capacitance requirements, thereby increasing the power noise. A methodology for efficiently placing on-chip distributed decoupling capacitors is the subject of Chapter 19. Design techniques to estimate the location and magnitude of a system of distributed decoupling capacitors are presented. Various tradeoffs in the design of a system of distributed on-chip decoupling capacitors are also investigated.

The effect of the inductance of on-chip interconnect on the high frequency impedance characteristics of a power distribution system is discussed in Chapter 20. Scaling trends of the chip-package resonance is described. The propagation of the power noise through an on-chip power distribution network is discussed. The significance of the inductance of on-chip power lines in sub-100 nanometer circuits is demonstrated.

The conclusions are summarized in Chapter 21. The design of power distribution networks, particularly for high complexity, high performance integrated circuits, will remain a particularly challenging task. The efficient integration of diverse circuit structures into complex digital circuits and systems-on-a-chip (SoC) requires a thorough understanding of the electrical phenomena in on-chip power distribution networks. On-chip decoupling capacitors will remain an efficient solution for reducing power/ground voltage fluctuations in nanoscale ICs. As technologies continue to scale, determining the proper amount of on-chip decoupling capacitance will become increasingly important to reduce leakage power. Due to higher frequencies, a hierarchical system of distributed on-chip decoupling capacitors should be utilized to provide the required decoupling capacitance under increasingly challenging technology constraints. Determining the specific location and magnitude of the on-chip decoupling capacitors is therefore a primary task in high complexity, high performance ICs.

2

Inductive Properties of Electric Circuits

Characterizing the inductive properties of the power and ground interconnect is essential in determining the impedance characteristics of a power distribution system. Several of the following chapters are dedicated to the inductive properties of on-chip power distribution networks. The objective of this chapter is to introduce the concepts used in these chapters to describe the inductive characteristics of complex interconnect structures.

The magnetic properties of circuits are typically introduced using circuits with inductive coils. The inductive characteristics of such circuits are dominated by the self and mutual inductances of these coils. The inductance of a coil is well described by the classical definition of inductance based on the magnetic flux through a current loop. The situation is more complex in circuits with no coils where no part of the circuit is inductively dominant and the circuit elements are strongly inductively coupled. The magnetic properties in this case are determined by the physical structure of the entire circuit, resulting in complex inductive behavior. The loop inductance formulation is inconvenient to represent the inductive characteristics of these circuits. The objective of this chapter is to describe various ways to represent a circuit inductance, highlighting specific assumptions. Intuitive interpretations are offered to develop a deeper understanding of the limitations and interrelations of these approaches. The variation of inductance with frequency and the relationship between the absolute inductance and the inductive behavior are also discussed in this chapter.

These topics are discussed in the following order. Several formulations of the circuit inductive characteristics as well as advantages and limitations of these formulations are described in Section 2.1.

Mechanisms underlying the variation of inductance with frequency are examined in Section 2.2. The relationship between the absolute inductance and the inductive behavior of circuits is discussed in Section 2.3. The inductive properties of on-chip interconnect structures are analyzed in Section 2.4. The chapter is summarized in Section 2.5.

2.1 Definitions of inductance

There are several ways to represent the magnetic characteristics of a circuit. Understanding the advantages and limitations of these approaches presents the opportunity to choose the approach most suitable for a particular task. Several representations of the inductive properties of a circuit are presented in this section. The field energy formulation of inductive characteristics is described in Section 2.1.1. The loop flux definition of inductance is discussed in Section 2.1.2. The concept of a partial inductance is introduced in Section 2.1.3. The net inductance formulation is introduced in Section 2.1.4.

2.1.1 Field energy definition

Inductance represents the capability of a circuit to store energy in the form of a magnetic field. Specifically, the inductance relates the electrical current to the magnetic flux and magnetic field energy. The magnetic field is interrelated with the electric field and current, as determined by Maxwell's equations and constitutive relations,[1]

$$\nabla D = \rho, \tag{2.1}$$

$$\nabla B = 0, \tag{2.2}$$

$$\nabla \times H = J + \frac{\partial D}{\partial t}, \tag{2.3}$$

$$\nabla \times E = -\frac{\partial B}{\partial t}, \tag{2.4}$$

$$D = \epsilon E, \tag{2.5}$$

$$B = \mu H, \tag{2.6}$$

$$J = \sigma E, \tag{2.7}$$

assuming a linear media. The domain of circuit analysis is typically confined to those operational conditions where the electromagnetic radiation phenomena are negligible. The direct effect of the displacement

---

[1] Vector quantities are denoted with bold italics, such as H.

current $\frac{\partial \boldsymbol{D}}{\partial t}$ on the magnetic field, as expressed by (2.3), can be neglected under these conditions (although the displacement current can
be essential to determine the current density $\boldsymbol{J}$). The magnetic field
is therefore determined only by the circuit currents. The local current
density determines the local behavior of the magnetic field, as expressed
by Ampere's law in the differential form,

$$\nabla \times \boldsymbol{H} = \boldsymbol{J}. \tag{2.8}$$

Equivalently, the elemental contribution to the magnetic field $d\boldsymbol{H}$
is expressed in terms of an elemental current $d\boldsymbol{J}$, according to the
Biot-Savart law,

$$d\boldsymbol{H} = \frac{d\boldsymbol{J} \times \boldsymbol{r}}{4\pi r^3}, \tag{2.9}$$

where $\boldsymbol{r}$ is the distance vector from the point of interest to the current
element $d\boldsymbol{J}$ and $r = |\boldsymbol{r}|$.

It can be demonstrated that the magnetic energy in a linear media
can be expressed as [40]

$$W_{\mathrm{m}} = \frac{1}{2} \int \boldsymbol{J} \cdot \boldsymbol{A} \, dr, \tag{2.10}$$

where $\boldsymbol{A}$ is the magnetic vector potential of the system, determined as

$$\boldsymbol{A}(\boldsymbol{r}) = \frac{\mu}{4\pi} \int \frac{\boldsymbol{J}(\boldsymbol{r}') \, dr'}{|\boldsymbol{r} - \boldsymbol{r}'|}. \tag{2.11}$$

Substituting (2.11) into (2.10) yields the expression of the magnetic
energy in terms of the current distribution in a system,

$$W_{\mathrm{m}} = \frac{\mu}{8\pi} \iint \frac{\boldsymbol{J}(\boldsymbol{r}) \cdot \boldsymbol{J}(\boldsymbol{r}')}{|\boldsymbol{r} - \boldsymbol{r}'|} \, dr \, dr'. \tag{2.12}$$

If the system is divided into several parts, each contained in a volume
V_i, the magnetic energy expression (2.12) can be rewritten as

$$W_{\mathrm{m}} = \frac{\mu}{8\pi} \sum_i \sum_j \int_{V_i} \int_{V_j} \frac{\boldsymbol{J}(\boldsymbol{r}) \cdot \boldsymbol{J}(\boldsymbol{r}')}{|\boldsymbol{r} - \boldsymbol{r}'|} \, dr \, dr'. \tag{2.13}$$

Assuming that the relative distribution of the current in each volume
V_i is independent of the current magnitude, the current density distribution $\boldsymbol{J}$ can be expressed in terms of the overall current magnitude

I and current distribution function $\boldsymbol{u}(\boldsymbol{r})$, so that $\boldsymbol{J}(\boldsymbol{r}) = I\boldsymbol{u}(\boldsymbol{r})$. The magnetic field energy can be expressed in terms of the overall current magnitudes I_i,

$$W_{\mathrm{m}} = \frac{1}{2}\sum_i \sum_j L_{ij} I_i I_j, \tag{2.14}$$

where

$$L_{ij} \equiv \frac{\mu}{4\pi} \int\limits_{V_i} \int\limits_{V_j} \frac{\boldsymbol{u}(\boldsymbol{r}) \cdot \boldsymbol{u}(\boldsymbol{r}')}{|\boldsymbol{r} - \boldsymbol{r}'|} \, d\boldsymbol{r}\, d\boldsymbol{r}' \tag{2.15}$$

is a mutual inductance between the system parts i and j. In a matrix formulation, the magnetic energy of a system consisting of N parts can be expressed as a positively defined binary form[2] $\mathbf{L}$ of a current vector $\boldsymbol{I} = \{I_1, \ldots, I_N\}$,

$$W_{\mathrm{m}} = \frac{1}{2}\boldsymbol{I}^T \mathbf{L} \boldsymbol{I} = \frac{1}{2}\sum_{i=1}^{N}\sum_{j=1}^{N} L_{ij} I_i I_j. \tag{2.16}$$

Each diagonal element L_{ii} of the binary form $\mathbf{L}$ is a self inductance of the corresponding current I_i and each non-diagonal element L_{ij} is a mutual inductance between currents I_i and I_j. Note that according to the definition of (2.15), the inductance matrix is symmetric, $i.e.$, $L_{ij} = L_{ji}$.

While the field approach is general and has no limitations, determining the circuit inductance through this approach is a laborious process, requiring numerical field analysis except for the simplest structures. The goal of circuit analysis is to provide an efficient yet accurate description of the system in those cases where the detail and accuracy of a full field analysis are unnecessary. Resorting to a field analysis to determine specific circuit characteristics greatly diminishes the efficiency of the circuit analysis formulation.

2.1.2 Magnetic flux definition

The concept of inductance is commonly introduced as a constant L relating a magnetic flux Φ through a circuit loop to a current I' in another loop,

$$\Phi = LI'. \tag{2.17}$$

---

[2] Matrix entities are denoted with bold roman symbols, such as $\mathbf{L}$.

In the special case where the two circuit loops are the same, the coefficient is referred to as a loop self inductance; otherwise, the coefficient is referred to as a mutual loop inductance.

For example, consider two isolated complete current loops ℓ and ℓ', as shown in Fig. 2.1. The mutual inductance M between these two

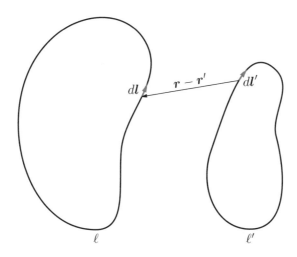

Fig. 2.1. Two complete current loops. The relative position of two differential loop elements dl and dl' is determined by the vector $\boldsymbol{r} - \boldsymbol{r'}$.

loops is a coefficient relating a magnetic flux Φ through a loop ℓ due to a current I' in loop ℓ',

$$\Phi = \iint_S \boldsymbol{B'} \cdot \boldsymbol{n}\, ds, \tag{2.18}$$

where S is a smooth surface bounded by the loop ℓ, $\boldsymbol{B'}$ is the magnetic flux created by the current in the loop ℓ', and $\boldsymbol{n}$ is a unit vector normal to the surface element ds. Substituting $\boldsymbol{B'} = \nabla \times \boldsymbol{A'}$ and using Stokes's theorem, the loop flux is expressed as

$$\Phi = \iint_S (\nabla \times \boldsymbol{A'}) \cdot \boldsymbol{n}\, ds = \oint_\ell \boldsymbol{A'}\, dl, \tag{2.19}$$

where $\boldsymbol{A'}$ is the vector potential created by the current $\boldsymbol{I'}$ in the loop ℓ'. The magnetic vector potential of the loop ℓ' $\boldsymbol{A'}$ is

$$A'(r) = \frac{\mu}{4\pi} \int_V \frac{J'(r')\, dr'}{|r - r'|} = I' \frac{\mu}{4\pi} \oint_{\ell'} \frac{dl'}{|r - r'|}, \qquad (2.20)$$

where $|r - r'|$ is the distance between the loop element dl' and the point of interest r. Substituting (2.20) into (2.19) yields

$$\Phi = I' \frac{\mu}{4\pi} \oint_\ell \oint_{\ell'} \frac{dl\, dl'}{|r - r'|} = MI', \qquad (2.21)$$

where

$$M \equiv \frac{\mu}{4\pi} \oint_\ell \oint_{\ell'} \frac{dl\, dl'}{|r - r'|} \qquad (2.22)$$

is a mutual inductance between the loops ℓ and ℓ'. As follows from the derivation, the integration in (2.20)–(2.22) is performed in the direction of the current flow. The mutual inductance (2.22) and associated magnetic flux (2.21) can therefore be either positive or negative, depending on the relative direction of the current flow in the two loops.

Note that the finite cross-sectional dimensions of the loop conductors are neglected in the transition between the general volume integration to a more constrained but simpler contour integration in (2.20). An entire loop current is therefore confined to an infinitely thin filament.

The thin filament approximation of a mutual inductance is acceptable where the cross-sectional dimensions of the conductors are much smaller than the distance $|r - r'|$ between any two points on loop ℓ and loop ℓ'. This approximation becomes increasingly inaccurate as the two loops are placed closer together. More importantly, the thin filament approach cannot be used to determine a self inductance by assuming ℓ to be identical to ℓ', as the integral (2.22) diverges at the points where $r = r'$.

To account for the finite cross-sectional dimensions of the conductors, both (2.19) and (2.20) are amended to include an explicit integration over the conductor cross-sectional area a,

$$\Phi = \frac{1}{I} \oint_\ell \int_a A'\, J dl\, da, \qquad (2.23)$$

and

$$A' = \frac{\mu}{4\pi} \oint_{\ell'} \int_{a'} \frac{J'\, dl'\, da'}{|r - r'|}, \qquad (2.24)$$

where a and a' are the cross sections of the elemental loop segments dl and dl', da and da' are the differential elements of the respective cross sections, $|\boldsymbol{r} - \boldsymbol{r}'|$ is the distance between da and da', and J is a current density distribution over the wire cross section a, $d\boldsymbol{J} = J\,dl\,da$, and $I = \int_a J\,da$. These expressions are more general than (2.19) and (2.20); the only constraint on the current flow imposed by formulations (2.23) and (2.24) is that the current flow has the same direction across the cross-sectional areas a and a'. This condition is satisfied in relatively thin conductors without sharp turns. Formulas (2.23) and (2.24) can be significantly simplified assuming a uniform current distribution (*i.e.*, $J = \text{const}$ and $I = aJ$) and a constant cross-sectional area a,

$$\Phi = \frac{1}{a} \oint_\ell \int_a \boldsymbol{A}'\,dl\,da, \tag{2.25}$$

and

$$\boldsymbol{A}' = \frac{\mu}{4\pi} \frac{I'}{a'} \oint_{\ell'} \int_{a'} \frac{dl'\,da'}{|\boldsymbol{r} - \boldsymbol{r}'|}. \tag{2.26}$$

The magnetic flux through loop ℓ is transformed into

$$\Phi = \frac{\mu}{4\pi} \frac{I'}{a\,a'} \oint_\ell \oint_{\ell'} \int_a \int_{a'} \frac{da\,da'\,dl\,dl'}{|\boldsymbol{r} - \boldsymbol{r}'|} = MI'. \tag{2.27}$$

The mutual loop inductance is therefore defined as

$$M_{\ell\ell'} \equiv \frac{\mu}{4\pi} \frac{1}{a\,a'} \oint_\ell \oint_{\ell'} \int_a \int_{a'} \frac{da\,da'\,dl\,dl'}{|\boldsymbol{r} - \boldsymbol{r}'|}. \tag{2.28}$$

The loop self inductance L_ℓ is a special case of the mutual loop inductance where the loop ℓ is the same as loop ℓ',

$$L_\ell \equiv M_{\ell\ell} = \frac{\mu}{4\pi} \frac{1}{a^2} \oint_\ell \oint_\ell \int_a \int_a \frac{da\,da'\,dl\,dl'}{|\boldsymbol{r} - \boldsymbol{r}'|}. \tag{2.29}$$

While straightforward and intuitive, the definition of the loop inductance as expressed by (2.17) cannot be applied to most practical circuits. Only the simplest circuits consist of a single current loop. In practical circuits with branch points, the current is not constant along the circumference of the conductor loops, as shown in Fig. 2.2. This difficulty can be circumvented by employing Kirchhoff's voltage law

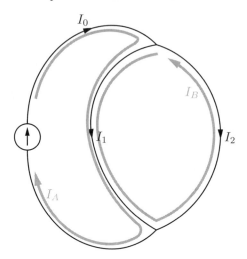

Fig. 2.2. A circuit with branch points. The current in each loop is not uniform along the circumference of the loop.

and including an inductive voltage drop within each loop equation. For example, two independent current loops carrying circular currents I_A and I_B can be identified in the circuit shown in Fig. 2.2. The inductive voltage drops V_A and V_B in loops A and B are

$$\begin{bmatrix} V_A \\ V_B \end{bmatrix} = \begin{bmatrix} L_{AA} & L_{AB} \\ L_{AB} & L_{BB} \end{bmatrix} \begin{bmatrix} I_A \\ I_B \end{bmatrix}. \tag{2.30}$$

The magnetic energy of the system is, analogous to (2.16),

$$W_{\mathrm{m}} = \frac{1}{2} \boldsymbol{I}^T \mathbf{L} \boldsymbol{I} = \frac{1}{2} \begin{bmatrix} I_A & I_B \end{bmatrix} \begin{bmatrix} L_{AA} & L_{AB} \\ L_{AB} & L_{BB} \end{bmatrix} \begin{bmatrix} I_A \\ I_B \end{bmatrix}. \tag{2.31}$$

Note that in a circuit with branch points, two current loops can share common parts, as illustrated in Fig. 2.2. The inductance between these two loops is therefore a hybrid between the mutual and self loop inductance, as defined by (2.28) and (2.29).

The flux formulation of the inductive characteristics, as expressed by (2.29) and (2.31), is a special case of the field formulation, as expressed by (2.15) and (2.16). The magnetic field expressions (2.16) and (2.31) are the same, while the definition of the loop inductance as expressed by (2.29) is obtained from (2.15) by assuming that the current flows in well formed loops; the thin filament definition of the mutual inductance (2.22) is the result of further simplification of (2.15). The magnetic energy and field flux derivations of the inductance are equivalent; both (2.15) and (2.29) can be obtained from either the energy formulation expressed by (2.31) or the flux formulation expressed by (2.22).

The loop inductance approach provides a more convenient description of the magnetic properties of the circuit with little loss of accuracy and generality, as compared to the field formulation as expressed by (2.16). Nevertheless, significant disadvantages remain. In the magnetic flux formulation of the circuit inductance, the basic inductive element is a closed loop. This aspect presents certain difficulties for a traditional circuit analysis approach. In circuit analysis, the impedance characteristics are described in terms of the circuit elements connecting two circuit nodes. Circuit analysis tools also use a circuit representation based on two-terminal elements. Few circuit elements are manufactured in a loop form. In a strict sense, a physical inductor is also a two terminal element. The current flowing through a coil does not form a complete loop, therefore, the definition of the loop inductance does not apply. The loop formulation does not provide a direct link between the impedance characteristics of the circuit and the impedance of the comprising two terminal circuit elements.

It is therefore of practical interest to examine how the inductive characteristics can be described by a network of two terminal elements with self and mutual impedances, without resorting to a multiple loop representation. This topic is the subject of the next section.

2.1.3 Partial inductance

The loop inductance, as defined by (2.28), can be deconstructed into more basic elements if the two loops are broken into segments, as shown in Fig. 2.3. The loop ℓ is broken into N segments $S_1, \ldots, S_N$ and loop ℓ' is broken into N' segments $S'_1, \ldots, S'_{N'}$. The definition of the loop inductance (2.28) can be rewritten as

$$M_{\ell\ell'} = \sum_{i=1}^{N} \sum_{j=1}^{N'} \frac{\mu}{4\pi} \frac{1}{a_i a'_j} \oint_{S_i} \oint_{S'_j} \int_{a_i} \int_{a'_j} \frac{da_i \, da'_j \, dl \, dl'}{|\boldsymbol{r} - \boldsymbol{r}'|} = \sum_{i=1}^{N} \sum_{j=1}^{N'} L_{ij}, \quad (2.32)$$

where

$$L_{ij} \equiv \frac{\mu}{4\pi} \frac{1}{a_i a'_j} \oint_{S_i} \oint_{S'_j} \int_{a_i} \int_{a'_j} \frac{da_i \, da'_j \, dl \, dl'}{|\boldsymbol{r} - \boldsymbol{r}'|}. \quad (2.33)$$

The integration along segments S_i and S'_j in (2.32) and (2.33) is performed in the direction of the current flow.

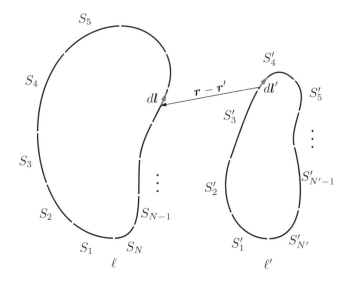

Fig. 2.3. Two complete current loops broken into segments.

Equation (2.33) defines the mutual partial inductance between two arbitrary segments S_i and S'_j. Similar to the loop inductance, the mutual partial inductance can be either positive or negative, depending on the direction of the current flow in the two segments. In the special case where S_i is identical to S'_j, (2.33) defines the partial self inductance of S_i. The partial self inductance is always positive.

The partial inductance formulation, as defined by (2.33), is more suitable for circuit analysis as the basic inductive element is a two terminal segment of interconnect. Any circuit can be decomposed into a set of interconnected two terminal elements. For example, the circuit shown in Fig. 2.2 can be decomposed into three linear segments instead of two loops as in the case of a loop analysis. The magnetic properties of the circuit are described by a partial inductance matrix $\mathbf{L} = \{L_{ij}\}$. Assigning to each element S_i a corresponding current I_i, the vector of magnetic electromotive forces $\boldsymbol{V}$ across each segment is

$$\boldsymbol{V} = \mathbf{L}\frac{d\boldsymbol{I}}{dt}. \qquad (2.34)$$

The magnetic energy of the circuit in terms of the partial inductance is determined, analogously to the loop inductance formulation (2.31), as

$$W_{\mathrm{m}} = \frac{1}{2}\boldsymbol{I}^T \mathbf{L} \boldsymbol{I} = \frac{1}{2}\sum_{i=1}^{N}\sum_{j=1}^{N} L_{ij} I_i I_j. \qquad (2.35)$$

The partial inductance has another practical advantage. If the self and mutual partial inductance of a number of basic segment shapes is determined as a function of the segment dimensions and orientations, the partial inductance matrix of any circuit composed of these basic shapes can be readily constructed according to the segment connectivity, permitting the efficient analysis of the magnetic properties of the circuit. In this regard, the partial inductance approach is more flexible than the loop inductance approach, as loops exhibit a greater variety of shapes and are difficult to precharacterize in an efficient manner.

For the purposes of circuit characterization, it is convenient to separate the sign and the absolute magnitude of the inductance. During precharacterization, the absolute magnitude of the mutual partial inductance L_{ij}^{abs} between basic conductor shapes (such as straight segments) is determined. During the process of analyzing a specific circuit structure, the absolute magnitude is multiplied by a sign function s_{ij}, resulting in the partial inductance as defined by (2.33), $L_{ij} = s_{ij} L_{ij}^{\mathrm{abs}}$. The sign function equals either 1 or -1, depending upon the sign of the scalar product of the segment currents: $s_{ij} = \mathrm{sign}\,(\boldsymbol{I}_i \cdot \boldsymbol{I}'_j)$.

The case of a straight wire is of particular practical importance. A conductor of any shape can be approximated by a number of short straight segments. The partial self inductance of a straight round wire is [41]

$$L_{\mathrm{line}} = \frac{\mu l}{2\pi}\left(\ln\frac{2l}{r} - \frac{3}{4}\right), \qquad (2.36)$$

where l is the length of the wire and r is the radius of the wire cross section, as shown in Fig. 2.4. The precise analytic expressions for the partial inductance are generally not available for straight conductors with a radially asymmetric cross section. The partial inductance of a straight line with a square cross section can be evaluated with good accuracy using approximate analytic expressions augmented with tables of correction coefficients [41], or expressions suitable for efficient numerical evaluation [42].

The partial self inductance, as expressed by (2.33), depends only on the shape of the conductor segment. It is therefore possible to assign a certain partial self inductance to an individual segment of the conductor. It should be stressed, however, that the partial self inductance

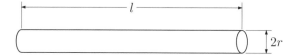

Fig. 2.4. A straight round wire.

of the comprising conductors by itself provides no information on the inductive properties of the circuit. For example, a loop of wire can have a loop inductance that is much greater than the sum of the partial self inductance of the comprising segments (where the wire is coiled) or much smaller than the sum of the comprising partial self inductances (where the wire forms a narrow long loop). The inductive properties of a circuit are described by *all* partial inductances in the circuit, necessarily including all mutual partial inductances between all pairs of elements, as expressed in (2.32) for the specific case of a current loop.

Unlike the loop inductance, the partial inductance cannot be measured experimentally. The partial inductance is, essentially, a convenient mathematical construct used to describe the inductive properties of a circuit. This point is further corroborated by the fact that the partial inductance is not uniquely defined. An electromagnetic field is described by an infinite number of vector potentials. If a specific field is described by a vector potential $\boldsymbol{A}$, any vector potential $\boldsymbol{A}'$ differing from $\boldsymbol{A}$ by a gradient of an arbitrary scalar function Ψ, *i.e.*, $\boldsymbol{A}' = \boldsymbol{A} + \boldsymbol{\nabla}\Psi$, also describes the field.[3] The magnetic field is determined through the curl operation of the vector potential and is not affected by the $\boldsymbol{\nabla}\Psi$ term, $\boldsymbol{\nabla} \times \boldsymbol{A} = \boldsymbol{\nabla} \times \boldsymbol{A}'$ as $\boldsymbol{\nabla} \times \boldsymbol{\nabla}\Psi = 0$. The choice of a specific vector potential is inconsequential. The vector potential definition (2.11) is therefore not unique. The choice of a specific vector potential is also immaterial in determining the loop inductance as expressed by (2.28), as the integration of a gradient of any function over a closed contour yields a null value. The choice of the vector potential, however, affects the value of the partial inductance, where the integration is performed over a conductor segment. Equation (2.33) therefore defines only one of many possible partial inductance matrices. This ambiguity does not present a problem as long as all of the partial inductances in the circuit are consistently determined using the same vector potential. The contributions of the function gradient to the partial inductance cancel

---

[3] This property of the electromagnetic field is referred to in electrodynamics as gauge invariance.

out, where the partial inductances are combined to describe the loop currents.

In the case of straight line segments, the partial inductance definition expressed by (2.33) has an intuitive interpretation. For a straight line segment, the partial self inductance is a coefficient of proportionality between the segment current and the magnetic flux through the infinite loop formed by a line segment S and two rays perpendicular to the segment, as illustrated in Fig. 2.5.

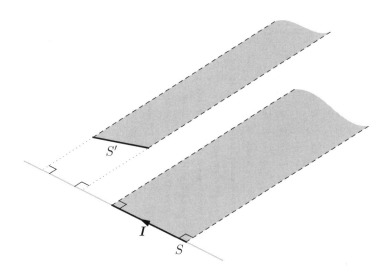

Fig. 2.5. Self and mutual partial inductance of a straight segment of wire. The partial self inductance of a segment S, as introduced by Rosa [43], is determined using the magnetic flux created by current I in segment S through an infinite contour formed by wire segment S (the bold arrow) and two rays perpendicular to the segment (the dashed lines). Similarly, the partial mutual inductance with another wire segment S' is determined using the flux created by current I through the contour formed by the segment S' and straight lines originating from the ends of the segment S' and perpendicular to segment S.

This flux is henceforth referred to as a partial flux. This statement can be proved as follows. The flux through the aforementioned infinite loop is determined by integrating the vector potential A along the loop contour, according to (2.25). The magnetic vector potential A of a straight segment, as determined by (2.11), is parallel to the segment. The integration of the vector potential along the rays perpendicular

to the segment is zero. The integration of the vector potential along
the segment completing the loop at infinity is also zero as the vector potential decreases inversely proportionally with distance. Similarly, the
mutual partial inductance between segments S and S' can be interpreted in terms of the magnetic flux through the infinite loop formed
by segment S' and two rays perpendicular to the segment S, as illustrated in Fig. 2.5.

This interpretation of the partial inductance in terms of the partial flux is in fact the basis for the original introduction of the partial
inductance by Rosa in 1908 in application to linear conductors [43].
Attempts to determine the inductance of a straight wire segment using
the total magnetic flux were ultimately unsuccessful as the total flux
of a segment is infinite. Rosa made an intuitive argument that only
the partial magnetic flux, as illustrated in Fig. 2.5, should be associated with the self inductance of the segment. The concept of partial
inductance proved useful and was utilized in the inductance calculation formulæ and tables developed by Rosa and Cohen [44], Rosa and
Grover [45], and Grover [41]. A rigorous theoretical treatment of the
subject was first developed by Ruehli in [42], where a general definition
of the partial inductance of an arbitrarily shaped conductor (2.33) is
derived. Ruehli also coined the term "partial inductance."

Connections between the loop and partial inductance can also be
established in terms of the magnetic flux. The magnetic flux through
a specific loop is a sum of all of the partial fluxes of the comprising
segments. The contribution of a magnetic field created by a specific
loop segment to the loop flux is also the sum of all of the partial inductances of this segment with respect to all segments of the loop. This
relationship is illustrated in Fig. 2.6.

2.1.4 Net inductance

The inductance of a circuit without branch points (*i.e.*, where the current flowing in all conductor segments is the same) can also be expressed
in a form with no explicit mutual inductances. Consider a current loop
consisting of N segments. As discussed in the previous section, the loop
inductance L_{loop} can be described in terms of the partial inductances
L_{ij} of the segments,

$$L_{\mathrm{loop}} = \sum_{i=1}^{N} \sum_{j=1}^{N} L_{ij}. \tag{2.37}$$

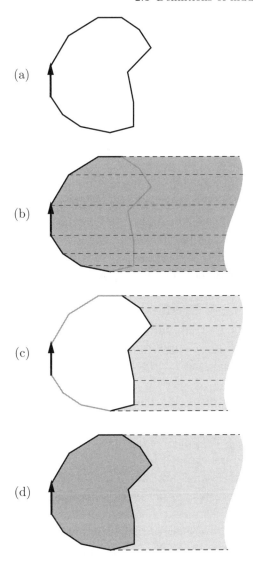

Fig. 2.6. The contribution of a current in a specific loop segment (shown with a bold arrow) to the total flux of the current loop is composed of partial fluxes of this segment with all segments of the loop. (a) A piecewise linear loop. (b) Partial flux of the segment with all other segments carrying current in the same direction (*i.e.*, the scalar product of the two segment vectors is positive). This flux is positive. (c) The partial flux of the segment with all other segments carrying current in the opposite direction (*i.e.*, the scalar product of the two segment vectors is negative). This flux is negative. (d) The sum of the positive and negative fluxes shown in (b) and (c) [*i.e.*, the geometric difference between the contours (b) and (c)] is the overall contribution of the segment to the magnetic flux of the loop. This contribution is expressed as the net inductance of the segment.

This sum can be rearranged as

$$L_{\text{loop}} = \sum_{i=1}^{N} L_i^{\text{eff}}, \tag{2.38}$$

where

$$L_i^{\text{eff}} \equiv \sum_{i,j=1}^{N} L_{ij}. \tag{2.39}$$

The inductance L_i^{eff}, as defined by (2.39), is often referred to as the *net* inductance [46], [47], [48]. The net inductance also has an intuitive interpretation in terms of the magnetic flux. As illustrated in Fig. 2.6, a net inductance (*i.e.*, the partial self inductance plus the partial mutual inductances with all other segments) of the segment determines the contribution of the segment current to the overall magnetic flux through the circuit.

The net inductance describes the behavior of coupled circuits without using explicit mutual inductance terms, simplifying the circuit analysis process. For example, consider a current loop consisting of a signal current path with inductance L_{sig} and return current path with inductance L_{ret}, as shown in Fig. 2.7. The mutual inductance between the two paths is M. The net inductance of the two paths is $L_{\text{sig}}^{\text{eff}} = L_{\text{sig}} - M$ and $L_{\text{ret}}^{\text{eff}} = L_{\text{ret}} - M$. The loop inductance in terms of the net inductance is $L_{\text{loop}} = L_{\text{sig}}^{\text{eff}} + L_{\text{ret}}^{\text{eff}}$. The inductive voltage drop along the return current path is $V_{\text{ret}} = L_{\text{ret}}^{\text{eff}} \frac{dI}{dt}$.

The net inductance has another desirable property. Unlike the partial inductance, the net inductance is independent of the choice of the magnetic vector potential, because, similar to the loop inductance, the integration of the vector potential is performed along a complete loop, as implicitly expressed by (2.39). The net inductance is therefore uniquely determined.

Note that the net inductance of a conductor depends on the structure of the overall circuit as indicated by the mutual partial inductance terms in (2.39). Modifying the shape of a single segment in a circuit changes the net inductance of *all* of the segments. The net inductance is, in effect, a specialized form of the partial inductance and should only be used in the specific circuit where the net inductance terms are determined according to (2.39).

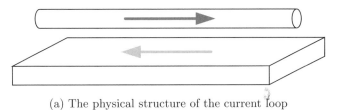

(a) The physical structure of the current loop

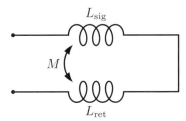

(b) The equivalent partial inductance model

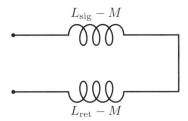

(c) The equivalent net inductance model

Fig. 2.7. The signal and return current paths.

2.2 Variation of inductance with frequency

A circuit inductance, either loop or partial, depends upon the current distribution across the cross section of the conductors, as expressed by (2.23) and (2.24). The current density is assumed constant across the conductor cross sections in the inductance formulas described in Section 2.1, as is commonly assumed in practice. This assumption is valid where the magnetic field does not appreciably change the path of the current flow. The conditions where this assumption is accurate are discussed in Section 2.2.1. Where the effect of the magnetic field on the current path is significant, the current density becomes non-uniform and the magnetic properties of the circuit vary significantly with frequency. The mechanisms causing the inductance to vary with frequency are described in Section 2.2.2. A circuit analysis of the variation of inductance with frequency is performed in Section 2.2.3 based on a simple circuit model. The section concludes with a discussion of the relative significance of the different inductance variation mechanisms.

2.2.1 Uniform current density approximation

The effect of the magnetic field on the current distribution can be neglected in two general cases. First, the current density is uniform where the magnetic impedance $L\frac{dI}{dt}$ is much smaller than the resistive impedance R of the interconnect structure. Under this condition, however, the magnetic properties of the circuit do not significantly affect the circuit behavior and are typically of little practical interest. The second case is of greater practical importance, where the magnetic impedance to the current flow, although greater than R, is uniform across the cross section of a conductor. This condition is generally satisfied where the separation between conductors is significantly greater that the cross-sectional dimensions. It can be shown by inspecting (2.11) that at a distance d much greater than the conductor cross-sectional dimension a, a non-uniform current distribution within the conductor contributes only a second order correction to the magnetic vector potential $\boldsymbol{A}$. The significance of this correction as compared to the primary term decreases with distance as a/d.

Where the separation of two conductors is comparable to the cross-sectional dimensions, the magnetic field significantly affects the current distribution within the conductors. The current density distribution across the cross section becomes non-uniform and varies with the signal frequency. In this case, the magnetic properties of an interconnect structure cannot be accurately represented by a constant value. Alternatively stated, the inductance varies with the signal frequency.

The frequency variation of the current density distribution and, consequently, of the conductor inductance can be explained from a circuit analysis point of view if the impedance characteristics of different paths *within the same conductor* are considered, as described in Section 2.2.2. The resistive properties of alternative parallel paths within the same conductors are identical, provided the conductivity of the conductor material is uniform. The magnetic properties of the paths however can be significantly different. At low frequencies, the impedance of the current paths is dominated by the resistance. The current density is uniform across the cross section, minimizing the overall impedance of the conductor. At sufficiently high frequencies, the impedance of the current paths is dominated by the inductive reactance. As the resistive impedance becomes less significant (as compared to the inductive impedance) at higher frequencies, the distribution of the current density asymptotically approaches the density profile that yields the minimum overall inductance of the interconnect structure. The inductance of the on-chip interconnect structures can therefore decrease significantly with signal frequency.

2.2.2 Inductance variation mechanisms

As discussed, the variation of inductance is the result of the variation of the current density distribution. The variation of the current distribution with frequency can be loosely classified into several categories.

Skin effect

With the onset of the skin effect, the current becomes increasingly concentrated near the line surface, causing a decrease in the magnetic field within the line core, as illustrated in Fig. 2.8. The magnetic field outside the conductor is affected relatively little. It is therefore convenient to divide the circuit inductance into "internal" and "external" parts, $L = L_{internal} + L_{external}$, where $L_{external}$ is the inductance associated with the magnetic field outside the conductors and $L_{internal}$ is the inductance associated with the magnetic field inside the conductors. In these terms, a well developed skin effect produces a significant decrease in the internal inductance $L_{internal}$. For a round wire at low frequency (where the current distribution is uniform across the line cross section), the internal inductance is $0.05 \frac{nH}{mm}$, independent of the radius (see the derivation in [49]). The external inductance of the round wire is unaffected by the skin effect.

Proximity effect

The current distribution also varies with frequency due to the proximity effect. At high frequencies, the current in the line concentrates along the side of the line facing an adjacent current return path, thereby reducing the effective area of the current loop and thus the loop inductance, as illustrated in Fig. 2.9.

The skin and proximity effects are closely related. These effects represent basically the same phenomenon — the tendency of the current to move closer to the current return path in order to minimize the

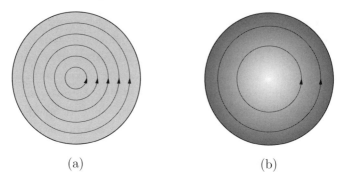

Fig. 2.8. Internal magnetic flux of a round conductor. (a) At low frequencies, the current density, as shown by the shades of gray, is uniform, resulting in the maximum magnetic flux inside the conductor, as shown by the circular arrows, and the associated internal inductance. (b) At high frequencies, the current flow is redistributed to the surface of the conductor, reducing the magnetic flux inside the conductor.

interconnect inductance at high frequencies. When a conductor is surrounded by several alternative current return paths, leading to a relatively symmetric current distribution at high frequency, the effect is typically referred to as the skin effect. The classical example of such an interconnect structure is a coaxial cable, where the shield provides equivalent current return paths along all sides of the core conductor. In the case where the current distribution is significantly asymmetric due to the close proximity of a dominant return path, the effect is referred to as the proximity effect.

Fig. 2.9. Proximity effect in two closely spaced lines. Current density distribution in the cross section of two closely spaced lines at high frequencies is shown in shades of gray. Darker shades of gray indicate higher current densities. In lines carrying current in the same direction (parallel currents), the current concentration is shifted away from the parallel current. In lines carrying current in opposite directions (antiparallel currents), the current concentrates toward the antiparallel current, minimizing the circuit inductance.

Multi-path current redistribution

The concept of current density redistribution within a conductor can be extended to redistribution of the current among several separate parallel conductors. This mechanism is henceforth referred to as *multi-path current redistribution*. For example, in standard single-ended digital logic, the forward current path is typically composed of a single line. No redistribution of the forward current occurs. The current return path, though, is not explicitly specified (although local shielding for particularly sensitive nets is becoming more common [50], [51]). Adjacent signal lines, power lines, and the substrate provide several alternative current return paths. A significant redistribution of the return current among these return paths can occur as signal frequencies increase. At low frequencies, the line impedance $Z(\omega) = R(\omega) + j\omega L(\omega)$ is dominated by the interconnect resistance R. In this case, the distribution of the return current over the available return paths is determined by the path resistance, as shown in Fig. 2.10(a). The return current spreads out far from the signal line to reduce the resistance of the return path and, consequently, the impedance of the current loop. At high frequencies, the line impedance $Z(\omega) = R(\omega) + j\omega L(\omega)$ is dominated by the reactive component $j\omega L(\omega)$. The minimum impedance path is primarily determined by the inductance $L(\omega)$, as shown in Fig. 2.10(b). Multi-path current redistribution, as described in Fig. 2.10, is essentially a proximity effect extended to several separate lines connected in parallel. In power grids, both the forward and return currents undergo multi-path redistribution as both the forward and return paths consist of multiple conductors connected in parallel.

The general phenomenon underlying these three mechanisms is, as viewed from a circuit perspective, the same. Where several parallel paths with significantly different electrical properties are available for current flow, the current is distributed among the paths so as to minimize the total impedance. As the frequency increases, the circuit inductance changes from the low frequency limit, determined by the ratio of the resistances of the parallel current paths, to the high frequency value, determined by the inductance ratios of the current paths. At high signal frequencies, the inductive reactance dominates the interconnect impedance; therefore, the path of minimum inductance carries the largest share of the current, minimizing the overall impedance (see Fig. 2.10). Note that parallel current paths can be formed either by several physically distinct lines, as in multi-path current redistribution, or

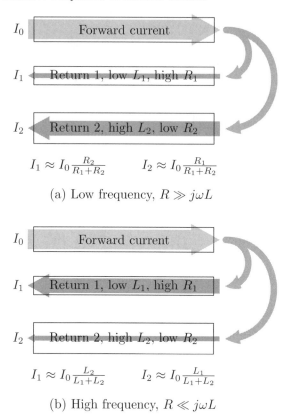

$$I_1 \approx I_0 \frac{R_2}{R_1+R_2} \qquad I_2 \approx I_0 \frac{R_1}{R_1+R_2}$$

(a) Low frequency, $R \gg j\omega L$

$$I_1 \approx I_0 \frac{L_2}{L_1+L_2} \qquad I_2 \approx I_0 \frac{L_1}{L_1+L_2}$$

(b) High frequency, $R \ll j\omega L$

Fig. 2.10. Current loop with two alternative current return paths. The forward current I_0 returns both through return path one with resistance R_1 and inductance L_1, and return path two with resistance R_2 and inductance L_2. In this structure, $L_1 < L_2$ and $R_1 > R_2$. At low frequencies (a), the path impedance is dominated by the line resistance and the return current is distributed between two return paths according to the resistance of the lines. Thus, at low frequencies, most of the return current flows through the return path of lower resistance, path two. At high frequencies (b), however, the path impedance is dominated by the line inductance and the return current is distributed between two return paths according to the inductance of the lines. Most of the return current flows through the path of lower inductance, path one, minimizing the overall inductance of the circuit.

by different paths within the same line, as in skin and proximity effects, as shown in Fig. 2.11. The difference is merely in the physical structure, the electrical behavior is fully analogous. A thick line can be thought of as being composed of multiple thin lines bundled together in parallel. The skin and proximity effects in such a thick line can be considered as a special case of current redistribution among multiple thin lines forming a thick line.

Fig. 2.11. A cross-sectional view of two parallel current paths (dark gray) sharing the same current return path (light gray circles). The path closest to the return path, path 1, has a lower inductance than the other path, path 2. The parallel paths can be either two physically distinct lines, as shown by the dotted line, or two different paths within the same line, as shown by the dashed line.

2.2.3 Simple circuit model

A simple model of current redistribution provides deeper insight into the process of inductance variation. This approach can be used to estimate the relative significance of the different current distribution mechanisms in various interconnect structures as well as the frequency characteristics of the inductance. Consider a simple case of two current paths with different inductive properties (for example, as shown in Fig. 2.11). The impedance characteristics are represented by the circuit diagram shown in Fig. 2.12, where the inductive coupling between the two paths is neglected for simplicity. Assume that $L_1 < L_2$ and $R_1 > R_2$.

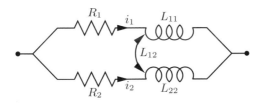

Fig. 2.12. A circuit model of two current paths with different inductive properties.

For the purpose of evaluating the variation of inductance with frequency, the electrical properties of the interconnect are characterized by the inductive time constant $\tau = L/R$. The impedance magnitude of these two paths is schematically shown in Fig. 2.13. The impedance of the first path is dominated by the inductive reactance above the frequency $f_1 = \frac{1}{2\pi}\frac{R_1}{L_1} = \frac{1}{2\pi\tau_1}$. The impedance of the second path is predominantly inductive above the frequency $f_2 = \frac{1}{2\pi}\frac{R_2}{L_2} = \frac{1}{2\pi\tau_2}$, such that $f_2 < f_1$. At low frequencies, *i.e.*, from DC to the frequency f_1, the ratio of the two impedances is constant. The effective inductance at low frequencies is therefore also constant, determining the low frequency inductance limit. At high frequencies, *i.e.*, frequencies exceeding f_2, the ratio of the impedances is also constant, determining the high frequency inductance limit, $\frac{L_1 L_2}{L_1+L_2}$. At intermediate frequencies from f_1 to f_2, the impedance ratio changes, resulting in a variation of the overall inductance from the low frequency limit to the high frequency limit. The frequency range of inductance variation is therefore determined by the two time constants, τ_1 and τ_2. The magnitude of the inductance variation depends upon both the difference between the time constants τ_1 and τ_2 and on the inductance ratio L_1/L_2. Analogously, in the case of multiple parallel current paths, the frequency range and the magnitude of the variation in inductance is determined by the minimum and maximum time constants as well as the difference in inductance among the current paths.

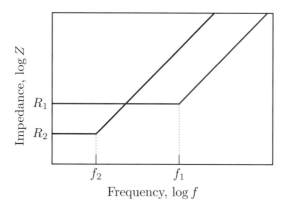

Fig. 2.13. Impedance magnitude versus frequency for two paths with dissimilar impedance characteristics.

The decrease in inductance begins when the inductive reactance $j\omega L$ of the path with the lowest R/L ratio becomes comparable to the path resistance R, $R \sim j\omega L$. The inductance, therefore, begins to decrease at a lower frequency if the minimum R/L ratio of the current paths is lower.

Due to this behavior, the proximity effect becomes significant at higher frequencies than the frequencies at which multi-path current redistribution becomes significant. Significant proximity effects occur in conductors containing current paths with significantly different inductive characteristics. That is, the inductive coupling of one edge of the line to the "return" current (*i.e.*, the current in the opposite direction) is substantially different from the inductive coupling of the other edge of the line to the same "return" current. In geometric terms, this characteristic means that the line width is larger than or comparable to the distance between the line and the return current. Consequently, the line with significant proximity effects is typically the immediate neighbor of the current return line. A narrower current loop is therefore formed with the current return path as compared to the other lines participating in the multi-path current redistribution. A smaller loop inductance L results in a higher R/L ratio. Referring to Fig. 2.10, current redistribution between paths one and two develops at frequencies lower than the onset frequency of the proximity effect in path one.

Efficient and accurate lumped element models are necessary to incorporate skin and proximity effects into traditional circuit simulation tools. Developing such models is an area of ongoing research [52], [53], [54], [55], [56], [57], [58]. The resistance and internal inductance of conductors are typically modeled with RL ladder circuits [52], as shown in Fig. 2.14. The sections of the RL ladder represent the resistance and inductance of the conductor parts at different distances from the current return path. Different methods for determining the value of the R and L elements have been proposed [53], [54], [55]. Analogously, RL ladders can also be extended to describe multi-path current redistribution [56], [57]. Techniques for reducing the order of a transfer function of an interconnect structure have also been proposed [58].

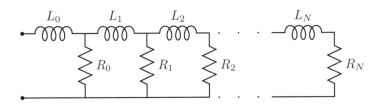

Fig. 2.14. An RL ladder circuit describing the variation of inductance with frequency.

2.3 Inductive behavior of circuits

The strict meaning of the term "inductance" is the *absolute inductance*, as defined in Section 2.1. The absolute inductance is measured in henrys. Sometimes, however, the same term "inductance" is loosely used to denote the *inductive behavior* of a circuit; namely, overshoots, ringing, signal reflections, *etc.* The inductive behavior of a circuit is characterized by such quantities as a damping factor and the magnitude of the overshoot and reflections of the signals. While any circuit structure carrying an electrical current has a finite absolute inductance, as defined in Section 2.1, not every circuit exhibits inductive behavior. Generally, a circuit exhibits inductive behavior if the absolute inductance of the circuit is sufficiently high. The relationship between the inductive behavior and the absolute inductance is, however, circuit specific and no general metrics for the onset of inductive behavior have been developed.

Specific metrics have been developed to evaluate the onset of inductive behavior in high speed digital circuits [59], [60], [61]. A digital signal that is propagating in an underdriven uniform lossy transmission line exhibits significant inductive effects if the line length l satisfies the following condition [60],

$$\frac{t_r}{2\sqrt{LC}} < l < \frac{2}{R}\sqrt{\frac{L}{C}}, \qquad (2.40)$$

where R, L, and C are the resistance, inductance, and capacitance per line length, respectively, and t_r is the rise time of the signal waveform.

The two inequalities comprising condition (2.40) have an intuitive circuit interpretation. The velocity of the electromagnetic signal propagation along a line is $v_c = \frac{1}{\sqrt{LC}}$. The left inequality of (2.40) therefore transforms into

$$t_r < \frac{2l}{v_c}, \qquad (2.41)$$

i.e., the signal rise time should be smaller than the round trip time of flight. Alternatively stated, the line length l should be a significant fraction of the shortest wavelength of significant signal frequencies λ_r. The spectral content of the signal with rise time t_r rolls off at -20 dB/decade above the frequency $f_r = 1/\pi t_r$. The shortest effective wavelength of the signal is therefore $\lambda_r = v_c/f_r = \pi v_c t_r$. The condition (2.41) can be rewritten as

$$\frac{l}{\lambda_r} > \frac{1}{2\pi}. \qquad (2.42)$$

The dimensionless ratio of the physical size of a circuit to the signal wavelength, l/λ, is referred to as the *electrical size* in high speed interconnect design [48], [62]. Circuits with an electrical size much smaller than unity are commonly called electrically small (or short), otherwise circuits are called electrically large (or long) [48], [62]. Electrically small circuits belong to the realm of classical circuit analysis and are well described by lumped circuits. Electrically large circuits require distributed circuit models and belong to the domain of high speed interconnect analysis techniques. The left inequality of condition (2.40) therefore restricts significant inductive effects to electrically long lines.

With the notion that the damping factor of the transmission line is $\zeta = \frac{R_0}{2}\sqrt{\frac{C_0}{L_0}}$, where $R_0 = Rl$, $L_0 = Ll$, and $C_0 = Cl$ are the total resistance, inductance, and capacitance of the line, respectively, the right inequality in condition (2.40) transforms into

$$\zeta < 1, \tag{2.43}$$

constraining the damping factor to be sufficiently small. Given a line with a specific R, L, and C, the inductive behavior is confined to a certain range of line length, as shown in Fig. 2.15. The upper bound of this range is determined by the damping factor of the line, while the lower bound is determined by the electrical size of the line.

Alternatively, condition (2.40) can be interpreted as a bound on the overall line inductance $L_0 = Ll$. The signal transmission exhibits inductive characteristics if the overall line inductance satisfies both of the following conditions,

$$L_0 > \frac{t_r^2}{4C_0} \tag{2.44}$$

and

$$L_0 > \frac{1}{4}R_0^2 C_0 . \tag{2.45}$$

Conditions (2.44) and (2.45) thereby quantify the term "inductance sufficiently large to cause inductive behavior" as applied to transmission lines. The design space for a line inductance with the region of inductive behavior, as determined by (2.44) and (2.45), is illustrated in Fig. 2.16.

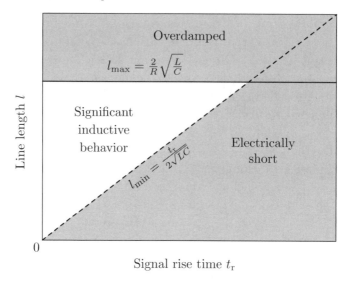

Fig. 2.15. The range of transmission line length where the signal propagation exhibits significant inductive behavior. The area of inductive behavior (the unshaded area) is bounded by the conditions of large electrical size (the dashed line) and insufficient damping (the solid line), as determined by (2.40). In the region where either of these conditions is not satisfied (the shaded area), the inductive effects are insignificant.

2.4 Inductive properties of on-chip interconnect

The distinctive feature of on-chip interconnect structures is the small cross-sectional dimensions and, consequently, a relatively high line resistance. For example, the resistance of a copper line with a $1\,\mu\text{m} \times 3\,\mu\text{m}$ cross section is approximately $7\,\Omega/\text{mm}$. The loop inductance of on-chip lines is typically between $0.4\,\text{nH/mm}$ and $1\,\text{nH/mm}$. At frequencies lower than several gigahertz, the magnetic characteristics do not significantly affect the behavior of on-chip circuits.

As the switching speed of digital integrated circuits increases with technology scaling, the magnetic properties have become essential for accurately describing on-chip circuit operation. The density and complexity of the on-chip interconnect structures preclude exploiting commonly assumed circuit simplifications, rendering the accurate analysis of inductive properties particularly challenging. Large integrated circuits contain many tens of millions of interconnect segments while the segment spacing is typically either equal to or less than the cross-sectional dimensions. Accurate treatment of magnetic coupling

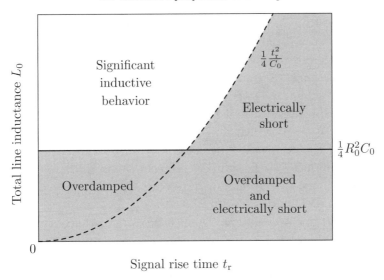

Fig. 2.16. The design space characterizing the overall transmission line inductance is divided into a region of inductive behavior and a region where inductive effects are insignificant. The region of inductive behavior (the unshaded area) is bounded by the conditions of large electrical size (the dashed line) and low damping (the solid line), as determined by (2.44) and (2.45). In the region where either of these conditions is not satisfied (the shaded area), the inductive effects are insignificant.

in these conditions is especially important. Neither the loop nor the partial inductance formulation can be directly applied to an entire circuit as the size of the resulting inductance matrices makes the process of circuit analysis computationally infeasible. Simplifying the inductive properties of a circuit is also difficult. Simply omitting relatively small partial inductance terms can significantly change the circuit behavior, possibly causing instability in an originally passive circuit. Techniques to simplify the magnetic characteristics so as to allow an accurate analysis of separate circuit parts is currently an area of focused research [63], [64], [65], [66].

The problem is further complicated by the significant variation of inductance with frequency. As discussed in Section 2.2, the inductance variation can be described in terms of the skin effect, proximity effect, and multi-path current redistribution. For a line with a rectangular cross section, the internal inductance is similar to the internal inductance of a round line, *i.e.*, 0.05 nH/mm, decreasing with the aspect ratio of the cross section. Over the frequency range of interest, up to 100 GHz, the skin effect reduces the internal inductance by only a small fraction. The reduction in the internal inductance due to the skin effect is, therefore, relatively insignificant, as compared to the overall inductance. Due to the relatively high resistance of on-chip interconnect, the proximity effect is significant only in immediately adjacent wide lines that carry high frequency current. Where several parallel lines are available for current flow, redistribution of the current among the lines is typically the primary cause in integrated circuits of the decrease in inductance with frequency. The proximity effect and multi-path current redistribution are therefore two mechanisms that can produce a significant change in the on-chip interconnect inductance with signal frequency.

Note that the statement "sufficiently high inductance causes inductive behavior" does not necessarily mean "any change in the interconnect physical structure that increases the line inductance increases the inductive behavior of the line." In fact, the opposite is often the case in an integrated circuit environment, where varying a single physical interconnect characteristic typically affects many electrical characteristics. The relationship between the physical structure of interconnect and the inductive behavior of a circuit is highly complex.

Consider a 3 mm long copper line with a $1\,\mu$m $\times$ $1\,\mu$m cross section. The resistance, inductance, and capacitance per length of the current loop (including both the line itself and the current return path) are $R = 25\,\Omega/$mm, $L = 0.8\,$nH/mm, and $C = 100\,$fF/mm, respectively. The velocity of the electromagnetic wave propagation along the line is 0.11 mm/ps. This velocity is somewhat smaller than the speed of light, 0.15 mm/ps, in the media with an assumed dielectric constant of 4 and is due to the additional capacitive load of the orthogonal lines in the lower layer. For a signal with a 30 ps rise time, the line is electrically long. The line damping factor, however, is $\zeta = \frac{Rl}{2}\sqrt{\frac{C}{L}} = 1.33 > 1$. The line is therefore overdamped and, according to the metrics expressed by (2.44) and (2.45), does not exhibit inductive behavior, as shown in Fig. 2.17(a).

Assume now that the line width is changed to $4\,\mu$m and the resistance, inductance, and capacitance of the line change to $R = 10\,\Omega/$mm, $L = 0.65\,$nH/mm, and $C = 220\,$fF/mm, respectively. The decrease in the loop resistance and inductance are primarily due to the smaller resistance and partial self inductance of the line. The increase in the line capacitance is primarily due to the greater parallel plate capacitance between the signal line and the perpendicular lines in the lower

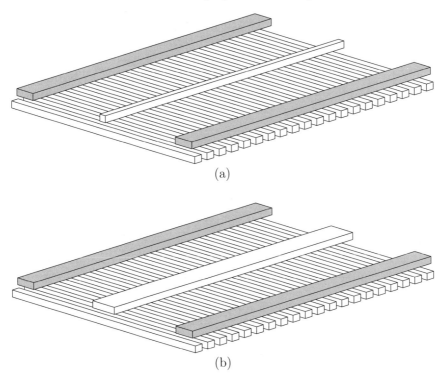

Fig. 2.17. A signal line in an integrated circuit environment. The power and ground lines (shaded gray) parallel to the signal line serve as a current return path. The lines in the lower metal layer increase the capacitive load of the line. The inductive behavior of a wide line, as shown in (b), is more significant as compared to a narrow line, as shown in (a).

layer. This capacitive load becomes more significant, as compared to the capacitance between the line and the return path, further slowing the velocity of the electromagnetic wave propagation to 0.084 mm/ps. For the same signal with a 30 ps transition time, the signal line becomes underdamped, $\zeta = 0.87 < 1$, and exhibits significant inductive behavior, as shown in Fig. 2.17(b).

The *inductive behavior* has become significant while the *absolute inductance* has *decreased* from $3\,\text{mm} \times 0.8\frac{\text{nH}}{\text{mm}} = 2.40\,\text{nH}$ to $1.95\,\text{nH}$. The reason for this seeming contradiction is that the inductance is a weak function of the cross-sectional dimensions, as compared to the resistance and capacitance. In integrated circuits, the signal lines that exhibit inductive behavior are the lowest resistance lines, *i.e.*, the wide lines in the thick upper metalization layers. These lines typically have a lower absolute inductance than other signal lines. It would therefore be misleading to state that the inductive behavior of on-chip interconnect has become important due to the increased inductance. This trend is due to the shorter signal transition times and longer line lengths, while maintaining approximately constant the resistive properties of the upper metal layers.

2.5 Summary

The preceding discussion of the inductive characteristics of electric circuits and different ways to represent these characteristics can be summarized as follows.

- The thin filament approximation is valid only for determining the mutual inductance of relatively thin conductors

- The partial inductance formulation is better suited to describe the inductive properties of circuits with branch points

- The partial inductance is a mathematical construct, not a physically observable property, and should only be used as part of a complete description of the circuit inductance

- The circuit inductance varies with frequency due to current redistribution within the circuit conductors. The current redistribution mechanisms can be classified as the skin effect, proximity effect, and multi-path current redistribution

- Signal propagation along a transmission line exhibits inductive behavior if the line is both electrically long and underdamped

- Characterizing on-chip inductance in both an efficient and accurate manner is difficult due to the density and complexity of on-chip interconnect structures

- The relationship between the physical structure of on-chip interconnect and the inductive behavior of a circuit is complex, as many electrical properties can be affected by changing a specific physical characteristic of an interconnect line

3

Properties of On-Chip Inductive Current Loops

The inductive characteristics of electric circuits are described in Chapter 2. Both accurate and efficient characterization of on-chip interconnect inductance is difficult, as discussed in the previous chapter. The objective of this chapter is to demonstrate that the task of inductance characterization, however, is considerably facilitated in certain interconnect structures. These results will be used in Chapter 9 to provide insight into the inductive properties of power distribution grids.

The chapter is organized as follows. A brief overview of the problem is presented in Section 3.1. The dependence of inductance on line length is discussed in Section 3.2. The inductive coupling of parallel conductors is described in Section 3.3. Application of these results to the circuit analysis process is discussed in Section 3.4. The conclusions are summarized in Section 3.5.

3.1 Introduction

IC performance has become increasingly constrained by on-chip interconnect impedances. Determining the electrical characteristics of the interconnect at early stages of the design process has therefore become necessary. The layout process is driven now by interconnect performance where the electrical characteristics of the interconnect are initially estimated and later refined. Impedance estimation is repeated throughout the circuit design and layout process and should, therefore, be computationally efficient.

The inductance of on-chip interconnect has become an important issue due to the increasing switching speeds of digital integrated circuits [61]. The inductive properties must, therefore, be effectively incorporated at all levels of the design, extraction, and simulation phases in the development of high speed ICs.

Operating an inductance extractor within a layout generator loop is computationally expensive. The efficiency of inductance estimation can be improved by extrapolating the inductive properties of a circuit from the properties of a precharacterized set of structures. Extrapolating inductance, however, is not straightforward as the inductance, in general, varies nonlinearly with the circuit dimensions. The partial inductance of a straight line, for example, is a nonlinear function of the line length. The inductive coupling of two lines decreases slowly with increasing line-to-line separation. The partial inductance representation of a circuit consists of strongly coupled elements with a nonlinear dependence on the geometric dimensions. The inductive properties of a circuit, therefore, do not, in general, vary linearly with the circuit geometric dimensions.

The objective of this chapter is to explore the dependence of inductance on the circuit dimensions, to provide insight from a circuit analysis perspective, and to determine the specific conditions under which the inductance properties scale linearly with the circuit dimensions with the objective of enhancing the layout extraction process. A comparison of the properties of the partial inductance with the properties of the loop inductance is shown to be effective for this purpose. The results of this investigation will be exploited in Chapter 9 to analyze the inductive properties of the power distribution grids.

The analysis is analogous to the procedure described in Section 9.3. The inductance extraction program FastHenry [67] is used to explore the inductive properties of interconnect structures. A conductivity of $58 \ \mathrm{S}/\mu\mathrm{m} \simeq (1.72 \ \mu\Omega \cdot \mathrm{cm})^{-1}$ is assumed for the interconnect material.

3.2 Dependence of inductance on line length

A non-intuitive property of the partial inductance is the characteristic that the partial inductance of a line is a nonlinear function of the line length. The partial inductance of a straight line is a superlinear function of length. The partial self inductance of a rectangular line at low frequency can be described by [41]

$$L_{\mathrm{part}} = 0.2l \left(\ln \frac{2l}{T+W} + \frac{1}{2} - \ln \gamma \right) \ \mu\mathrm{H}, \qquad (3.1)$$

where T and W are the thickness and width of the line, and l is the length of the line in meters. The $\ln\gamma$ term is a function of only the T/W ratio, is small as compared to the other terms (varying from 0 to 0.0025), and has a negligible effect on the dependence of the inductance with length. This expression is an approximation, valid for $l \gg T + W$; a precise formula for round conductors can be found in [43].

From a circuit analysis point of view, this nonlinearity is caused by the significant inductive coupling among the different segments of the same line. Consider a straight line; the corresponding circuit representation of the self inductance of the line is a single inductor, as shown in Fig. 3.1(a). Consider also the same line as two shorter lines connected in series. The corresponding circuit representation of the inductance of these two lines is two coupled inductors connected in series, as shown in Fig. 3.1(b).

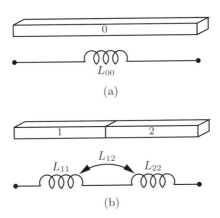

Fig. 3.1. Two representations of a straight line inductance. (a) The line can be considered as a single element, with a corresponding circuit representation as a single inductor. (b) The same line can also be considered as two lines connected in series with a corresponding circuit representation as two coupled inductors connected in series.

The inductance of this circuit is

$$L_{1+2} = L_{11} + L_{22} + 2L_{12}. \qquad (3.2)$$

If the partial inductance is a linear function of length, the inductance of the circuit is the sum of the inductance of its elements, *i.e.*,

$$L_{linear} = L_{11} + L_{22}. \qquad (3.3)$$

The difference between the nonlinear dependence [see (3.2)] and the linear dependence [see (3.3)] is the presence of the cross coupling term $2L_{12}$. This term increases the inductance beyond the linear value, *i.e.*, the sum $L_{11} + L_{22}$. The cross coupling term increases with line length and does not become small as compared to the self inductance of the lines L_{11} as the line length increases.

However, the loop inductance of a complete current loop formed by two parallel straight rectangular conductors, shown in Fig. 3.2, is given by [41]

$$L_{\text{loop}} = 0.4l \left(\ln \frac{P}{H + W} + \frac{3}{2} - \ln \gamma + \ln k \right) \mu H, \qquad (3.4)$$

where P is the distance between the center of the lines (the pitch) and $\ln k$ is a tabulated function of the H/W ratio. This expression is accurate for long lines (*i.e.*, for $l \gg P$). The expression is a linear function of the line length l.

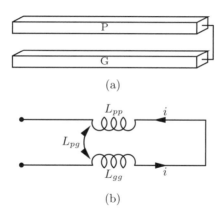

(a)

(b)

Fig. 3.2. A complete current loop formed by two straight parallel lines; (a) a physical structure and (b) a circuit diagram of the partial inductance.

To compare the dependence of inductance on length for both a single line and a complete loop and to assess the accuracy of the long line approximation assumed in (3.4), the inductance extractor FastHenry is used to evaluate the partial inductance of a single line and the loop inductance of two identical parallel lines forming forward and return current paths. The cross section of the lines is $1 \, \mu m \times 1 \, \mu m$. The spacing between the lines in the complete loop is $4 \, \mu m$. The length is varied from $10 \, \mu m$ to $10 \, mm$.

The linearity of a function is difficult to visualize when the function argument spans three orders of magnitude (particularly when plotted in a semi-logarithmic coordinate system). The inductance per length, alternatively, is a convenient measure of the linearity of the inductance. The inductance per length (the inductance of the structure divided by the length of the structure) is independent of the length if the inductance is linear with length, and varies with length otherwise.

The inductance per length is therefore determined for a single line and a two-line loop of various length. The results are shown in Fig. 3.3. The linearity of the data rather than the absolute magnitude of the

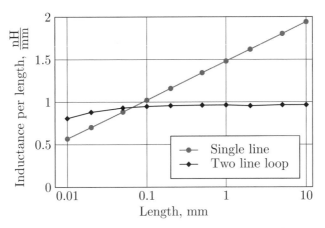

Fig. 3.3. Inductance per length versus line length for a single line and for a loop formed by two identical parallel lines. The line cross section is $1\,\mu$m $\times\ 1\,\mu$m.

data is of primary interest here. To facilitate the assessment of the inductance per length in relative terms, the data shown in Fig. 3.3 are recalculated as a per cent deviation from the reference value and are shown in Fig. 3.4. The inductance per length at a length of 1 mm is chosen as a reference. Thus, the inductance per length versus line length is plotted in Fig. 3.4 as a per cent deviation from the magnitude of the inductance per length at a line length of 1 mm. As shown in Fig. 3.4, the inductance per length of a single line changes significantly with length. The inductance per length of a complete loop is practically constant for a wide range of lengths [68] (varying approximately 4% over the range from 50 μm to 10,000 μm). The inductance of a complete loop can, therefore, be considered linear when the length of a loop exceeds the loop width by approximately a factor of ten. Note that in the case of a simple structure, such as the two line loop shown in Fig. 3.2, it is not necessary to use FastHenry to produce the data shown in Fig. 3.4. The formulæ for inductance in [41] can be applied to derive the data shown in Fig. 3.4 with sufficient accuracy.

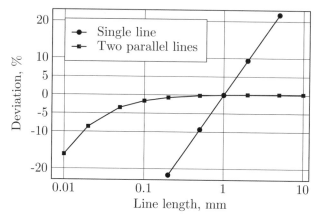

Fig. 3.4. Inductance per length versus line length in terms of the per cent difference from the inductance per length at a 1 mm length. The data is the same as shown in Fig. 3.3 but normalized to the value of inductance per length at 1 mm (0% deviation). The (loop) inductance per length of a two line loop is virtually constant over a wide range of length, while the (partial) inductance per length of a single line varies linearly with length.

To gain further insight into why the inductance of a complete loop increases linearly with length while the inductance of a single line increases nonlinearly, consider a complete loop as two loop segments connected in series, as shown in Fig. 3.5. The inductance of the left side loop segment (formed by line segments one and two) is

$$L_{1+2} = L_{11} + L_{22} - 2L_{12}, \qquad (3.5)$$

while the inductance of the right side segment of the loop (formed by line segments three and four) is, analogously,

$$L_{3+4} = L_{33} + L_{44} - 2L_{34}. \qquad (3.6)$$

The inductance of the entire loop is

$$L_{\text{loop}} = \sum_{i,j=1}^{4} L_{ij}$$
$$= L_{11} + L_{22} + L_{33} + L_{44} - 2L_{12} - 2L_{34}$$
$$+2L_{13} - 2L_{14} + 2L_{24} - 2L_{23}. \tag{3.7}$$

Considering that $L_{13} = L_{24}$ and $L_{14} = L_{23}$ due to the symmetry of the structure and substituting (3.5) and (3.6) into (3.7), this expression reduces to

$$L_{\text{loop}} = L_{1+2} + L_{3+4} + 2M, \tag{3.8}$$

where $M = L_{13} - L_{14} + L_{24} - L_{23} = 2(L_{13} - L_{14})$ is the coupling between the two loop segments. This expression for a complete loop is completely analogous to (3.2) for a single line. Similar to (3.2), the nonlinear term within the parenthesis augments the inductance beyond a linear value of $L_{1+2}+L_{3+4}$. An important difference, however, is that the nonlinear part is a difference of two terms close in value, because $L_{13} \approx L_{14}$ and $L_{24} \approx L_{23}$.

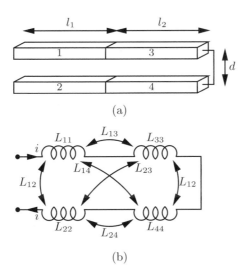

(a)

(b)

Fig. 3.5. A complete current loop formed by two straight parallel lines consists of two loop segments in series; (a) a physical structure and (b) a circuit diagram of the partial inductance.

This behavior can be intuitively explained as follows. Segments three and four are physically (and inductively) much closer to each other than these segments are to segment one. The effective distance (the distance to move one segment so as to completely overlap the other segment) between line segment three and line segment four is small as compared to the effective distance between segment three and segment one. The inductive coupling is a smooth function of distance. The magnitude of the inductive coupling of segment one to segment three is, therefore, quite close to the magnitude of the coupling of segment one to segment four (but of opposite sign due to the opposite direction of the current flow).

A mathematical treatment of this phenomenon confirms this intuitive insight. The mutual partial inductance of two parallel line segments, for example, segments one and four shown in Fig. 3.5(b), is

$$L_M = 0.1 \left(l_1 \ln \frac{l_1 + l_2}{l_1} + l_2 \ln \frac{l_1 + l_2}{l_2} - d \right) \mu\text{H}, \qquad (3.9)$$

where l_1 and l_2 are the segment lengths, and d is the distance between the center axes of the segments, as shown in Fig. 3.5(a). Consider, for example, the case where the line segments are of equal length, $l_1 = l_2 = l/2$. The mutual inductance as a function of the axis distance d is

$$L_M(d) = (l \ln 2 - d) \ \mu\text{H}, \qquad (3.10)$$

where $L_M(d)$ is a weak function of d if $d \ll l$. The mutual inductance of two segments forming a straight line, i.e., L_{12} in Fig. 3.1 and L_{13} and L_{24} in Fig. 3.2, is $L_M(0)$. The mutual inductance L_{14} and L_{23} is $L_M(d)$. The effective inductive coupling of two loop segments, therefore, simplifies to the following expression,

$$4(L_{13} - L_{14}) = 4(L_M(0) - L_M(d)) = 0.4d \ \mu\text{H}. \qquad (3.11)$$

Note that this coupling is much smaller than the coupling of two straight segments, $L_M(0)$, and is independent of the loop segment length l. As the loop length l exceeds several loop widths d (as $l \gg d$), the coupling becomes negligible as compared to the self inductance of the loop segments. As compared to the coupling between two single line segments, the effective coupling is reduced by a factor of

$$\frac{M_{\text{line}}}{M_{\text{loop}}} = \frac{L_M(0)}{2(L_M(0) - L_M(d))} = \frac{\ln 2}{2} \frac{l}{d}. \qquad (3.12)$$

In general, it can be stated that at distances much larger than the effective separation between the forward and return currents, the inductive coupling is dramatically reduced as the coupling of the forward current and return current is mutually cancelled. Hence, the inductance of a long loop ($l \gg d$) depends linearly upon the length of the loop.

3.3 Inductive coupling between two parallel loop segments

As shown in the previous section, the relative proximity of the forward and return current paths is the reason for the cancellation of the inductive coupling between two loop segments connected in series (*i.e.*, different sections of the same current loop).

The same argument can be applied to show that the effective inductive coupling between two sections of parallel current loops is also reduced [68]. As in the case of the collinear loop segments considered above, this behavior is due to cancellation of the coupling if the distance between the forward and return current paths (the loop width) is much smaller than the distance between the parallel loop segments. The physical structure and circuit diagram of two parallel loop segments are shown in Fig. 3.6.

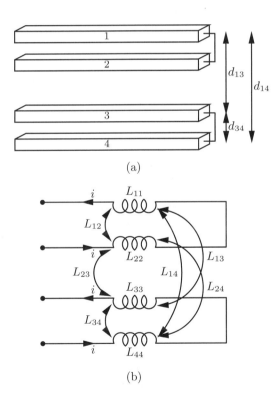

(a)

(b)

Fig. 3.6. Two parallel loop segments where each loop segment is formed by two straight parallel lines; (a) a physical structure and (b) a circuit diagram of the partial inductance.

The mutual loop inductance of the two loop segments is

$$M_{\text{loop}} = L_{13} - L_{14} + L_{24} - L_{23}. \qquad (3.13)$$

The mutual inductance between two parallel straight lines of equal length is [43]

$$M_{\text{loop}} = 0.2l \left(\ln \frac{2l}{d} - 1 + \frac{d}{l} - \ln \gamma + \ln k \right) \mu\text{H}, \qquad (3.14)$$

where l is the line length, and d is the distance between the line centers. This expression is an approximation for the case where $l \gg d$. The mutual inductance of two straight lines is a weak function of the distance between the lines. Therefore, if the distance between lines one and three d_{13} is much greater than the distance between lines three and four d_{34}, such that $d_{13} \approx d_{14}$, then $L_{13} \approx L_{14}$. Analogously, $L_{23} \approx L_{24}$. The coupling of the loops M_{loop} is a difference of two similar values, which is small as compared to the self inductance of the loop segments. The loop segments can be considered weakly coupled in this case. This effective decoupling means that the reluctance of the loop segments wired in parallel is the sum of the reluctances of the individual loop segments. Alternatively, the inductance of the parallel connection is the inductance of two parallel uncoupled inductors, $L_{\|} = \frac{L_{11}L_{22}}{L_{11}+L_{22}}$, similar to the resistance of parallel elements. The circuit inductance, therefore, varies linearly with the circuit "width": as more identical circuit elements (*i.e.*, loop segments) are connected in parallel, the inductance of the circuit decreases inversely linearly with the number of parallel elements. This linear property of inductance is demonstrated in [69], [70] with specific application to high performance power grids.

3.4 Application to circuit analysis

Although rectangular lines with a unity height to width aspect ratio are used in the case studies described above, the conclusions are quite general and hold for different wire shapes and aspect ratios. At low frequencies, where the current density is uniform throughout a wire cross section, the self inductance of a wire is determined by the *geometric mean distance* of the wire cross section and the mutual inductance of two wires is determined by the *geometric mean distance* between two cross sections, as first described by Maxwell [71]. Rosa and Grover systematized and tabulated the geometric mean distance data for several practically important cases [41], [45]. Both the self and mutual inductance of a conductor is moderately dependent on the perimeter length of the inductor cross section and is virtually independent of the conductor cross-sectional shape. For example, the self inductance of rectangular conductors with aspect ratios of 1 and 10, but with the same perimeter length-to-line length ratio of 40, differ by only 0.012%, according to (3.1).

Skin effects reduce the current density as well as the magnetic field within the core of the conductor. This effect slightly reduces the self inductance of a wire and has virtually no effect on the mutual inductance. Proximity effects in two neighboring wires carrying current in the opposite directions (as in a loop formed by two parallel wires) can only reduce the effective distance between the two currents, making the loop effectively "longer." Therefore, a uniform current density distribution is a conservative assumption regarding the linearity of inductance with loop length.

The particular on-chip current return path is rarely known before analysis of the circuit. Nevertheless, if an approximate but conservative estimate of the distance d between the signal wire and the current return path can be made which satisfies the long loop condition $l \gg d$, the inductance can be considered to vary linearly with the length of the structure. Similar to the resistance and capacitance, the inductance of such a structure is effectively a local characteristic, independent of the length of the structure. The analysis of a large interconnect structure is therefore not necessary to obtain the local inductive characteristics, greatly simplifying the circuit analysis process. This property is demonstrated in Chapter 9 on an example of regularly structured power distribution grids.

3.5 Summary

The variation of the partial and loop inductance with line length is analyzed in this chapter. The primary conclusions are summarized as follows.

- The nonlinear variation of inductance with length is due to inductive coupling between circuit segments

- In long loops, the effective coupling between loop segments is small as compared to the self inductance of the segments due to the mutual cancellation of the forward current coupling with the return current coupling

- The inductance of long loops increases virtually linearly with length

- In a similar manner, the effective inductive coupling between two parallel current loops is greatly reduced due to coupling cancellation as compared to the coupling between line segments of the same length

- As a general rule, the inductance of circuits scales *linearly* with the circuit dimensions where the distance between the forward and return currents is much smaller than the dimensions of the circuit

- The linear variation of inductance with the circuit dimensions can be exploited to simplify the inductance extraction process and the related circuit analysis of on-chip interconnect

4

Electromigration

The power current requirements of integrated circuits are rapidly rising, as discussed in Chapter 1. The current density in on-chip power and ground lines can reach several hundred thousands of amperes per square centimeter. At these current densities, electromigration becomes significant. Electromigration is the transport of metal atoms under the force of an electron flux. The depletion and accumulation of the metal material resulting from the atomic flow can lead to the formation of extrusions (or hillocks) and voids in the metal structures. The hillocks and voids can lead to short circuit and open circuit faults [72], respectively, as shown in Fig. 4.1, degrading the reliability of an integrated circuit.

The significance of electromigration has been established early in the development of integrated circuits [73], [74]. Electromigration should be considered in the design process of an integrated circuit to ensure reliable operation over the target lifetime. Electromigration reliability and related design implications are the subject of this chapter. A more detailed discussion of the topic of electromigration can be found in the literature [75], [76].

The chapter is organized as follows. The physical mechanism of electromigration is described in Section 4.1. The role of mechanical stress in electromigration reliability is discussed in Section 4.2. The steady state conditions of electromigration damage in interconnect lines are established in Section 4.3. The dependence of electromigration reliability on the dimensions of the interconnect line is discussed in Section 4.4. The statistical distribution of the electromigration lifetime is presented in Section 4.5. Electromigration reliability under an AC current is discussed in Section 4.6. Electromigration reliability in novel interconnect

(a)   (b)

Fig. 4.1. Electromigration induced circuit faults. (a) Line extrusions formed due to metal accumulation can short circuit the adjacent line. (b) The voids formed due to metal depletion increase the line resistance and can lead to an open circuit.

technologies using copper metalization and low-k dielectrics is discussed in Section 4.7. Certain approaches to designing for electromigration reliability are briefly reviewed in Section 4.8. The chapter concludes with a summary.

4.1 Physical mechanism of electromigration

Electromigration is a microscopic mass transport of metal ions through diffusion under an electrical driving force. An atomic flux under a driving force F is

$$J_{\mathrm{a}} = C_a \mu F, \tag{4.1}$$

where C_a is the atomic concentration and μ is the mobility of the atoms. From the Einstein relationship, the mobility can be expressed in terms of the atomic diffusivity D_{a} and the thermal energy kT,

$$\mu = \frac{D_{\mathrm{a}}}{kT}. \tag{4.2}$$

Two forces act on metal ions: an electric field force and an electron wind force. The electric field force is proportional to the electric field E and acts in the direction of the field. The electric field also accelerates the conduction electrons in the direction opposite to the electric field. The electrons transfer the momentum to the metal ions in the course of scattering, exerting the force in the direction opposite to the field E. This force is commonly referred to as the electron wind force. The electron wind force equals the force exerted by the electric field on the conduction electrons, which is proportional to the electric field intensity E.

In the metals of interest, such as aluminum and copper, the electron wind force dominates and the net force acts in the direction opposite to the electric current. The resulting atomic flux is therefore in the direction opposite to the electric current j, as shown in Fig. 4.2. The net

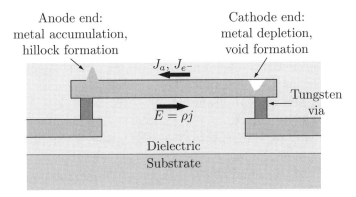

Fig. 4.2. Electromigration mass transport in an interconnect line. An electron flux J_{e^-} flowing in the opposite direction to the electric field $E = \rho j$ induces an atomic flow J_a in the direction of the electron flow. Diffusion barriers, such as tungsten vias, create an atomic flux divergence, leading to electromigration damage in the form of voids and hillocks.

force acting on the metal atoms is proportional to the electric field and is typically expressed as $-Z^*eE$ or $-Z^*e\rho j$, where Z^*e is the effective charge of the metal ions, j is the current density, and ρ is the resistivity of the metal. The electromigration atomic flux is therefore

$$J_a = -C_a \frac{D_a Z^* e\rho j}{kT}, \tag{4.3}$$

where the minus sign indicates the direction opposite to the field E. The diffusivity D_a has an Arrhenius dependence on temperature, $D_a = D_0 \exp(-Q/kT)$, where Q is the activation energy of diffusion. Substituting this relationship into (4.3), the atomic flux becomes

$$J_a = -C_a \frac{Z^* e\rho j}{kT} D_0 \exp\left(-\frac{Q}{kT}\right). \tag{4.4}$$

The material properties C_a, ρ, and Z^*e are difficult to change. The diffusion coefficient D_0 and the activation energy Q, although also material specific, vary significantly depending on the processing conditions and the resulting microstructure of the metal film. There are several paths for electromigration in interconnect lines, such as diffusion through the lattice, along the grain boundary, and along the metal surface. Each path is characterized by the individual diffusion coefficient and activation energy. The atomic flux depends exponentially on the activation energy Q and is therefore particularly sensitive to variations in Q. The diffusion path with the lowest activation energy dominates the overall atomic flux.

The on-chip metal films are polycrystalline, consisting of individual crystals of various size. The individual crystals are commonly referred to as grains. The activation energy Q is relatively low for diffusion along the grain boundary, as low as 0.5 eV for aluminum, while the activation energy for diffusion through the lattice is the highest, up to 1.4 eV in aluminum. Diffusion along the grain boundary is the primary path of electromigration in relatively wide aluminum interconnect lines, as discussed in Section 4.4.

An atomic flux does not cause damage to on-chip metal structures where the influx of the atoms is balanced by the atom outflow. The depletion and accumulation of metal (and the associated damage) develop at those sites where the influx and outflux are not equal, $i.e.$, the flux divergence is not zero, $\nabla \cdot J_a \neq 0$. Flux divergence can be caused by several factors, including an interface between materials with dissimilar diffusivity and resistivity, inhomogeneity in microstructure, and a temperature gradient.

Consider, for example, the current flow in an aluminum metal line segment connected to two tungsten vias at the ends, as shown in Fig. 4.2. At the current densities of interest, tungsten is not susceptible to electromigration and can be considered as an ideal barrier for the diffusion of aluminum atoms. The tungsten-aluminum interface at the anode line end, where the electric current enters the line ($i.e.$, the electron flux exits the line), prevents the outflow of the aluminum atoms from the line. The incoming atomic flux causes an accumulation of aluminum atoms. At the cathode end of the line, the tungsten-aluminum interface blocks the atomic flux from entering the line. The electron flux enters the line at this end and carries away aluminum atoms, leading to metal depletion and, potentially, formation of a void.

In a similar manner, an inhomogeneity in the microstructure can cause flux divergence. As has been discussed, grain boundaries have significantly lower activation energy, facilitating atomic flow. The electromigration atomic flux is enhanced at the locations with small grains, where the grain boundaries provide numerous paths of facilitated diffusion. At those sites where the grain size changes abruptly, the high atomic flux in the region of the smaller grain size is mismatched with the relatively low atomic flux in the larger grain region. The atomic flux divergence leads to a material depletion or accumulation, depending upon the direction of current flow. These sites are particularly susceptible to electromigration damage [77].

4.2 Electromigration-induced mechanical stress

The depletion and accumulation of material at the sites of atomic flux divergence induce a mechanical stress gradient in the metal structures. At the sites of metal accumulation the stress is compressive, while at the sites of metal depletion the stress is tensile. The resulting stress gradient in turn induces a flux of metal atoms J_a^{stress},

$$J_a^{\text{stress}} = C_a \frac{D_a}{kT} \Omega \frac{\partial \sigma}{\partial l}, \qquad (4.5)$$

where σ is the mechanical stress, assumed positive in tension, Ω is the atomic volume, and l is the line length. The atomic flux due to the stress gradient is opposite in direction to the electromigration atomic flux, counteracting the electromigration mass transport. This flux is therefore often referred to as an atomic backflow. The distribution of the mechanical stress and the net atomic flux are therefore interrelated. Accurate modeling of the mechanical stress is essential in predicting electromigration reliability. Mechanical stress can also have components unrelated to electromigration atomic flux, such as a difference in the thermal expansion rates of the materials.

    The on-chip metal structures are encapsulated in a dielectric material, typically silicon dioxide. The stiffness of the dielectric material significantly affects the electromigration reliability. A rigid dielectric, such as silicon dioxide, limits the variation in metal volume at the sites of the metal depletion and accumulation, resulting in greater electromigration-induced mechanical stress as compared to a metal line in a less rigid environment. The greater mechanical stress induces a greater atomic flux in the direction opposite to the electromigration atomic flux, limiting the net atomic flux and the related structural damage. Rigid encapsulation therefore significantly improves the electromigration reliability of the metal interconnect.

A rigid dielectric can however lead to structural defects due to a mismatch in the thermal expansion rate of the materials. As the silicon wafer is cooled from the temperature of silicon dioxide deposition, the rigid and well adhering silicon dioxide prevents the aluminum lines from contracting according to the thermal expansion rate of aluminum. The resulting tensile stress in the aluminum lines can cause void formation [78]. This effect is exacerbated in narrow lines.

4.3 Steady state limit of electromigration damage

Under certain conditions, the stress induced atomic flux can fully compensate the atomic flux due to electromigration, preventing further damage. In this case, the atomic concentration along a metal line is stationary, $\frac{\partial C_a}{\partial t} = 0$. The net atomic flow is related to the atomic concentration by the continuity equation,

$$\frac{\partial C_a}{\partial t} = -\nabla J_a = \frac{\partial J_a}{\partial l} = 0. \tag{4.6}$$

The atomic flow is uniform along the line length l, $J_a(l) = \text{const}$. In a line segment confined by diffusion barriers, the steady state atomic flux is zero. Under this condition, the diffusion due to the electrical driving force is compensated by the diffusion due to the mechanical driving force. The net atomic flux J_a is the sum of the electromigration and stress induced fluxes,

$$J_a = J_a^{\text{em}} + J_a^{\text{stress}} = C_a \frac{D_a}{kT} \left(\Omega \frac{\partial \sigma}{\partial l} - Z^* e\rho j \right). \tag{4.7}$$

The steady state condition is established where

$$\Omega \frac{\partial \sigma}{\partial l} = Z^* e\rho j. \tag{4.8}$$

High mechanical stress gradients are required to compensate the electromigration atomic flow at high current densities. The magnitude of the stress gradient depends upon the formation of voids and hillocks.

Consider a line segment of length l_0 between two sites of flux divergence, such as tungsten vias or severe microstructural irregularities. The compressive stress at the anode end should reach a certain yield stress σ_y to develop the extrusion damage [79]. Assuming a near zero stress at the cathode end of the segment, the damage critical stress gradient is $\left(\frac{\partial \sigma}{\partial l}\right)_{\text{max}} = \sigma_y/l_0$. Substituting this expression for the stress gradient into (4.8) yields the maximum current density,

$$j_{\text{max}}^{\text{hillock}} = \frac{\Omega \sigma_y}{Z^* e \rho} \frac{1}{l_0}. \tag{4.9}$$

If the current density is lower than the limit determined by (4.9), the steady state condition is reached before formation of the hillocks at the anode, and the interconnect line is highly resistive to electromigration damage. Resistance to electromigration damage below a certain current threshold was first experimentally observed by Blech [80]

Void formation also causes mechanical stress that counteracts electromigration atomic flow. As the stress becomes sufficiently high to fully compensate the electromigration flow, the void stops growing and remains at the steady state size [81]. The steady state void size is well described by [82], [83]

$$\Delta l = \frac{Z^* e \rho}{2B\Omega} j l_0^2, \tag{4.10}$$

where B is the elastic modulus relating the line strain to the line stress. Equation (4.10) can be obtained from (4.8) by assuming that a void of size Δl induces a line stress $2B\frac{\Delta l}{l_0}$.

If the steady state void does not lead to a circuit failure, the electromigration lifetime of the line is practically unlimited. A formation of a void in a line does not necessarily lead to a circuit failure. Metal lines in modern semiconductor processes are covered with a thin refractory metal film, such as Ti, TiN, or TiAl$_3$, in the case of aluminum interconnect. These metal films are highly resistant to electromigration damage, providing an alternative current path in parallel to the metal core of the line. A void spanning the line width will therefore increase the resistance of the line, rather than lead to an open circuit fault. The increase in line resistance is proportional to the void size Δl. An increase in the line resistance from 10% to 20% is typically considered critical, leading to a circuit failure. A critical increase in the resistance occurs when the void size reaches a certain critical value Δl_{crit}. The current density resulting in a void of critical steady state size is determined from (4.10),

$$j_{\text{max}}^{\text{void}} = \frac{2B\Omega}{Z^* e \rho} \frac{\Delta l_{\text{crit}}}{l_0^2}. \tag{4.11}$$

If the current density of the line is lower than $j_{\mathrm{max}}^{\mathrm{void}}$, the void damage saturates at a subcritical size and the lifetime of the line becomes practically unlimited.

The critical current density as determined by (4.9) for hillock formation and (4.11) for void formation increases with shorter line length l_0. For a given current density there exists a certain critical line length l_{crit}. Lines shorter than l_{crit} are highly resistant to electromigration damage. This phenomenon is referred to as the electromigration threshold or Blech effect.

4.4 Dependence of electromigration lifetime on the line dimensions

While the electromigration reliability depends upon many parameters, as expressed by (4.4), most of these parameters cannot be varied due to material properties and manufacturing process characteristics. The parameters that can be flexibly varied at the circuit design phase are the line width, length, and current. Varying the current is often restricted by circuit performance considerations. The dependence of electromigration reliability on the line width and length are discussed in this section.

The dependence of the electromigration lifetime on the width of a line is relatively complex [84], [85], as illustrated in Fig. 4.3. In relatively wide lines, *i.e.*, where the average grain size is much smaller than the line width, the grain boundaries form a continuous diffusion path along the line length, as shown in Fig. 4.4(b). Although the grain boundary diffusion path enhances the electromigration atomic flow due to a higher diffusion coefficient along the grain boundary, the probability of abrupt microstructural inhomogeneity along the line length is small, due to a large number of grains spanning the width of the line. The probability of an atomic flux divergence in these polygranular lines is relatively low and the susceptibility of the line to electromigration damage is moderate.

As the line width approaches the average grain size, the polygranular line structure is likely to be interrupted by grains spanning the entire width of the line, disrupting the boundary diffusion path, as shown in Fig. 4.4(c). Electromigration transport in the spanning grain can occur only through the lattice or along the surface of the line. The diffusivity of these paths is significantly lower than the diffusivity of

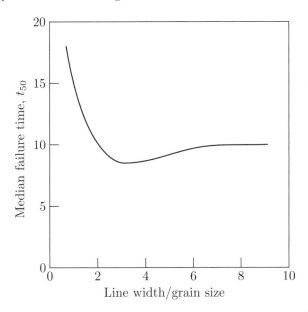

Fig. 4.3. Representative variation of the median electromigration lifetime with line width (based on data obtained from [85]).

the polycrystalline segments. The spanning grains therefore present a barrier to an atomic flux in relatively long polygranular segments of the line. The atomic flux discontinuity at the boundary of the spanning grains renders such lines more susceptible to electromigration damage, shortening the electromigration lifetime.

As the line width is reduced below the average grain size, the spanning grains dominate the line microstructure, forming the so-called "bamboo" pattern, as shown in Fig. 4.4(d). The polygranular segments become sparse and short. The shorter polygranular segments are resistant to electromigration damage, increasing the expected lifetime of the narrow lines.

The electromigration lifetime also varies with line length. Shorter lines have a longer lifetime than longer lines [84]. The longer lines are more likely to contain a significant microstructural discontinuity, such as a spanning grain in wide lines or a long polygranular segment in narrow lines. The longer lines are therefore more susceptible to electromigration damage.

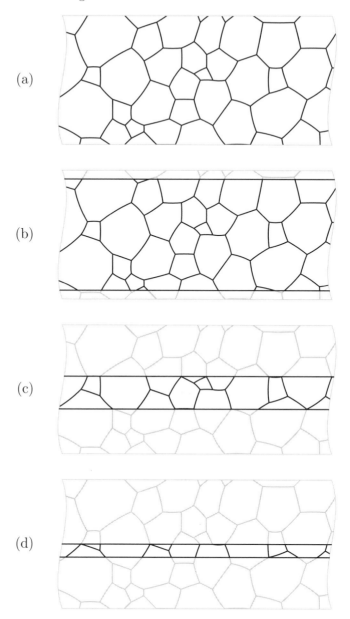

Fig. 4.4. Grain structure of interconnect lines. (a) Grain structure of an unpatterned thin metalization film. (b) The structure of the wide lines is polygranular along the entire line length. (c) In lines with a width close to the average grain size, the polygranular segments are interrupted by the grains spanning the entire line width. (d) In narrow lines, most of the grains span the entire line width, forming a "bamboo" pattern.

4.5 Statistical distribution of electromigration lifetime

Electromigration failure is a statistical process. Identically designed
interconnect structures fail at different times due to variations in the
microstructure. Failure times are typically described by a log-normal
distribution. A variable distribution is log-normal if the distribution
of the logarithm of the variable is normal. The log-normal probability
density function $p(t)$ is

$$p(t) = \frac{1}{\sqrt{2\pi}\sigma t} \exp\left(-\frac{(\ln(t/t_{50}))^2}{2\sigma^2}\right). \tag{4.12}$$

The log-normal distribution is characterized by the median time to
failure t_{50} and the shape factor σ. The probability density function for
several values of σ are shown in Fig. 4.5.

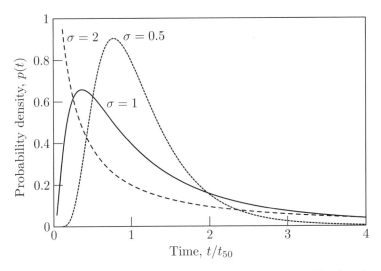

Fig. 4.5. Log-normal distribution of electromigration failures. The distribution is
unimodal and is determined by the median time to failure t_{50} and the shape parameter σ.

Accelerated electromigration testing is performed to evaluate the
lifetime distribution parameters for a specific manufacturing process.
The interconnect structures are subjected to a current and temperature significantly greater than the target specifications to determine
the statistical characteristics of the interconnect failures within a limited time period. The following relationship, first proposed by Black in

1967 [74], [86], is commonly used to estimate the median time to failure at different temperatures and current densities,

$$t_{50} = \frac{A}{j^n} \exp\left(\frac{Q}{kT}\right),$$ (4.13)

where A and n are empirically determined parameters. In the absence of Joule heating effects, the value of n varies from one to two, depending on the characteristics of the manufacturing process [72]. As demonstrated by models of the electromigration process, $n = 1$ corresponds to the case of void induced failures, and $n = 2$ represents the case of hillock induced failures [87].

4.6 Electromigration lifetime under AC current

On-chip interconnect lines typically carry time-varying AC current. It is necessary to determine the electromigration lifetime under AC conditions based on accelerated testing, which is typically performed under DC current.

Consider a train of current square pulses with a duty ratio $d = t_{on}/T$, as shown in Fig. 4.6, versus a DC current of the same magnitude. Assuming that a linear accumulation of the electromigration damage

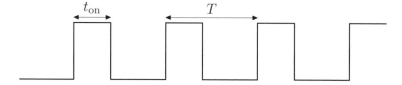

Fig. 4.6. A train of current pulses.

during the active phase t_{on} of the pulses results in the pulsed current lifetime t_{50}^{pulsed} that is $1/d$ times longer than the DC current lifetime t_{50}^{dc}, i.e., $t_{50}^{dc}/t_{50}^{pulsed} = d$. This estimate is, however, overly conservative, suggesting a certain degree of electromigration damage repair during the quiet phase of the pulses. This "self-healing" effect significantly extends the lifetime of a line. Experimental studies [88], [89], [90] have demonstrated that, in the absence of Joule heating effects, the pulsed current lifetime is determined by the average current j_{avg},

$$t_{50}^{\text{pulsed}} = \frac{A}{j_{\text{avg}}^n} \exp\left(\frac{Q}{kT}\right). \tag{4.14}$$

As $j_{\text{avg}} = dj_{\text{dc}}$, the pulsed current lifetime is related to the DC current lifetime as $t_{50}^{\text{dc}}/t_{50}^{\text{pulsed}} = d^n$.

Electromigration reliability is greatly enhanced under bidirectional current flow. Accurate characterization of electromigration reliability becomes difficult due to the long lifetimes. Available experimental data are in agreement with the average current model, as expressed by (4.14). According to (4.14), the lifetime becomes infinitely long as the DC component of the current approaches zero. The current density these lines can carry, however, is also limited. As the magnitude of the bidirectional current becomes sufficiently high, the Joule heating becomes significant, degrading the self-healing process and, consequently, the electromigration lifetime.

Clock and data lines in integrated circuits are usually connected to a single driver. The average current in these lines is zero and the lines are typically highly resistant to electromigration failure in the absence of significant Joule heating. Power and ground lines carry a high unidirectional current. Power and ground lines are therefore particularly susceptible to electromigration damage.

4.7 Electromigration in novel interconnect technologies

Heretofore, this discussion is largely specific to standard interconnect technologies based on aluminum and silicon dioxide. Electromigration effects in these technologies have been extensively studied and are relatively well understood. Advanced interconnect technologies using novel materials, such as copper and low-k dielectrics, have been recently introduced to enhance the electrical characteristics of interconnect structures. The electromigration characteristics of these advanced interconnect processes are different as compared to aluminum based technologies, as demonstrated by early studies [91]. The electromigration characteristics of advanced interconnect technologies are currently an area of intensive study [91].

The electromigrational atomic flux in copper interconnect is significantly lower than in aluminum interconnect mainly due to the higher activation energies of the diffusion paths. The threshold effect has been observed in copper lines [92], [93], [94], with threshold current densities several times higher than in aluminum lines. Experimental results suggest that surface diffusion and diffusion along the interface between copper and silicon nitride covering the top of the line dominate electromigration transport. The silicon nitride film serves as a diffusion barrier at the upper surface of the line. Several features of the copper interconnect structure, however, exacerbate the detrimental effects of electromigration [91].

The sides and bottom of a copper line are covered with a refractory metal film (Ta, Ti, TaN, or TiN) to prevent copper from diffusing into the dielectric. The thickness of this film is much smaller than the thickness of the film covering the aluminum interconnect. The formation of a void in a copper line therefore leads to a higher relative increase in the line resistance. Thin redundant layers can also lead to an abrupt open-circuit failure rather than a gradual increase in the line resistance [95].

The via structure in dual-damascene copper interconnect is susceptible to failure due to void formation [96]. The refractory metal film lining the bottom of the via forms a diffusion barrier between the via and the lower metal line, as shown in Fig. 4.7. The resistance of dual-damascene interconnect is particularly sensitive to void formation at the bottom of the via. The structural characteristics of the via contact are crucial to interconnect reliability [97].

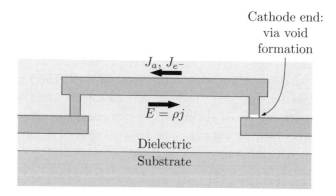

Cathode end:
via void
formation

J_a, J_{e^-}

$E = \rho j$

Dielectric

Substrate

Fig. 4.7. A dual-damascene interconnect structure. A diffusion barrier is formed by the refractory metal film lining the side and lower surfaces of the lines and vias. The via bottom at the cathode end is the site of metal depletion and is susceptible to void formation. The resistance of the line is particularly sensitive to void formation at the bottom of the via.

Low-k dielectric materials decrease the distributed capacitance of the metal line and are a desirable complement to low resistivity copper metalization. Low-k dielectrics are significantly less rigid than silicon dioxide. Metal depletion and accumulation therefore result in a smaller mechanical stress, decreasing stress-induced backflow and electromigration reliability [98], [99], [100]. Low-k dielectrics also have a lower thermal conduction coefficient, exacerbating detrimental Joule heating effects.

The dominant failure mechanisms of advanced copper interconnect are therefore different than in aluminum based technologies. A statistical distribution of failure times is also more complex and cannot be described by a unimodal probability density distribution [96], [101].

The accumulated experience in understanding the electromigration characteristics of aluminum based interconnect structures therefore has limited applicability to novel interconnect technologies. Further investigation is required to ensure high yield and reliability of circuits with advanced interconnect technologies.

4.8 Designing for electromigration reliability

The electromigration reliability of integrated circuits has traditionally been ensured by requiring the average current density of each interconnect line to be below a predetermined design rule specified threshold. The cross-sectional dimensions of on-chip interconnect decrease with technology scaling, while the current increases. Simply limiting the line current density by a design rule threshold becomes increasingly restrictive under these conditions.

To ensure a target reliability, the current density threshold is selected under the implicit assumption that the current density threshold is reached with a certain number of interconnect lines. If the number of lines with a critical current density is fewer than the assumed estimate, the design rule is overly conservative. Alternatively, if the number of critical lines is larger than the estimate, the design rule does not guarantee the target reliability characteristics.

The "one size fits all" threshold approach can be replaced with a more flexible statistical electromigration budgeting methodology [102]. If the failure probability of the ith line is estimated as $p_i(t)$ based on the line dimensions and average current, the probability that none of the lines in a circuit fails at time t is $\prod_i (1 - p_i(t))$. The failure probability of the overall system $P(t)$ is therefore

$$P(t) = 1 - \prod_i (1 - p_i(t)). \tag{4.15}$$

A system with a few lines carrying current exceeding the threshold design rule [and therefore exhibiting a relatively high failure probability $p(t)$] can be as reliable as a system with a larger number of lines carrying current at the threshold level.

This statistical approach permits individual budgeting of the line failure probabilities $p_i(t)$, while maintaining the target reliability of the overall system $P(t)$. This flexibility supports more efficient use of interconnect resources, particularly in congested areas, reducing circuit area.

4.9 Summary

The electromigration reliability of integrated circuits has been discussed in this chapter. The primary conclusions of the chapter are the following.

- Electromigration damage develops near the sites of atomic flux divergence, such as vias and microstructural discontinuities

- Electromigration reliability depends upon the mechanical properties of the interconnect structures

- Electromigration reliability of short lines is greatly enhanced due to stress gradient induced backflow, which compensates the electromigration atomic flow

- Power and ground lines are particularly susceptible to electromigration damage as these lines carry a unidirectional current

5

High Performance Power Distribution Systems

Supplying power to high performance integrated circuits has become a challenging task. The system supplying power to an integrated circuit greatly affects the performance, size, and cost characteristics of the overall electronic system. This system is comprised of interconnect networks with decoupling capacitors on a printed circuit board, an integrated circuit package, and a circuit die. The entire system is henceforth referred to as the *power distribution system*. The design of power distribution systems is described in this chapter. The focus of the discussion is the overall structure and interaction among the various parts of the system. The impedance characteristics and design of on-chip power distribution networks, the most complex part of the power distribution system, are discussed in greater detail in the following chapters.

The chapter is organized as follows. A typical power distribution system for a high power, high speed integrated circuit is described in Section 5.1. A circuit model of a power distribution system is presented in Section 5.2. The output impedance characteristics are discussed in Section 5.3. The effect of a shunting capacitance on the impedance of a power distribution system is considered in Section 5.4. The hierarchical placement of capacitors to satisfy target impedance requirements is described in Section 5.5. Design strategies to control the resonant effects in power distribution networks are discussed in Section 5.6. A purely resistive impedance characteristic of power distribution systems can be achieved using an impedance compensation technique, as described in Section 5.7. A case study of a power distribution system is presented in Section 5.8. Design techniques to enhance the impedance characteristics of power distribution networks are discussed in Section 5.9. The limitations of the analysis presented herein are discussed in Section 5.10. The chapter concludes with a summary.

5.1 Physical structure of a power distribution system

A cross-sectional view of a power distribution system for a high performance integrated circuit is shown in Fig. 5.1. The power supply system

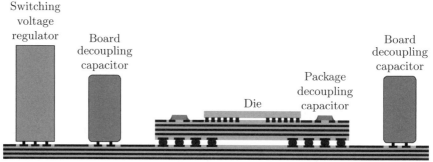

Fig. 5.1. A cross-sectional view of a power distribution system of a high performance integrated circuit.

spans several levels of packaging hierarchy. It consists of a switching voltage regulator module (VRM), the power distribution networks on a printed circuit board (PCB), on an integrated circuit package, and on-chip, plus the decoupling capacitors connected to these networks. The power distribution networks at the board, package, and circuit die levels form a conductive path between the power source and the power load. A switching voltage regulator converts the DC voltage level provided by a system power supply unit to a voltage V_{dd} required for powering an integrated circuit. The regulator serves as a power source, effectively decoupling the power distribution system of an integrated circuit from the system level power supply. The power and ground planes of the printed circuit board connect the switching regulator to the integrated circuit package. The board-level decoupling capacitors are placed across the power and ground planes to provide charge storage for current transients faster than the response time of the regulator. The board power and ground interconnect is connected to the power and ground networks of the package through a ball grid array

(BGA) or pin grid array (PGA) contacts. Similar to a printed circuit board, the package power and ground distribution networks are typically comprised of several metal planes. High frequency (*i.e.*, with low parasitic impedance) decoupling capacitors are mounted on the package, electrically close to the circuit die, to ensure the integrity of the power supply during power current surges drawn by the circuit from the package level power distribution network. The package power and ground distribution networks are connected to the on-chip power distribution grid through a flip-chip array of bump contacts or alternative bonding technologies [103]. On-chip decoupling capacitors are placed across the on-chip power and ground networks to maintain a low impedance at signal frequencies comparable to the switching speed of the on-chip devices.

The physical structure of the power distribution system is hierarchical. Each tier of the power distribution system typically corresponds to a tier of packaging hierarchy and consists of a power distribution network and associated decoupling capacitors. The hierarchical structure of the power distribution system permits the desired impedance characteristics to be obtained in a cost effective manner, as described in the following sections.

5.2 Circuit model of a power distribution system

A simplified circuit model of a power supply system[1] is shown in Fig. 5.2. This lumped circuit model is effectively one-dimensional, where each level of the packaging hierarchy is modeled by a pair of power and ground conductors and a decoupling capacitor across these conductors. A one-dimensional model accurately describes the impedance characteristics of a power distribution system over a wide frequency range [104], [105], [106]. The conductors are represented by the parasitic resistive and inductive impedances; the decoupling capacitors are represented by a series RLC circuit reflecting the parasitic impedances of the capacitors. The italicized superscripts "p" and "g" refer to the power and ground conductors, respectively; superscript "C" refers to the parasitic impedance characteristics of the capacitors. The subscripts "r," "b," "p," and "c" refer to the regulator, board, package, and on-chip

---

[1] The magnetic coupling of the power and ground conductors is omitted in the circuit diagrams of a power distribution system. The inductive elements shown in the diagrams therefore denote a net inductance, as described in Section 2.1.4.

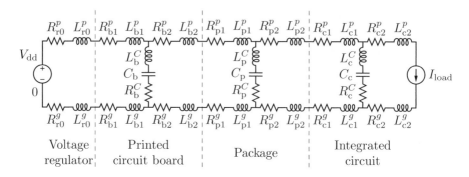

Fig. 5.2. A one-dimensional circuit model of the power supply system shown in Fig. 5.1.

conductors. Subscript 1 refers to the conductors upstream of the respective decoupling capacitor, with respect to the flow of energy from the power source to the load. That is, the conductors denoted with subscript 1 are connected to the appropriate decoupling capacitor at the voltage regulator. Subscript 2 refers to the conductors downstream of the appropriate decoupling capacitor. For example, R_{b1}^g refers to the parasitic resistance of the ground conductors connecting the board capacitors to the voltage regulator, while L_{p2}^p refers to the parasitic inductance of the power conductors connecting the package capacitors to the on-chip I/O contacts. Similarly, L_p^C refers to the parasitic series inductance of the package capacitors.

The model shown in Fig. 5.2 can be reduced by combining the circuit elements connected in series. The reduced circuit model is shown in Fig. 5.3. The circuit characteristics of the circuit shown in Fig. 5.3 are related to the characteristics of the circuit shown in Fig. 5.2 as

$$R_r^p = R_{r0}^p + R_{b1}^p \qquad\qquad L_r^p = L_{r0}^p + L_{b1}^p \qquad (5.1a)$$
$$R_b^p = R_{b2}^p + R_{p1}^p \qquad\qquad L_b^p = L_{b2}^p + L_{p1}^p \qquad (5.1b)$$
$$R_p^p = R_{p2}^p + R_{c1}^p \qquad\qquad L_p^p = L_{p2}^p + L_{c1}^p \qquad (5.1c)$$
$$R_c^p = R_{c2}^p \qquad\qquad\qquad L_c^p = L_{c2}^p \qquad\qquad (5.1d)$$

for conductors carrying power current and, analogously, for the ground conductors,

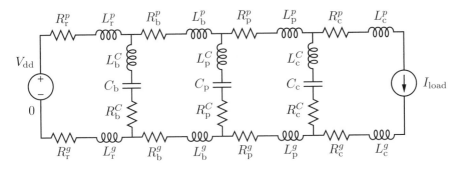

Fig. 5.3. A reduced version of the circuit model characterizing a power supply system.

$$R_r^g = R_{r2}^g + R_{b1}^g \qquad\qquad L_r^g = L_{r2}^g + L_{b1}^g \qquad (5.1e)$$
$$R_b^g = R_{b2}^g + R_{p1}^g \qquad\qquad L_b^g = L_{b2}^g + L_{p1}^g \qquad (5.1f)$$
$$R_p^g = R_{p2}^g + R_{c1}^g \qquad\qquad L_p^g = L_{p2}^g + L_{c1}^g \qquad (5.1g)$$
$$R_c^g = R_{c2}^g \qquad\qquad\qquad L_c^g = L_{c2}^g. \qquad (5.1h)$$

As defined, the circuit elements with subscript "r" represent the output impedance of the voltage regulator and the impedance of the on-board current path from the regulator to the board decoupling capacitors. The elements denoted with subscript "b" represent the impedance of the current path from the board capacitors to the package, the impedance of the socket and package pins (or solder bumps in the case of a ball grid array mounting solution), and the impedance of the package lines and planes. Similarly, the elements with subscript "p" signify the impedance of the current path from the package capacitors to the die mounting site, the impedance of the solder bumps or bonding wires, and, partially, the impedance of the on-chip power distribution network. Finally, the resistors and inductors with subscript "c" represent the impedance of the current path from the on-chip capacitors to the on-chip power load.

The board, package, and on-chip power distribution networks have significantly different electrical characteristics due to the different physical properties of the interconnect at the various tiers of the packaging hierarchy. From the board level to the die level, the cross-sectional dimensions of the interconnect lines decrease while the aspect ratio increases, producing a dramatic increase in interconnect density. The inductance of the board level power distribution network, *i.e.*, L_{b1}^p, L_{b1}^g, L_{b2}^p, and L_{b2}^g, is large as the power and ground planes are typically separated by tens or hundreds of micrometers. The effective output impedance of the voltage switching regulator over a wide range of frequencies can be described as $R + j\omega L$ [104], [107], [108], hence the presence of R_{r2}^{p} and R_{r2}^{g}, and L_{r2}^{p} and L_{r2}^{g} in the model shown in Fig. 5.2. The inductance of the on-chip power distribution network, L_{c1}^{p}, L_{c1}^{g}, L_{c2}^{p}, and L_{c2}^{g}, is comparatively low due to the high interconnect density. The inductance L_{p1}^{p}, L_{p1}^{g}, L_{p2}^{p}, and L_{p2}^{g} of the package level network is of intermediate magnitude, larger than the inductance of the on-chip network but smaller than the inductance of the board level network. The inductive characteristics of the circuit shown in Fig. 5.3 typically exhibit a hierarchical relationship: $L_{b}^{p} > L_{p}^{p} > L_{c}^{p}$ and $L_{b}^{g} > L_{p}^{g} > L_{c}^{g}$. The resistance of the power distribution networks follows the opposite trend. The board level resistances R_{b1}^{p}, R_{b1}^{g}, R_{b2}^{p}, and R_{b2}^{g} are small due to the large cross-sectional dimensions of the relatively thick board planes, while the on-chip resistances R_{c1}^{p}, R_{c1}^{g}, R_{c2}^{p}, and R_{c2}^{g} are large due to the small cross-sectional dimensions of the on-chip interconnect. The resistive characteristics of the power distribution system are therefore reciprocal to the inductive characteristics, $R_{b}^{p} < R_{p}^{p} < R_{c}^{p}$ and $R_{b}^{g} < R_{p}^{g} < R_{c}^{g}$. The physical hierarchy is thus reflected in the electrical hierarchy: the progressively finer physical features of the conductors typically result in a higher resistance and a lower inductance.

5.3 Output impedance of a power distribution system

To ensure a small variation in the power supply voltage under a significant current load, the power distribution system should exhibit a small impedance at the terminals of the load within the frequency range of interest, as shown in Fig. 5.4. The impedance of the power distribution system as seen from the terminals of the load circuits is henceforth referred to as the impedance of the power distribution system. In order to ensure correct and reliable operation of an integrated circuit, the impedance of the power distribution system is specified to be lower than a certain upper bound Z_0 in the frequency range from DC to the maximum frequency f_0 [107], as illustrated in Fig. 5.6. Note that the maximum frequency f_0 is determined by the switching time of the on-chip signal transients, rather than by the clock frequency f_{clk}. The shortest signal switching time is typically smaller than the clock period by at least an order of magnitude; therefore, the maximum frequency of interest f_0 is considerably higher than the clock frequency f_{clk}.

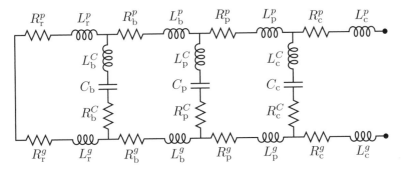

Fig. 5.4. A circuit network representing a power distribution system impedance as seen from the terminals of the power load. The power source at the left side of the network has been replaced with a short circuit. The terminals of the power load are shown at the right side of the network.

The objective of designing a power distribution system is to ensure a target output impedance characteristic. It is therefore important to understand how the output impedance of the circuit shown in Fig. 5.4 depends on the impedance of the comprising circuit elements. The impedance characteristics of a power distribution system are analyzed in the following sections. The impedance characteristics of a power distribution system with no capacitors are considered first. The effect of the decoupling capacitors on the impedance characteristics is described in the following sections.

A power distribution system with no decoupling capacitors is shown in Fig. 5.5. The power source and load are connected by interconnect

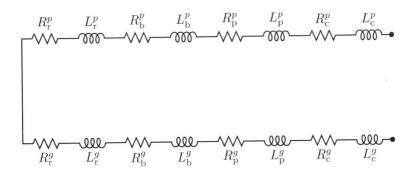

Fig. 5.5. A circuit network representing a power distribution system without decoupling capacitors.

with resistive and inductive parasitic impedances. The magnitude of
the impedance of this network is

$$|Z_{\text{tot}}(\omega)| = |R_{\text{tot}} + j\omega L_{\text{tot}}|, \tag{5.2}$$

where R_{tot} and L_{tot} are the total series resistance and inductance of
the power distribution system, respectively,

$$R_{\text{tot}} = R_{\text{tot}}^p + R_{\text{tot}}^g, \tag{5.3}$$

$$R_{\text{tot}}^p = R_{\text{r}}^p + R_{\text{b}}^p + R_{\text{p}}^p + R_{\text{c}}^p, \tag{5.4}$$

$$R_{\text{tot}}^g = R_{\text{r}}^g + R_{\text{b}}^g + R_{\text{p}}^g + R_{\text{c}}^g, \tag{5.5}$$

$$L_{\text{tot}} = L_{\text{tot}}^p + L_{\text{tot}}^g, \tag{5.6}$$

$$L_{\text{tot}}^p = L_{\text{r}}^p + L_{\text{b}}^p + L_{\text{p}}^p + L_{\text{c}}^p, \tag{5.7}$$

$$L_{\text{tot}}^g = L_{\text{r}}^g + L_{\text{b}}^g + L_{\text{p}}^g + L_{\text{c}}^g. \tag{5.8}$$

The variation of the impedance with frequency is illustrated in
Fig. 5.6. To satisfy the specification at low frequency, the resistance
of the power distribution system should be sufficiently low, $R_{\text{tot}} < Z_0$.
Above the frequency $f_{L_{\text{tot}}} = \frac{1}{2\pi}\frac{R_{\text{tot}}}{L_{\text{tot}}}$, however, the impedance of the
power distribution system is dominated by the inductive reactance
$j\omega L_{\text{tot}}$ and increases linearly with frequency, exceeding the target specification at the frequency $f_{\max} = \frac{1}{2\pi}\frac{Z_0}{L_{\text{tot}}}$.

The high frequency impedance should be reduced to satisfy the target specifications. Opportunities for reducing the inductance of the
interconnect structures comprising a power system are limited. The
feature size of the board and package level interconnect, which largely
determines the inductance, depends on the manufacturing technology.
Furthermore, the output impedance of the voltage regulator is highly
inductive and difficult to reduce.

The high frequency impedance is effectively reduced by placing
capacitors across the power and ground conductors. These shunting
capacitors terminate the high frequency current loop, permitting the
current to bypass the inductive interconnect, such as the board and
package power distribution networks. The high frequency impedance
of the system as seen from the load terminals is thereby reduced. The
capacitors effectively "decouple" the high impedance parts of the power
distribution system from the load at high frequencies. These capacitors
are therefore commonly referred to as decoupling capacitors. Several
stages of decoupling capacitors are typically used to confine the output
impedance within the target specifications.

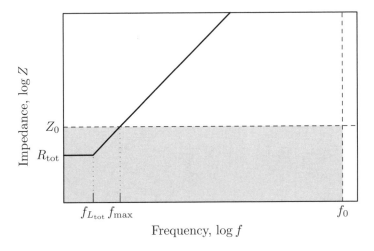

Fig. 5.6. Impedance of the power distribution system with no decoupling capacitors. The shaded area denotes the target impedance specifications of the power distribution system.

In the following sections, the impedance characteristics of a power distribution system with several stages of decoupling capacitors are described in several steps. The effect of a single decoupling capacitor on the impedance of a power distribution system is considered in Section 5.4. The hierarchical placement of the decoupling capacitors is described in Section 5.5. The impedance characteristics near the resonant frequencies are examined in Section 5.6.

5.4 A power distribution system with a decoupling capacitor

The effects of a decoupling capacitor on the output impedance of a power distribution system are the subject of this section. The impedance characteristics of a power distribution system with a decoupling capacitor is described in Section 5.4.1. The limitations of using a single stage decoupling scheme are discussed in Section 5.4.2.

5.4.1 Impedance characteristics

A power distribution system with a shunting capacitance is shown in Fig. 5.7. The shunting capacitance can be physically realized with a single capacitor or, alternatively, with a bank of several identical capacitors connected in parallel. Similar to a single capacitor, the

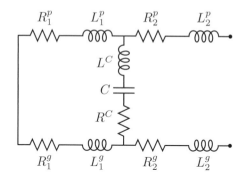

Fig. 5.7. A circuit model of a power distribution network with a decoupling capacitor C.

impedance of a bank of identical capacitors is accurately described by a series RLC circuit. The location of the capacitors partitions the power network into upstream and downstream sections, with respect to the flow of energy (or power current) from the power source to the load. The upstream section is referred to as stage 1 in Fig. 5.7; the downstream section is referred to as stage 2. Also note that the overall series inductance and resistance of the power distribution network remains the same as determined by (5.3) to (5.8): $L_1^p + L_2^p = L_{\text{tot}}^p$, $L_1^g + L_2^g = L_{\text{tot}}^g$, $R_1^p + R_2^p = R_{\text{tot}}^p$, and $R_1^g + R_2^g = R_{\text{tot}}^g$. The impedance of the power distribution network shown in Fig. 5.7 is

$$Z(\omega) = R_2 + j\omega L_2 + (R_1 + j\omega L_1) \,\|\, \left(R^C + j \left(\omega L^C - \frac{1}{\omega C} \right) \right), \quad (5.9)$$

where $R_1 = R_1^p + R_1^g$ and $L_1 = L_1^p + L_1^g$; analogously, $R_2 = R_2^p + R_2^g$ and $L_2 = L_2^p + L_2^g$.

The impedance characteristics of a power distribution network with a shunting capacitor are illustrated in Fig. 5.8. At low frequencies, the impedance of the capacitor is greater than the impedance of the upstream section. The power current loop extends to the power source, as illustrated by the equivalent circuit shown in Fig. 5.9(a). Assuming that the parasitic inductance of the decoupling capacitor is significantly smaller than the upstream inductance, i.e., $L^C < L_1$, the capacitor has a lower impedance than the impedance of the upstream section

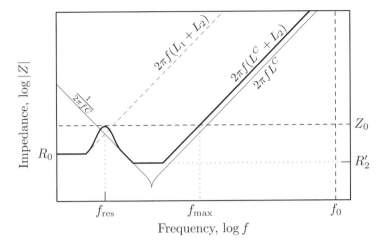

Fig. 5.8. Impedance of the power distribution system with a decoupling capacitor as shown in Fig. 5.7. The impedance of the power distribution system with a decoupling capacitor (the black line) exhibits an extended region of low impedance as compared to the impedance of the power distribution system with no decoupling capacitors, shown for comparison with the dashed gray line. The impedance of the decoupling capacitor is shown with a thin solid line.

$R_1 + j\omega L_1$ above the frequency $f_{\text{res}} \approx \frac{1}{2\pi} \frac{1}{\sqrt{L_1 C}}$. Above the frequency f_{res}, the bulk of the power current bypasses the impedance $R_1 + j\omega L_1$ through the capacitor, shrinking the size of the power current loop, as shown in Fig. 5.9(b). Note that the decoupling capacitor and the upstream stage form an underdamped LC tank circuit, resulting in a resonant[2] peak in the impedance at frequency f_{res}. The impedance at the resonant frequency f_{res} is *increased* by Q_{tank}, the quality factor of the tank circuit, as compared to the impedance of the power network at the same frequency without the capacitor. Between f_{res} and $f_R = \frac{1}{2\pi} \frac{1}{(R_2 + R^C)C}$, the impedance of the power network is dominated by the capacitive reactance $\frac{1}{\omega C}$ and decreases with frequency, reaching $R_2' = R_2 + R^C$. The parasitic resistance of the decoupling capacitor R^C should therefore also be sufficiently low, such that $R_2 + R^C < Z_0$, in order to satisfy the target specification. At frequencies greater than f_R, the

---

[2] The circuit impedance drastically increases near the resonant frequency in parallel (tank) resonant circuits. The parallel resonance is therefore often referred to as the *antiresonance* to distinguish this phenomenon from the series resonance, where the circuit impedance *decreases*. Correspondingly, peaks in the impedance are often referred to as antiresonant.

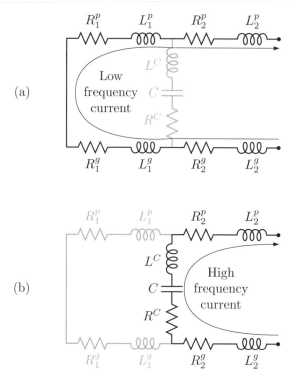

Fig. 5.9. The path of current flow in a power distribution system with a decoupling capacitor. (a) At frequencies below $f_{\text{res}} = \frac{1}{2\pi}\frac{1}{\sqrt{L_1 C}}$, the capacitive impedance is relatively high and the current loop extends throughout an entire network. (b) At frequencies above f_{res}, the capacitive impedance is lower than the impedance of the upstream section and the bulk of the current bypasses the upstream inductance $L_1^p + L_1^g$ through the decoupling capacitor C.

network impedance increases as

$$Z(\omega) = R_2 + R^C + j\omega(L_2 + L^C). \tag{5.10}$$

A comparison of (5.10) with (5.2) indicates that placing a decoupling capacitor reduces the high frequency inductance of the power distribution network, as seen by the load, from $L_{\text{tot}} = L_1 + L_2$ to $L_2' = L_2 + L^C$.

The output resistance R_2' and inductance L_2' as seen from the load are henceforth referred to as the *effective* resistance and inductance of the power distribution system to distinguish these quantities from the *overall series* resistance R_{tot} and inductance L_{tot} of the system (which are the effective resistance and inductance at DC). The shunting capacitor "decouples" the upstream portion of the power distribution system from the power current loop at high frequencies, decreasing the maximum impedance of the system. The frequency range where the impedance specification is satisfied is correspondingly increased to $f_{\max} = \frac{1}{2\pi}\frac{Z_0}{L_2'}$.

5.4.2 Limitations of a single-tier decoupling scheme

It follows from the preceding discussion that the impedance of a power distribution system can be significantly reduced by a single capacitor (or a group of capacitors placed at the same location) over a wide range of frequencies. This solution is henceforth referred to as *single-tier decoupling*. A single-tier decoupling scheme, however, is difficult to realize in practice as this solution imposes stringent requirements on the performance characteristics of the decoupling capacitor, as discussed below.

To confine the network impedance within the specification boundaries at the highest required frequencies, *i.e.*, the $2\pi f_0(L_2 + L^C) < Z_0$, a low inductance path is required between the decoupling capacitance and the load,

$$L_2 + L^C < \frac{1}{2\pi}\frac{Z_0}{f_0}. \tag{5.11}$$

Simultaneously, the capacitance should be sufficiently high to bypass the power current at sufficiently low frequencies, thereby preventing the violation of target specifications above $f_{\max}$ due to a high inductive impedance $j\omega(L_1 + L_2)$,

$$\frac{1}{2\pi f_{\max}C} < 2\pi f_{\max}(L_1 + L_2). \tag{5.12}$$

Condition (5.12) is, however, insufficient. As mentioned above, the decoupling capacitor C and the inductance of the upstream section L_1 form a resonant tank circuit. The impedance reaches the maximum at the resonant frequency $\omega_{\mathrm{res}} \approx \frac{1}{\sqrt{L_1C}}$, where the inductive impedance of the tank circuit complements the capacitive impedance, $j\omega_{\mathrm{res}}L_1 = -\frac{1}{j\omega_{\max}C}$. Where the quality factor of the tank circuit Q_{tank} is sufficiently larger than unity (*i.e.*, $Q_{\mathrm{tank}} \gtrsim 3$), the impedance of the tank circuit at the resonant frequency ω_{res} is purely resistive and is larger than the characteristic impedance $\sqrt{\frac{L_1}{C}} = \omega_{\mathrm{res}}L_1$ by the quality factor Q_{tank}, $Z_{\mathrm{tank}}(\omega_{\mathrm{res}}) = Q_{\mathrm{tank}}\sqrt{\frac{L_1}{C}}$. The quality factor of the tank circuit is

$$Q_{\text{tank}} = \frac{1}{R_1 + R^C} \sqrt{\frac{L_1}{C}}, \tag{5.13}$$

yielding the magnitude of the peak impedance,

$$Z_{\text{peak}} = \frac{1}{R_1 + R^C} \frac{L_1}{C}. \tag{5.14}$$

To limit the impedance magnitude below the target impedance Z_0, the decoupling capacitance should satisfy

$$C > \frac{L_1}{Z_0(R_1 + R^C)}. \tag{5.15}$$

A larger upstream inductance L_1 (*i.e.*, the inductance that is decoupled by the capacitor) therefore requires a larger decoupling capacitance C to maintain the target network impedance Z_0. The accuracy of the simplifications used in the derivation of (5.14) decreases as the factor Q_{tank} approaches unity. A detailed treatment of the case where $Q \approx 1$ is presented in Section 5.6.

These impedance characteristics have an intuitive physical interpretation. From a physical perspective, the decoupling capacitor serves as an intermediate storage of charge and energy. To be effective, such an energy storage device should possess two qualities. First, the device should have a high capacity to store a sufficient amount of energy. This requirement is expressed in terms of the impedance characteristics as the minimum capacitance constraint (5.15). Second, to supply sufficient power at high frequencies, the device should be able to release and accumulate energy at a sufficient rate. This quality is expressed as the maximum inductance constraint (5.11).

Constructing a device with both high energy capacity and high power capability is, however, challenging. The conditions of low inductance (5.11) and high capacitance (5.15) cannot be simultaneously satisfied in a cost effective manner. In practice, these conditions are contradictory. The physical realization of a large decoupling capacitance as determined by (5.15) requires the use of discrete capacitors with a large nominal capacity, which, consequently, have a large form factor. The large physical dimensions of the capacitors have two implications. The parasitic series inductance of a physically large capacitor L^C is relatively high due to the increased area of the current loop within the capacitors, contradicting requirement (5.11). Furthermore, the large physical size of the capacitors prevents placing the capacitors sufficiently close to the power load. Greater physical separation increases the inductance L_2 of the current path from the capacitors to the load, also contradicting requirement (5.11). The available component technology therefore imposes a tradeoff between the high capacity and low parasitic inductance of a capacitor.

Gate switching times of a few tens of picoseconds are common in contemporary integrated circuits, creating high transients in the power load current. Only on-chip decoupling capacitors have a sufficiently low parasitic inductance to maintain a low impedance power distribution system at high frequencies. Placing a sufficiently large on-chip decoupling capacitance, as determined by (5.15), requires a die area many times greater than the area of the load circuit. Therefore, while technically feasible, the single-tier decoupling solution is prohibitively expensive. A more efficient approach to the problem, a multi-stage hierarchical placement of decoupling capacitors, is described in the following section.

5.5 Hierarchical placement of decoupling capacitance

Low impedance, high frequency power distribution systems are realized in a cost effective way by using a hierarchy of decoupling capacitors. The capacitors are placed in several stages: on the board, package, and circuit die. The impedance characteristics of a power distribution system with several stages of decoupling capacitors are described in this section. The evolution of the system output impedance is described as the decoupling stages are consecutively placed across the network. Arranging the decoupling capacitors in several stages eliminates the need to satisfy both the high capacitance and low inductance requirements, as expressed by (5.11) and (5.15), in the same decoupling capacitor stage.

Board decoupling capacitors

Consider a power distribution system with decoupling capacitors placed on the board, as shown in Fig. 5.10. The circuit is analogous to the network shown in Fig. 5.7. Similar to the single-tier decoupling scheme, the board decoupling capacitance should satisfy the following condition in order to meet the target specification at low frequencies,

$$C_{\mathrm{b}} > \frac{L_{\mathrm{r}}}{Z_0(R_{\mathrm{r}} + R_{\mathrm{b}}^C)} \ . \tag{5.16}$$

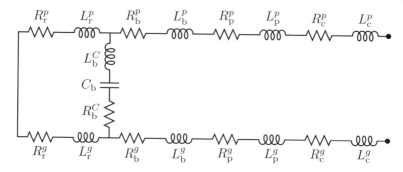

Fig. 5.10. A circuit model of a power distribution system with a board decoupling capacitance.

$L_r = L_r^p + L_r^g$ and $R_r = R_r^p + R_r^g$ are the resistance and inductance, respectively, of the network upstream of the board capacitance, including the output inductance of the voltage regulator and the impedance of the current path between the voltage regulator and the board capacitors, as defined by (5.1). At frequencies greater than f_{res}^B, the power current flows through the board decoupling capacitors, bypassing the voltage regulator. Condition (5.11), however, is not satisfied due to the relatively high inductance $L_b + L_p + L_c$ resulting from the large separation from the load and the high parasitic series inductance of the board capacitors L_b^C. Above the frequency $f_{R_B} = \frac{1}{2\pi} \frac{1}{R_B C_b}$, the impedance of the power distribution system is

$$Z_B = R_B + j\omega L_B, \tag{5.17}$$

where $L_B = L_b^C + L_b + L_p + L_c$ and $R_B = R_b^C + R_b + R_p + R_c$. Although the high frequency inductance of the network is reduced from L_{tot} to L_B, the impedance exceeds the target magnitude Z_0 above the frequency $f_{max}^B = \frac{1}{2\pi} \frac{Z_0}{L_B}$, as shown in Fig. 5.11.

Package decoupling capacitors

The excessive high frequency inductance L_B of the power distribution system with board decoupling capacitors is further reduced by placing an additional decoupling capacitance physically closer to the power load, i.e., on the integrated circuit package. The circuit model of a power distribution system with board and package decoupling capacitors is shown in Fig. 5.12. The impedance of the network with board and package decoupling capacitors is illustrated in Fig. 5.13. The package decoupling capacitance decreases the network inductance L_B in

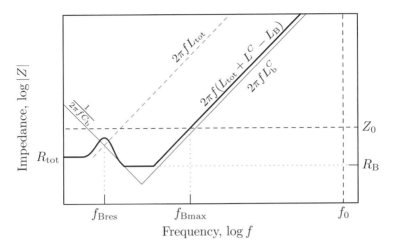

Fig. 5.11. Impedance of the power distribution system with the board decoupling capacitance shown in Fig. 5.7. The impedance characteristic is shown with a black line. The impedance of the power distribution system with no decoupling capacitors is shown, for comparison, with a gray dashed line. The impedance of the board decoupling capacitance is shown with a thin solid line.

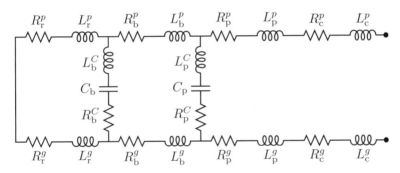

Fig. 5.12. A circuit model of a power distribution system with board and package decoupling capacitances.

a fashion similar to the board decoupling capacitance. A significant difference is that for the package decoupling capacitance the capacity requirement (5.15) is relaxed to

$$C_p > \frac{L_b'}{Z_0(R_b' + R_p^C)} \,, \tag{5.18}$$

where $L_b' = L_b + L_b^C$ and $R_b' = R_b + R_b^C$ are the effective inductance and resistance in parallel with the package decoupling capacitance. The upstream inductance L_{b}' as seen by the package capacitance at high frequencies (*i.e.*, $f > f_{R_{\mathrm{B}}}$) is significantly lower than the upstream inductance L_{r} as seen by the board capacitors. The package capacitance requirement (5.18) is therefore significantly less stringent than the board capacitance requirement (5.16). The lower capacitance require-

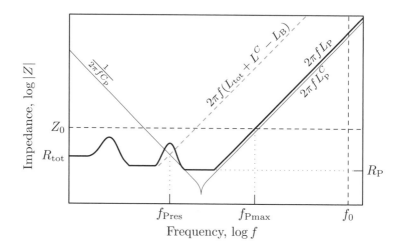

Fig. 5.13. Impedance of the power distribution system with board and package decoupling capacitances as shown in Fig. 5.7. The impedance characteristic is shown with a black line. The impedance of the power distribution system with only the board decoupling capacitance is shown with a dashed gray line for comparison. The impedance of the package decoupling capacitance is shown with a thin solid line.

ment is satisfied by using several small form factor capacitors mounted onto a package. As shown in Fig. 5.13, the effect of the package decoupling capacitors on the system impedance is analogous to the effect of the board capacitors, but occurs at a higher frequency. With package decoupling capacitors, the impedance of the power distribution system above a frequency $f_{R_{\mathrm{P}}} = \frac{1}{2\pi}\frac{1}{R_{\mathrm{P}}C_{\mathrm{p}}}$ becomes

$$Z_{\mathrm{P}} = R_{\mathrm{P}} + j\omega L_{\mathrm{P}}, \tag{5.19}$$

where $L_{\mathrm{P}} = L_{\mathrm{p}}^{C} + L_{\mathrm{p}} + L_{\mathrm{c}}$ and $R_{\mathrm{P}} = R_{\mathrm{p}}^{C} + R_{\mathrm{p}} + R_{\mathrm{c}}$. Including a package capacitance therefore lowers the high frequency inductance from L_{B} to L_{P}. The reduced inductance L_{P}, however, is not sufficiently low to satisfy (5.11) in high speed circuits. The impedance of the power system exceeds the target magnitude Z_0 at frequencies above $f_{\max}^{\mathrm{P}} = \frac{1}{2\pi}\frac{Z_0}{L_{\mathrm{P}}}$.

On-chip decoupling capacitors

On-chip decoupling capacitors are added to the power distribution system in order to extend the frequency range of the low impedance characteristics to f_0. A circuit model characterizing the impedance of a power distribution system with board, package, and on-chip decoupling capacitors is shown in Fig. 5.14. The impedance characteristics of a

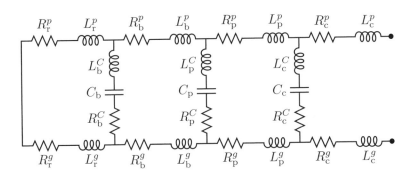

Fig. 5.14. A circuit network characterizing the impedance of a power distribution system with board, package, and on-chip decoupling capacitances.

power distribution system with all three types of decoupling capacitors are illustrated in Fig. 5.15. To ensure that the network impedance does not exceed Z_0 due to the inductance L_P of the current loop terminated by the package capacitance, the on-chip decoupling capacitance should satisfy

$$C_c > \frac{L'_p}{Z_0(R'_p + R_c^C)} , \qquad (5.20)$$

where $L'_p = L_p + L_p^C$ is the effective inductance in parallel with the package decoupling capacitance. The capacity requirement for the on-chip capacitance is further reduced as compared to the package capacitance due to lower effective inductances, $L_p < L_b$ and $L'_p < L'_b$.

Advantages of hierarchical decoupling

Hierarchical placement of decoupling capacitors exploits the trade-off between the capacity and the parasitic series inductance of a capacitor to achieve an economically effective solution. The total decoupling capacitance of a hierarchical scheme $C_b + C_p + C_c$ is larger than the total decoupling capacitance of a single-tier solution C as determined

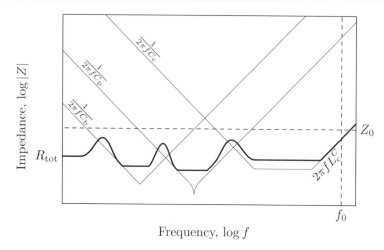

Fig. 5.15. Impedance of a power distribution system with board, package, and on-chip decoupling capacitances, as shown in Fig. 5.7. The impedance is shown with a black line. The impedance characteristic is within the target specification outlined by the dashed lines. The impedance characteristics of the decoupling capacitances are shown with thin solid lines.

by (5.15). The advantage of the hierarchical scheme is that the inductance limit (5.11) is imposed only on the final stage of decoupling capacitors which constitute a small fraction of the total decoupling capacitance. The constraints on the physical dimensions and parasitic impedance of the capacitors in the remaining stages are dramatically reduced, permitting the use of cost efficient electrolytic and ceramic capacitors.

The decoupling capacitance at each tier is effective within a limited frequency range, as determined by the capacitance and inductance of the capacitor. The range of effectiveness of the board, package, and on-chip decoupling capacitances overlaps each other, as shown in Fig. 5.15, spanning an entire frequency region of interest from DC to f_0 (the maximum operating frequency).

As described in Section 5.4, a decoupling capacitor lowers the high frequency impedance by allowing the power current to bypass the inductive interconnect upstream of the capacitor. In a power distribution system with several stages of decoupling capacitors, the inductive interconnect is excluded from the power current loop in several steps, as illustrated in Fig. 5.16. At near-DC frequencies, the power current loop extends through an entire power supply system to the power source,

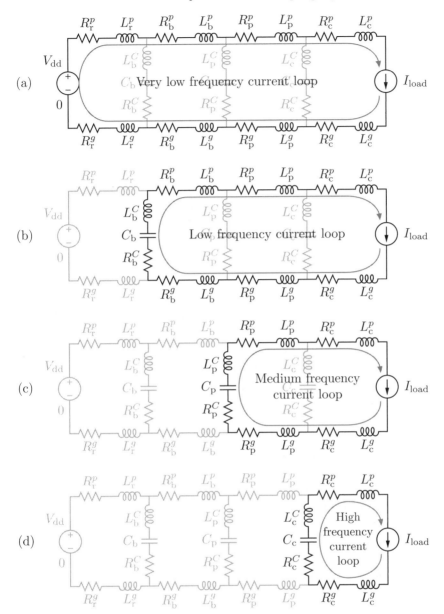

Fig. 5.16. Variation of the power current path with frequency. (a) At low frequencies the current loop extends through an entire system to the power source; as frequencies increase, the current loop is terminated by the board, package, and on-chip decoupling capacitors, as shown in (b), (c), and (d), respectively.

as shown in Fig. 5.16(a). As frequencies increase, the power current is shunted by the board decoupling capacitors, shortening the current loop as shown in Fig. 5.16(b). At higher frequencies, the package capacitors terminate the current loop, further reducing the loop size, as depicted in Fig. 5.16(c). Finally, at the highest frequencies, the power current is terminated by the on-chip capacitors, as illustrated in Fig. 5.16(d). The higher the frequency, the shorter the distance between the shunting capacitor and the load and the smaller the size of the current loop. At transitional frequencies where the bulk of the current is shifted from one decoupling stage to the next, both decoupling stages carry significant current, giving rise to resonant behavior.

Each stage of decoupling capacitors determines the impedance characteristics over a limited range of frequencies, where the bulk of the power current flows through the stage. Outside of this frequency range, the stage capacitors have an insignificant influence on the impedance characteristics of the system. The lower frequency impedance characteristics are therefore determined by the upstream stages of decoupling capacitors, while the impedance characteristics at the higher frequencies are determined by the capacitors closer to the load. For example, the board capacitors determine the low frequency impedance characteristics but do not affect the high frequency response of the system, which is determined by the on-chip capacitors.

5.6 Resonance in power distribution networks

Using decoupling capacitors is a powerful technique to reduce the impedance of a power distribution system within a significant range of frequencies, as discussed in preceding sections. A decoupling capacitor, however, *increases* the network impedance in the vicinity of the resonant frequency f_{res}. Controlling the impedance behavior near the resonant frequency is therefore essential to effectively use decoupling capacitors. A relatively high quality factor, $Q \gtrsim 3$, is assumed in the expression for the peak impedance of the parallel resonant circuit described in Section 5.4. Maintaining low impedance characteristics in power distribution systems necessitates minimizing the quality factor of all of the resonant modes; relatively low values of the quality factor are therefore common in power distribution systems. The impedance characteristics of parallel resonant circuits with a low quality factor, $Q \approx 1$, are the focus of this section.

The purpose of the decoupling capacitance is to exclude the high impedance upstream network from the load current path, as discussed in Section 5.4. The decoupling capacitor and the inductive upstream network form a parallel resonant circuit with a significant resistance in both the inductive and capacitive branches. An equivalent circuit diagram is shown in Fig. 5.17.

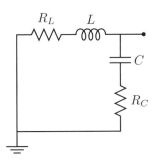

Fig. 5.17. A parallel resonant circuit with a significant parasitic resistance in both branches. L and R_L represent the effective impedance of the upstream network; C and R_C represent the impedance characteristics of the decoupling capacitor.

Near the resonant frequency, the impedance of the upstream network exhibits an inductive-resistive nature, $R_L + j\omega L$. The impedance of the decoupling capacitor is well described near the resonant frequency as $R_C + j\omega C$. The effective series inductance of the capacitor is significantly lower than the upstream network inductance (otherwise the capacitor would be ineffective, as discussed in Section 5.4) and can be typically neglected near the resonant frequency. The impedance characteristics of the capacitor and upstream network are schematically illustrated in Fig. 5.18.

The impedance of the tank circuit shown in Fig. 5.17 is

$$
\begin{aligned}
Z_{\text{tank}} &= (R_L + j\omega L) \,\|\, \left(R_C + \frac{1}{j\omega C} \right) \\
&= \frac{(R_L + j\omega L)\left(R_C + \frac{1}{j\omega C} \right)}{(R_L + j\omega L) + \left(R_C + \frac{1}{j\omega C} \right)} \\
&= R_C + \frac{s(L - CR_C^2) + (R_L - R_C)}{s^2 LC + sC(R_C + R_L) + 1}.
\end{aligned}
\tag{5.21}
$$

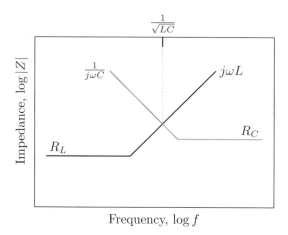

Fig. 5.18. An asymptotic plot of the impedance magnitude of the inductive and capacitive branches forming the tank circuit shown in Fig. 5.17.

The poles of the system described by (5.21) are determined by the characteristic equation,

$$s^2 + 2\zeta\omega_n + \omega_n^2 = 0, \tag{5.22}$$

$$\text{where} \quad \omega_n = \frac{1}{\sqrt{LC}} \tag{5.23}$$

$$\text{and} \quad \zeta = \frac{R_L + R_C}{2}\sqrt{\frac{C}{L}} \, . \tag{5.24}$$

Note that the magnitude of the circuit impedance varies from R_L at low frequencies ($\omega \ll \frac{1}{\sqrt{LC}}$) to R_C at high frequencies ($\omega \gg \frac{1}{\sqrt{LC}}$). It is desirable that the impedance variations near the resonant frequency do not significantly increase the maximum impedance of the network, *i.e.*, the variations do not exceed or only marginally exceed $\max(R_L, R_C)$, the maximum resistive impedance. Different constraints can be imposed on the parameters of the tank circuit to ensure this behavior [109]. Several cases of these constraints are described below.

CASE I: The tank circuit has a monotonic (*i.e.*, overshoot- and oscillation-free) response to a step current excitation if the circuit damping is critical or greater: $\zeta \leqslant 1$. Using (5.24), sufficient damping is achieved when

$$R_L + R_C \geqslant 2R_0 = 2\sqrt{\frac{L}{C}}, \tag{5.25}$$

where $R_0 = \sqrt{\frac{L}{C}}$ is the characteristic impedance of the tank circuit. Therefore, if the network impedance $R_L + j\omega L$ is to be reduced to the target impedance Z_0 at high frequencies (meaning that $R_C = Z_0$), the required decoupling capacitance is

$$C \geqslant \frac{L}{Z_0^2} \frac{4}{(1 + R_L/Z_0)^2}. \tag{5.26}$$

CASE II: Alternatively, the monotonicity of the impedance variation with frequency can be ensured. It can be demonstrated [109] that the magnitude of the impedance of the tank circuit varies monotonically from R_L at low frequencies, $\omega \ll \omega_n$, to R_C at high frequencies, $\omega \gg \omega_n$, if

$$R_L R_C \geqslant R_0^2 = \frac{L}{C}. \tag{5.27}$$

The required decoupling capacitance in this case is

$$C \geqslant \frac{L}{Z_0^2} \frac{1}{R_L/Z_0}. \tag{5.28}$$

Note that condition (5.27) is stronger than condition (5.25), i.e., satisfaction of (5.27) ensures satisfaction of (5.25). Correspondingly, the capacitance requirement described by (5.28) is always greater or equal to the capacitance requirement described by (5.26).

CASE III: The capacitance requirement determined in Case II, as described by (5.28), increases rapidly when either R_L or R_C is small. In this situation, condition (5.28) is overly conservative and restrictive. The capacitive requirement is significantly relaxed if a small peak in the impedance characteristics is allowed. To restrict the magnitude of the impedance peak to approximately 1% (in terms of the resistive baseline), the following condition must be satisfied [109],

$$(b_2 r^2 + b_1 r + b_0) R_{\max}^2 \geqslant R_0^2 = \frac{L}{C}, \tag{5.29}$$

where $R_{\max} = \max(R_L, R_C)$, $r = \min\left(\frac{R_L}{R_C}, \frac{R_C}{R_L}\right)$, $b_0 = 0.4831$, $b_1 = 0.4907$, and $b_2 = -0.0139$. The decoupling capacitance requirement is relaxed in this case to

$$C \geqslant \frac{L}{Z_0^2} \frac{1}{b_2 r^2 + b_1 r + b_0}. \tag{5.30}$$

Case III is the least constrained as compared to Cases I and II.

The design space of the resistances R_L and R_C as determined by (5.25), (5.27), and (5.29) is illustrated in Fig. 5.19 in terms of the circuit characteristic impedance $R_0 = \sqrt{\frac{L}{C}}$. In the unshaded region near the origin, the impedance exhibits significant resonant behavior. The rest of the space is partitioned into three regions. The more constrained regions are shaded in progressively darker tones. In the lightest shaded region, only the Case III condition is satisfied. The Case I and III conditions are satisfied in the adjacent darker region. All three conditions are satisfied in the darkest region.

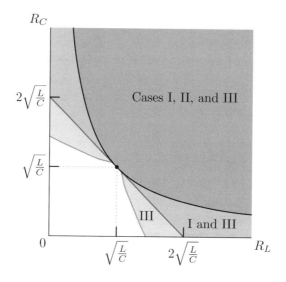

Fig. 5.19. Design space of the resistances R_L and R_C for the tank circuit shown in Fig. 5.17. All three design constraints, as determined by (5.25), (5.27), and (5.29), are satisfied in the darkest area. The constraints of Cases I and III are satisfied in the medium gray area, while only the Case III constraint is satisfied in the lightly shaded area.

Alternatively, conditions (5.25), (5.27), and (5.29) can be used to determine the decoupling capacitance requirement as a function of the circuit inductance L and resistances R_L and R_C. The normalized capacitance $C\frac{R_L^2}{L}$ versus the normalized resistance R_C/R_L is shown in Fig. 5.20. The shading is analogous to the scheme used in Fig. 5.19. The darkest region boundary is determined by condition (5.28), the boundary of the intermediate region is determined by condition (5.26), and the boundary of the lightest region is determined by condition

(5.30). Similar to the resistance design space illustrated in Fig. 5.19, Case II is the most restrictive, while Case III is the least restrictive. The capacitance requirements decrease in all three cases as the normalized resistance $\frac{R_C}{R_L}$ increases.

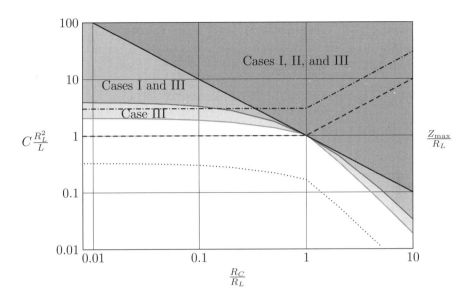

Fig. 5.20. Decoupling capacitance requirements as determined by Cases I, II, and III, *i.e.*, by (5.26), (5.28), and (5.30), respectively. Case II requires the greatest amount of capacitance while Case III requires the lowest capacitance.

The normalized maximum impedance of the network $Z_0 = \frac{R_{\max}}{R_L} = \frac{\max(R_L, R_C)}{R_L}$ is also shown in Fig. 5.20 by a dashed line. For $\frac{R_C}{R_L} \leqslant 1$, the maximum network impedance is $Z_0 = R_L \left(\frac{Z_0}{R_L} = 1 \right)$. Increasing resistance R_C to R_L therefore reduces the required decoupling capacitance without increasing the maximum impedance of the network. Increasing R_C beyond R_L further reduces the capacitance requirement, but at a cost of increasing the maximum impedance. For $\frac{R_C}{R_L} \geqslant 1$, the maximum impedance becomes $Z_0 = R_C \left(\frac{Z_0}{R_L} = \frac{R_C}{R_L} \right)$.

The requirement for the decoupling capacitance can be further reduced if the resonant impedance is designed to significantly exceed $R_{\max}$. For comparison, the capacitance requirement as determined by (5.15) for the case $Z_0 = 3R_{\max}$ ($Q \approx 3$) is also shown in Fig. 5.20 by the dotted line. This significantly lower requirement, as compared to

Cases I, II, and III, however, comes at a cost of a higher maximum impedance, as shown by the dashed line with dots.

5.7 Full impedance compensation

A special case of both (5.26) and (5.28) is the condition where

$$R_L = R_C = R_0 = \sqrt{\frac{L}{C}}, \tag{5.31}$$

as shown in Figures 5.19 and 5.20. Under condition (5.31), the zeros of the tank circuit impedance (5.21) cancel the poles and the impedance becomes purely resistive and independent of frequency,

$$Z_{\text{tank}}(\omega) = R_0. \tag{5.32}$$

This specific case is henceforth referred to as fully compensated impedance. As shown in Fig. 5.20, choosing the capacitive branch resistance R_C in accordance with (5.31) results in the lowest decoupling capacitance requirements for a given maximum circuit impedance.

The impedance of the inductive and capacitive branches of the tank circuit under condition (5.31) is illustrated in Fig. 5.21. The condition of full compensation (5.31) is equivalent to two conditions: $R_L = R_C$, i.e., the impedance at the lower frequencies is matched to the impedance at the higher frequencies, and $\frac{L}{R_L} = R_C C$, i.e., the time constants of the inductor and capacitor currents are matched, as shown in Fig. 5.21.

A constant, purely resistive impedance is achieved across the entire frequency range of interest, as illustrated in Fig. 5.22, if each decoupling stage is fully compensated. The resistance and capacitance of the decoupling capacitors in a fully compensated system are completely determined by the impedance characteristics of the power and ground interconnect and the location of the capacitors. The overall capacitance and resistance of the board decoupling capacitors are, according to (5.31),

$$R_b^C = R_r, \tag{5.33}$$

$$C_b = \frac{L_r}{R_r^2}. \tag{5.34}$$

The inductance of the upstream part of the power distribution system as seen at the terminals of the package capacitors is $L_b^C + L_b$. The capacitance and resistance of the package capacitors are

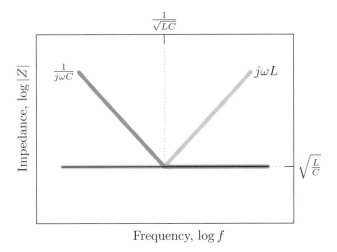

Fig. 5.21. Asymptotic impedance of the inductive (light gray line) and capacitive (dark gray line) branches of a tank circuit under the condition of full compensation (5.31). The overall impedance of the tank circuit is purely resistive and does not vary with frequency, as shown by the black line.

$$R_{\mathrm{p}}^{C} = R_{\mathrm{b}}^{C} + R_{\mathrm{b}} = R_{\mathrm{r}} + R_{\mathrm{b}}, \tag{5.35}$$

$$C_{\mathrm{p}} = \frac{L_{\mathrm{b}}^{C} + L_{\mathrm{b}}}{(R_{\mathrm{r}} + R_{\mathrm{b}})^2}. \tag{5.36}$$

Analogously, the resistance and capacitance of the on-chip decoupling capacitors are

$$R_{\mathrm{c}}^{C} = R_{\mathrm{p}}^{C} + R_{\mathrm{p}} = R_{\mathrm{r}} + R_{\mathrm{b}} + R_{\mathrm{p}}, \tag{5.37}$$

$$C_{\mathrm{c}} = \frac{L_{\mathrm{p}}^{C} + L_{\mathrm{p}}}{(R_{\mathrm{r}} + R_{\mathrm{b}} + R_{\mathrm{p}})^2}. \tag{5.38}$$

Note that the effective series resistance of the decoupling capacitors increases toward the load, $R_{\mathrm{b}}^{C} < R_{\mathrm{p}}^{C} < R_{\mathrm{c}}^{C}$, such that the resistance of the load current loop, no matter which decoupling capacitor terminates this loop, remains the same as the total resistance R_{tot} of the system without decoupling capacitors,

$$R_{\mathrm{b}}^{C} + R_{\mathrm{b}} + R_{\mathrm{p}} + R_{\mathrm{c}} =$$
$$R_{\mathrm{p}}^{C} + R_{\mathrm{p}} + R_{\mathrm{c}} = \tag{5.39}$$
$$R_{\mathrm{c}}^{C} + R_{\mathrm{c}} = R_{\mathrm{tot}}.$$

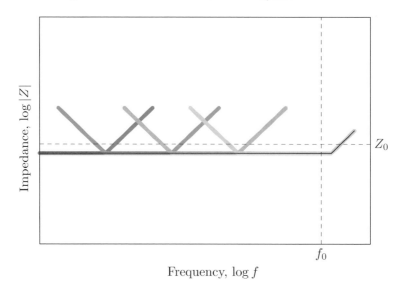

Fig. 5.22. Impedance diagram of a power distribution system with full compensation at each decoupling stage. The impedance of the decoupling stages is shown in shades of gray. The resulting system impedance is purely resistive and constant over the frequency range of interest, as shown by the black line.

If the resistance of the decoupling capacitors is reduced below the values determined by (5.33), (5.35), and (5.37), the resonance behavior degrades the impedance characteristics of the power distribution system.

5.8 Case study

A case study of a power distribution system for a high performance integrated circuit is presented in this section. This case study is intended to provide a practical perspective to the analytic description of the impedance characteristics of the power distribution systems developed in the previous section.

Consider a microprocessor consuming 60 watts from a 1.2 volt power supply. The average power current of the microprocessor is 50 amperes. The maximum frequency of interest is assumed to be 20 GHz, corresponding to the shortest gate switching time of approximately 17 ps. The objective is to limit the power supply variation to approximately 8% of the nominal 1.2 volt power supply level under a 50 ampere load. This objective results in a target impedance specification of

Table 5.1. Parameters of a case study power distribution system

Circuit parameter	Initial system	Near-critical damping	Fully compensated
R_r	$1\,\text{m}\Omega$	$1\,\text{m}\Omega$	$1\,\text{m}\Omega$
L_r	$10\,\text{nH}$	$10\,\text{nH}$	$10\,\text{nH}$
C_b	$5\,\text{mF}$	$5\,\text{mF}$	$10\,\text{mF}$
R_b^C	$0.1\,\text{m}\Omega$	$1\,\text{m}\Omega$	$1\,\text{m}\Omega$
L_b^C	$0.3\,\text{nH}$	$0.3\,\text{nH}$	$0.3\,\text{nH}$
R_b	$0.3\,\text{m}\Omega$	$0.3\,\text{m}\Omega$	$0.3\,\text{m}\Omega$
L_b	$0.2\,\text{nH}$	$0.2\,\text{nH}$	$0.2\,\text{nH}$
C_p	$250\,\mu\text{F}$	$250\,\mu\text{F}$	$296\,\mu\text{F}$
R_p^C	$0.2\,\text{m}\Omega$	$1.5\,\text{m}\Omega$	$1.3\,\text{m}\Omega$
L_p^C	$1\,\text{pH}$	$1\,\text{pH}$	$1\,\text{pH}$
R_p	$0.1\,\text{m}\Omega$	$0.1\,\text{m}\Omega$	$0.1\,\text{m}\Omega$
L_p	$1\,\text{pH}$	$1\,\text{pH}$	$1\,\text{pH}$
C_c	$500\,\text{nF}$	$500\,\text{nF}$	$1020\,\text{nF}$
R_c^C	$0.4\,\text{m}\Omega$	$1.5\,\text{m}\Omega$	$1.4\,\text{m}\Omega$
L_c^C	$1\,\text{fH}$	$1\,\text{fH}$	$1\,\text{fH}$
R_c	$0.05\,\text{m}\Omega$	$0.05\,\text{m}\Omega$	$0.05\,\text{m}\Omega$
L_c	$4\,\text{fH}$	$4\,\text{fH}$	$4\,\text{fH}$

0.08×1.2 volts/50 amperes $= 2$ milliohms over the frequency range from DC to 20 GHz. Three stages of decoupling capacitors are assumed. The parameters of the initial version of the power distribution system are displayed in Table 5.1.

The impedance characteristics of the overall power distribution system are shown in Fig. 5.23. The resonant modes of this system are significantly underdamped. The resulting impedance peaks exceed the target impedance specifications. The impedance characteristics improve significantly if the damping of the resonant mode is increased to the near-critical level. The greater damping can be achieved by only manipulating the effective series resistance of the decoupling capacitors. The impedance characteristics approach the target specifications, as shown in Fig. 5.23, as the resistance of the board, package, and on-chip capacitors is increased from initial values of 0.1, 0.2, and 0.4 milliohms to 1, 1.5, and 1.5 milliohms, respectively. Further improvements in the system impedance characteristics require increasing the decoupling capacitance. Fully compensating each decoupling stage renders the impedance purely resistive. The magnitude of the fully compensated

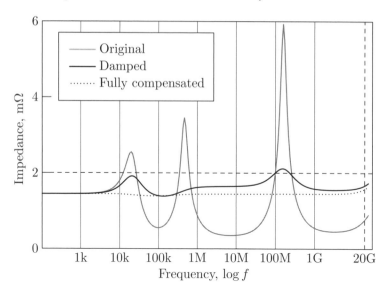

Fig. 5.23. Impedance characteristics of a power distribution system case study. The initial system is significantly underdamped. The resonant impedance peaks exceed the target impedance specification, as shown with a gray line. As the effective series resistance of the decoupling capacitors is increased, the damping of the resonant modes also increases, reducing the magnitude of the peak impedance. The damped system impedance effectively satisfies the target specifications, as shown with a solid black line. The impedance characteristics can be further improved if the impedance of each decoupling stage is fully compensated, as shown by the dotted line.

impedance equals the total resistance of the power distribution system, 1.45 milliohms ($1 + 0.3 + 0.1 + 0.05$), as described in Section 5.7.

The resonant frequencies of the case study are representative of typical resonant frequencies encountered in power distribution systems. The board decoupling stage resonates in the kilohertz range of frequencies, while the frequency of the resonant peak due to the package decoupling stage is in the low megahertz frequencies [108], [110]. The frequency of the chip capacitor resonance, often referred to as the chip-package resonance as this effect includes the inductance of the package interconnect, typically varies from tens of megahertz to low hundreds of megahertz [105], [111], [112].

5.9 Design considerations

The power current requirements of high performance integrated circuits are rapidly growing with technology scaling. These requirements necessitate a significant reduction in the output impedance of the power distribution system over a wider range of frequencies for new generations of circuits. Design approaches to improve the impedance characteristics of next generation power distribution systems are discussed in this section.

As described in previous sections, the inductance of the power distribution interconnect structures is efficiently excluded from the path of high frequency current by the hierarchical placement of decoupling capacitors. The inductance of the power and ground interconnect, however, greatly affects the decoupling capacitance requirements. As indicated by (5.16), (5.18), and (5.20), a lower inductance current path connecting the individual stages of decoupling capacitors relaxes the requirements placed on the decoupling capacitance at each stage of the power distribution network. Lowering the interconnect inductance decreases both the overall cost of the decoupling capacitors and the effective impedance of the power distribution system. It is therefore desirable to reduce the inductance of the power distribution interconnect.

As indicated by (5.18) and (5.20), the lower bound on the capacitance at each decoupling stage is determined by the effective inductance of the upstream current path $L^C + L^{\text{int}}$, which consists of the inductance of the previous stage decoupling capacitors L^C and the inductance of the interconnect connecting the two stages L^{int}. The impedance characteristics are thereby improved by lowering both the effective series inductance of the decoupling capacitors, as described in Section 5.9.1, and the interconnect inductance, as described in Section 5.9.2.

5.9.1 Inductance of the decoupling capacitors

The series inductance of the decoupling capacitance can be decreased by using a larger number of lower capacity capacitors to realize a specific decoupling capacitance rather than using fewer capacitors of greater capacity. Assume that a specific type of decoupling capacitor is used to realize a decoupling capacitance C, as shown in Fig. 5.7. Each capacitor has a capacity C_1 and a series inductance L_1^C. Placing $N_1 = \frac{C}{C_1}$ capacitors in parallel realizes the desired capacitance C with an effective series inductance $L^{C_1} = \frac{L_1^C}{N_1}$. Alternatively, using capacitors of lower capacity C_2 requires a larger number of capacitors $N_2 = \frac{C}{C_2}$ to realize the same capacitance C, but results in a lower overall inductance of the capacitor bank $L^{C_2} = \frac{L_2^C}{N_2}$. The efficacy of this approach is enhanced if the lower capacity component C_2 has a smaller form factor than the components of capacity C_1 and therefore has a significantly smaller series inductance L_2^C. Using a greater number of capacitors, however, requires additional board area and incurs higher component and assembly costs. Furthermore, the efficacy of the technique is diminished if the larger area required by the increased number of capacitors necessitates placing some of the capacitors at a greater distance from the load, increasing the inductance of the downstream current path L_2.

The series inductance of the decoupling capacitance L^C, however, constitutes only a portion of the overall inductance $L_2 + L^C$ of the current path between two stages of the decoupling capacitance, referring again to the circuit model shown in Fig. 5.7. Once the capacitor series inductance L^C is much lower than L_2, any further reduction of L^C has an insignificant effect on the overall impedance. The inductance of the current path between the decoupling capacitance and the load therefore imposes an upper limit on the frequency range of the capacitor efficiency. A reduction of the interconnect inductance is therefore necessary to improve the efficiency of the decoupling capacitors.

5.9.2 Interconnect inductance

Techniques for reducing interconnect inductance can be divided into *extensive* and *intensive* techniques. Extensive techniques lower inductance by using additional interconnect resources to form additional parallel current paths. Intensive techniques lower the inductance of existing current paths by modifying the structure of the power and ground interconnect so as to minimize the area of the current loop. The inductance of a power distribution system can be extensively decreased by allocating more metal layers on a printed circuit board and circuit package for power and ground distribution, increasing the number of package power and ground pins (solder balls in the case of ball grid array mounting) connecting the package to the board, and increasing the number of power and ground solder bumps or bonding wires connecting the die to the package. Extensive methods are often constrained by technological and cost considerations, such as the number and thickness of the metal and isolation layers in a printed circuit board, the total number of pins in a specific package, and the die bonding technology. Choosing a solution with greater interconnect capacity typically incurs a significant cost penalty.

Intensive methods reduce the interconnect inductance without incurring higher manufacturing costs. These methods alter the interconnect structure in order to decrease the area of the power current loop. For example, the inductance of a current path formed by two square parallel power and ground planes is proportional to the separation of the planes[3] h,

$$L_{\text{plane}} = \mu_0 h. \tag{5.40}$$

Thus, parallel power and ground planes separated by $1\,\text{mil}$ ($25.4\,\mu\text{m}$) have an inductance of $32\,\text{pH}$ per square. A smaller separation between the power and ground planes results in a proportionally smaller inductance of the power current loop. Reducing the thickness of the dielectric between the power and ground planes at the board and package levels is an effective means to reduce the impedance of the power distribution network [113], [114].

The same strategy of placing the power and ground paths in close proximity can be exploited to lower the effective inductance of the "vertical conductors," *i.e.*, the conductors connecting the parallel power and ground planes or planar networks. Examples of such connections are pin grid arrays and ball grid arrays connecting a package to a board, a flip-chip area array of solder balls connecting a die to a package, and an array of vias connecting the power and ground planes within a package. In regular pin and ball grid arrays, this strategy leads to a so-called checkerboard pattern [115], as shown in Fig. 5.24.

5.10 Limitations of the one-dimensional circuit model

The one-dimensional lumped circuit model shown in Fig. 5.3 has been used to describe the frequency dependent impedance characteristics of

---

[3] Equation (5.40) neglects the internal inductance of the metal planes, *i.e.*, the inductance associated with the magnetic flux within the planes. Omission of the internal inductance is a good approximation where the thickness of the planes is much smaller than the separation between the planes and at high signal frequencies, where current flow is restricted to the surface of the planes due to a pronounced skin (proximity) effect.

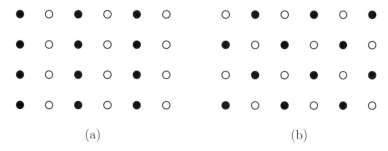

Fig. 5.24. Placement of area array connections for low inductance. The inductance of the area array interconnect strongly depends on the pattern of placement of the power and ground connections. The inductance of pattern (a) is relatively high, while pattern (b) has the lowest inductance. The power and ground connections are shown in black and white, respectively.

power distribution systems. This model captures the essential characteristics of power distribution systems over a wide range of frequencies. Due to the relative simplicity, the model is an effective vehicle for demonstrating the primary issues in the design of power distribution systems and related design challenges and tradeoffs. The use of this model for other purposes, however, is limited due to certain simplifications present in the model. Several of these limitations are summarized below.

A capacitor at each decoupling stage is modeled by a single RLC circuit. In practice, however, the decoupling capacitance on the board and package is realized by a large number of discrete capacitors. Each capacitor has a distinct physical location relative to the load. The parasitic inductance of the current path between a specific capacitor and the power load is somewhat different for each of the capacitors. The frequency of the resonant impedance peaks of an individual capacitor therefore varies depending upon the location of the capacitor. The overall impedance peak profile of a group of capacitors is therefore spread, resulting in a lower and wider resonant peak.

More significantly, it is common to use two different types of capacitors for board level decoupling. High capacity electrolytic capacitors, typically ranging from a few hundreds to a thousand microfarads, are used to obtain the high capacitance necessary to decouple the high inductive impedance of a voltage regulator. Electrolytic capacitors, however, have a relatively high series inductance of several nanohenrys. Ceramic capacitors with a lower capacity, typically tens of microfarads, and a lower parasitic inductance, a nanohenry or less, are used to extend the frequency range of the low impedance region to higher frequencies. Effectively, there are two decoupling stages on the board, with a very small parasitic impedance between the two stages.

Using lumped circuit elements implies that the wavelength of the signals of interest is much larger than the physical dimensions of the circuit structure, permitting an accurate representation of the impedance characteristics with a few lumped elements connected in series. The current transitions at the power load are measured in tens of picoseconds, translating to thousands of micrometers of signal wavelength. This wavelength is much smaller than the size of a power distribution system, typically of several inches, making the use of a lumped model at first glance unjustified. The properties of power distribution systems, however, support a lumped circuit representation over a wider range of conditions as suggested by the aforementioned simple size criterion.

The design of a power distribution system is intended to restrict the flow of the power current as close to the load as possible. Due to the hierarchy of the decoupling capacitors, the higher the current frequency, the shorter the current loop, confining the current flow closer to the load. As seen from the terminals of the power source, a power distribution system is a multi-stage low pass filter, each stage having a progressively lower cut-off frequency. The spectral content of the current in the on-board power distribution system is limited to several megahertz. The corresponding wavelength is much larger than the typical system dimensions of a few inches, making a lumped model sufficiently accurate. The spectral content of the power current within the package network extends into the hundreds of megahertz frequency range. The signal wavelength remains sufficiently large to use a lumped model; however, a more detailed network may be required rather than a one-dimensional model to accurately characterize the network impedances.

A one-dimensional model is inadequate to describe an on-chip power distribution network. The on-chip power distribution network is the most challenging element of the power delivery system design problem. The board and package level power distribution networks consist of several metal planes and thousands of vias, pins, and traces as well as several dozen of decoupling capacitors. In comparison, the on-chip power distribution network in a high complexity integrated circuit typically consists of millions of line segments, dramatically exacerbating the complexity of the design and analysis process. The design and analysis of on-chip power distribution networks is the focus of the Chapter 7.

5.11 Summary

The impedance characteristics of power distribution systems with multiple stages of decoupling capacitances have been described in this chapter. These impedance characteristics can be briefly summarized as follows.

- The significant inductance of the power and ground interconnect is the primary obstacle to achieving a low output impedance power distribution system

- The hierarchical placement of decoupling capacitors achieves a low output impedance in a cost effective manner by terminating the power current loop progressively closer to the load as the frequency increases

- The capacitance and effective series inductance determine the frequency range where the decoupling capacitor is effective

- Resonant circuits are formed within the power distribution networks due to the placement of the decoupling capacitors, increasing the output impedance near the resonant signal frequencies

- The effective series resistance of the decoupling capacitors is a critical factor in controlling the resonant phenomena

- The lower the inductance of the power interconnect and decoupling capacitors, the lower the decoupling capacitance necessary to achieve the target impedance characteristics

6

Decoupling Capacitance

The on-going miniaturization of integrated circuit feature sizes has placed significant requirements on the on-chip power and ground distribution networks. Circuit integration densities rise with each VDSM technology generation due to smaller devices and larger dies. The on-chip current densities and the total current also increase. Simultaneously, the higher switching speed of smaller transistors produces faster current transients in the power distribution network. Supplying high average currents and continuously increasing transient currents through the high impedance on-chip interconnects results in significant fluctuations of the power supply voltage in scaled CMOS technologies.

Such a change in the supply voltage is referred to as power supply noise. Power supply noise adversely affects circuit operation through several mechanisms, as described in Chapter 1. Supplying sufficient power current to high performance ICs has therefore become a challenging task. Large average currents result in increased IR noise and fast current transients result in increased $L\frac{dI}{dt}$ voltage drops (ΔI noise) [21].

Decoupling capacitors are often utilized to manage this power supply noise. Decoupling capacitors can have a significant effect on the principal characteristics of an integrated circuit, i.e., speed, cost, and power. Due to the importance of decoupling capacitors in current and future ICs, significant research has been developed over the past several decades, covering different areas such as hierarchical placement of decoupling capacitors, sizing and placing of on-chip decoupling capacitors, resonant phenomenon in power distribution systems with decoupling capacitors, and static on-chip power dissipation due to leakage current through the gate oxide.

In this chapter, a brief review of the background of decoupling capacitance is provided. In Section 6.1, the concept of a decoupling capacitance is introduced and an historical retrospective is described. A practical model of a decoupling capacitor is also introduced. In Section 6.2, the impedance of a power distribution system with decoupling capacitors is presented. Target specifications of the impedance of a power distribution system are reviewed. Antiresonance phenomenon in a system with decoupling capacitors is intuitively explained. A hydraulic analogy of the hierarchical placement of decoupling capacitors is also presented. Intrinsic and intentional on-chip decoupling capacitances are discussed and compared in Section 6.3. Different types of on-chip decoupling capacitors are qualitatively analyzed in Section 6.4. The advantages and disadvantages of several types of widely used on-chip decoupling capacitors are also discussed in Section 6.4. Enhancing the efficiency of on-chip decoupling capacitors with a switching voltage regulator is presented in Section 6.5. Finally, some conclusions are offered in Section 6.6.

6.1 Introduction to decoupling capacitance

Decoupling capacitors are often used to maintain the power supply voltage within specification so as to provide signal integrity while reducing electromagnetic interference (EMI) radiated noise. In this book, the use of decoupling capacitors to mitigate power supply noise is investigated. The concept of a decoupling capacitor is introduced in this section. An historical retrospective is presented in Section 6.1.1. A description of a decoupling capacitor as a reservoir of charge is discussed in Section 6.1.2. Decoupling capacitors are shown to be an effective way to provide sufficient charge to a switching current load within a short period of time. A practical model of a decoupling capacitor is presented in Section 6.1.3.

6.1.1 Historical retrospective

About 600 BC, Thales of Miletus recorded that the ancient Greeks could generate sparks by rubbing balls of amber on spindles [116]. This is the triboelectric effect [117], the mechanical separation of charge in a dielectric (insulator). This effect is the basis of the capacitor.

In October 1745, Ewald Georg von Kleist of Pomerania invented the first recorded capacitor: a glass jar coated inside and out with metal.

The inner coating was connected to a rod that passed through the lid and ended in a metal sphere, as shown in Fig. 6.1 [118]. By layering the insulator between two metal plates, von Kleist dramatically increased the charge density. Before Kleist's discovery became widely known, a Dutch physicist, Pieter van Musschenbroek, independently invented a similar capacitor in January 1746 [119]. It was named the Leyden jar, after the University of Leyden where van Musschenbroek worked.

Benjamin Franklin investigated the Leyden jar and proved that the charge was stored on the glass, not in the water as others had assumed [120]. Originally, the units of capacitance were in "jars." A jar is equivalent to about 1 nF. Early capacitors were also known as *condensors*, a term that is still occasionally used today. The term condensor was coined by Alessandro Volta in 1782 (derived from the Italian *condensatore*), referencing the ability of a device to store a higher density of electric charge than a normal isolated conductor [120].

6.1.2 Decoupling capacitor as a reservoir of charge

A capacitor consists of two electrodes, or plates, each of which stores an equal amount of opposite charge. These two plates are conductive and are separated by an insulator (dielectric). The charge is stored on the surface of the plates at the boundary with the dielectric. Since each plate stores an equal but opposite charge, the net charge across the capacitor is always zero.

The capacitance C of a capacitor is a measure of the amount of charge Q stored on each plate for a given potential difference (voltage V) which appears between the plates,

$$C = \frac{Q}{V}. \tag{6.1}$$

The capacitance is proportional to the surface area of the conducting plate and inversely proportional to the distance between the plates [121]. The capacitance is also proportional to the permittivity of the dielectric substance that separates the plates. The capacitance of a parallel plate capacitor is

$$C \approx \frac{\epsilon A}{d}, \tag{6.2}$$

where ϵ is the permittivity of the dielectric, A is the area of the plates, and d is the spacing between the plates. Equation (6.2) is only accurate for a plate area much greater than the spacing between the plates,

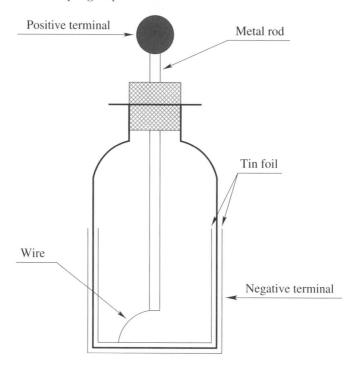

Fig. 6.1. Leyden jar originally developed by Ewald Georg von Kleist in 1745 and independently invented by Pieter van Musschenbroek in 1746. The charge is stored on the glass between two tin foils (capacitor plates) [118].

$A \gg d^2$. In general, the capacitance of the metal interconnects placed over the substrate is composed of three primary components: a parallel plate capacitance, fringe capacitance, and lateral flux (side) capacitance [122], as shown in Fig. 6.2. Accurate closed-form expressions have been developed by numerically fitting a model that describes parallel lines above the plane or between two parallel planes [123], [124], [125], [126], [127], [128].

As opposite charge accumulates on the plates of a capacitor across an insulator, a voltage develops across the capacitor due to the electric field formed by the opposite charge. Work must be done against this electric field as more charge is accumulated. The energy stored in a capacitor is equal to the amount of work required to establish the voltage across the capacitor. The energy stored in the capacitor is

$$E_{\text{stored}} = \frac{1}{2}CV^2 = \frac{1}{2}\frac{Q^2}{C} = \frac{1}{2}VQ. \tag{6.3}$$

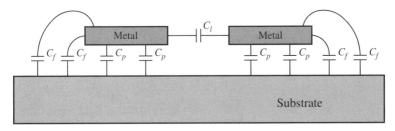

Fig. 6.2. Capacitance of two metal lines placed over a substrate. Three primary components compose the total capacitance of the on-chip metal interconnects. C_l denotes the lateral flux (side) capacitance, C_f denotes the fringe capacitance, and C_p denotes the parallel plate capacitance.

From a physical perspective, a decoupling capacitor serves as an intermediate storage of charge and energy. The decoupling capacitor is located between the power supply and current load, *i.e.*, electrically closer to the switching circuit. The decoupling capacitor is therefore more efficient in terms of supplying charge as compared to a remote power supply. The amount of charge stored on the decoupling capacitor is limited by the voltage and the capacitance. Unlike a decoupling capacitor, the power supply can provide an almost infinite amount of charge. A hydraulic model of a decoupling capacitor is illustrated in Fig. 6.3. Similar to water stored in a water tank and connected to the consumer through a system of pipes, the charge on the decoupling capacitor stored between the conductive plates is connected to the current load through a hierarchical interconnect system. To be effective, the decoupling capacitor should satisfy two requirements. First, the capacitor should have sufficient capacity to store a significant amount of energy. Second, to supply sufficient power at high frequencies, the capacitor should be able to release and accumulate energy at a high rate.

6.1.3 Practical model of a decoupling capacitor

Decoupling capacitors are often used in power distribution systems to provide the required charge in a timely manner and to reduce the output impedance of the overall power delivery network [48]. An ideal decoupling capacitor is effective over the entire frequency range: from DC to the maximum operating frequency of a system. Practically, a decoupling capacitor is only effective over a certain frequency range. The impedance of a practical decoupling capacitor decreases linearly with

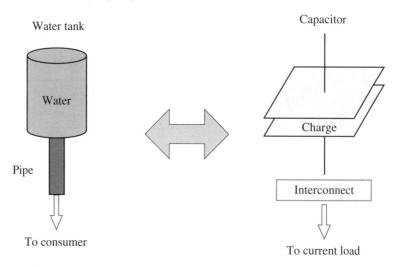

Water tank

Capacitor

Water

Charge

Pipe

Interconnect

To consumer

To current load

Fig. 6.3. Hydraulic model of a decoupling capacitor as a reservoir of charge. Similar to water stored in a water tank and connected to the consumer through a system of pipes, charge on the decoupling capacitor is stored between the conductive plates connected to the current load through a hierarchical interconnect system.

frequency at low frequencies (with a slope of -20 dB/dec in a logarithmic scale). As the frequency increases, the impedance of the decoupling capacitor increases linearly with frequency (with a slope of 20 dB/dec in a logarithmic scale), as shown in Fig. 6.4. This increase in the impedance of a practical decoupling capacitor is due to the parasitic inductance of the decoupling capacitor. The parasitic inductance is referred to as the effective series inductance (ESL) of a decoupling capacitor [103].The impedance of a decoupling capacitor reaches the minimum impedance at the frequency $\omega = \frac{1}{\sqrt{LC}}$. This frequency is known as the resonant frequency of a decoupling capacitor. Observe that the absolute minimum impedance of a decoupling capacitor is limited by the parasitic resistance, *i.e.*, the effective series resistance (ESR) of a decoupling capacitor.The parasitic resistance of a decoupling capacitor is due to the resistance of the metal leads and conductive plates and the dielectric losses of the insulator. The ESR and ESL of an on-chip metal-oxide-semiconductor (MOS) decoupling capacitor are illustrated in Fig. 6.5. Note that the parasitic inductance of the decoupling capacitor is determined by the area of the current loops, decreasing with smaller area, as shown in Fig. 6.5(b) [129].

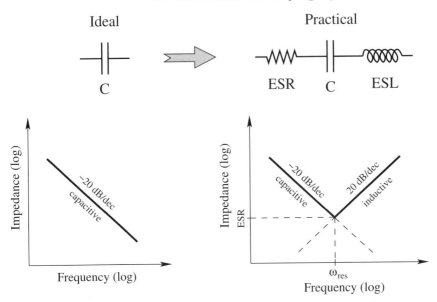

Fig. 6.4. Practical model of a decoupling capacitor. The impedance of a practical decoupling capacitor decreases linearly with frequency, reaching the minimum at a resonant frequency. Beyond the resonant frequency, the impedance of the decoupling capacitor increases linearly with frequency due to the ESL. The minimum impedance is determined by the ESR of the decoupling capacitor.

The impedance of a decoupling capacitor depends upon a number of characteristics. For instance, as the capacitance is increased, the capacitive curve moves down and to the right (see Fig. 6.4). Since the parasitic inductance for a particular capacitor is fixed, the inductive curve remains unaffected. As different capacitors are selected, the capacitive curve moves up and down relative to the fixed inductive curve. The primary way to decrease the total impedance of a decoupling capacitor for a specific semiconductor package is to increase the value of the capacitor [131]. Note that to move the inductive curve down, lowering the total impedance characteristics, a number of decoupling capacitors should be connected in parallel. In the case of identical capacitors, the total impedance is reduced by a factor of two for each doubling in the number of capacitors [107].

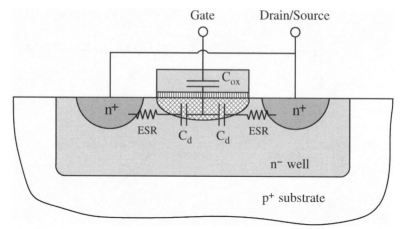

(a) ESR of a MOS-based decoupling capacitor. The ESR of an on-chip MOS decoupling capacitor is determined by the doping profiles of the n^+ regions and n^- well, the size of the capacitor, and the impedance of the vias and gate material [130].

(b) ESL of a MOS-based decoupling capacitor. The ESL of an on-chip MOS decoupling capacitor is determined by the area of the current return loops. The parasitic inductance is lowered by shrinking the area of the current return loops.

Fig. 6.5. Physical structure of an on-chip MOS decoupling capacitor.

6.2 Impedance of power distribution system with decoupling capacitors

As described in Section 6.1.2, a decoupling capacitor serves as a reservoir of charge, providing the required charge to the switching current load. Decoupling capacitors are also used to lower the impedance of the power distribution system. The impedance of a decoupling capacitor decreases rapidly with frequency, shunting the high frequency currents and reducing the effective current loop of a power distribution network. The impedance of the overall power distribution system with decoupling capacitors is the subject of this section. In Section 6.2.1, the target impedance of a power distribution system is introduced. It is shown that the impedance of a power distribution system should be maintained below a target level to guarantee fault-free operation of the entire system. The antiresonance phenomenon is presented in Section 6.2.2. A hydraulic analogy of a system of decoupling capacitors is described in Section 6.2.3. The analogy is drawn between a water supply system and the hierarchical placement of decoupling capacitors at different levels of a power delivery network.

6.2.1 Target impedance of a power distribution system

To ensure a small variation in the power supply voltage under a significant current load, the power distribution system should exhibit a small impedance as seen from the current load within the frequency range of interest [132]. A circuit network representing the impedance of a power distribution system as seen from the terminals of the current load is shown in Fig. 6.6.

The impedance of a power distribution system is with respect to the terminals of the load circuits. In order to ensure correct and reliable operation of an IC, the impedance of a power distribution system should be maintained below a certain upper bound Z_{target} in the frequency range from DC to the maximum operating frequency f_0 of the system [133], [134], [135]. The maximum tolerable impedance of a power distribution system is henceforth referred to as the target impedance. Note that the maximum operating frequency f_0 is determined by the switching time of the on-chip signal transients, rather than by the clock frequency. The shortest signal switching time is typically an order of magnitude smaller than the clock period. The maximum operating frequency is therefore considerably higher than the clock frequency.

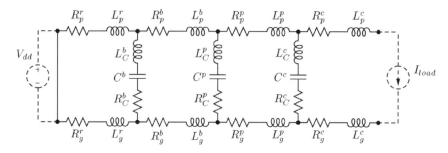

Fig. 6.6. A circuit network representing the impedance of a power distribution system with decoupling capacitors as seen from the terminals of the current load. The ESR and ESL of the decoupling capacitors are also included. Subscript p denotes the power paths and subscript g denotes the ground path. Superscripts r, b, p, and c refer to the voltage regulator, board, package, and on-chip power delivery networks, respectively.

One primary design objective of an effective power distribution system is to ensure that the output impedance of the network is below a target output impedance level. It is therefore important to understand how the output impedance of the circuit, shown schematically in Fig. 6.6, depends upon the impedance of the comprising circuit elements. A power distribution system with no decoupling capacitors is shown in Fig. 6.7. The power source and load are connected by interconnect with resistive and inductive parasitic impedances. The magnitude of the impedance of this network is

$$|Z_{\text{tot}}(\omega)| = |R_{\text{tot}} + j\omega L_{\text{tot}}|, \tag{6.4}$$

where R_{tot} and L_{tot} are the total resistance and inductance of the power distribution system, respectively,

$$R_{\text{tot}} = R_{\text{tot}}^p + R_{\text{tot}}^g, \tag{6.5}$$
$$R_{\text{tot}}^p = R_p^r + R_p^b + R_p^p + R_p^c, \tag{6.6}$$
$$R_{\text{tot}}^g = R_g^r + R_g^b + R_g^p + R_g^c, \tag{6.7}$$
$$L_{\text{tot}} = L_{\text{tot}}^p + L_{\text{tot}}^g, \tag{6.8}$$
$$L_{\text{tot}}^p = L_p^r + L_p^b + L_p^p + L_p^c, \tag{6.9}$$
$$L_{\text{tot}}^g = L_g^r + L_g^b + L_g^p + L_g^c. \tag{6.10}$$

The variation of the impedance with frequency is illustrated in Fig. 6.8. To satisfy a specification at low frequency, the resistance of

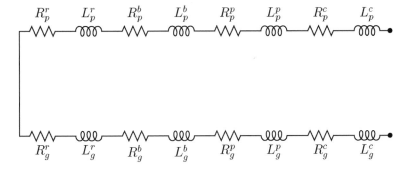

Fig. 6.7. A circuit network representing the impedance of a power distribution system without decoupling capacitors.

the power delivery network should be sufficiently low, $R_{\text{tot}} < Z_{\text{target}}$. Above the frequency $f_{L_{\text{tot}}} = \frac{1}{2\pi} \frac{R_{\text{tot}}}{L_{\text{tot}}}$, however, the impedance of the power delivery network is dominated by the inductive reactance $j\omega L_{\text{tot}}$ and increases linearly with frequency, exceeding the target impedance at the frequency $f_{\text{max}} = \frac{1}{2\pi} \frac{Z_{\text{target}}}{L_{\text{tot}}}$.

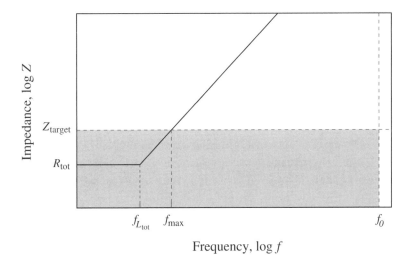

Fig. 6.8. Impedance of a power distribution system without decoupling capacitors. The shaded area denotes the target impedance specifications of the overall power distribution system.

The high frequency impedance should be reduced to satisfy the target specifications. Opportunities for reducing the inductance of the power and ground paths of a power delivery network are limited [23], [136], [137], [138], [139]. The inductance of the power distribution system is mainly determined by the board and package interconnects [140], [141], [142]. The feature size of the board and package level interconnect depends upon the manufacturing technology. The output impedance of a power distribution system is therefore highly inductive and is difficult to lower [105].

The high frequency impedance is effectively reduced by placing capacitors across the power and ground interconnections. These shunting capacitors effectively terminate the high frequency current loop, permitting the current to bypass the inductive interconnect, such as the board and package power delivery networks [143], [144], [145], [146]. The high frequency impedance of the system as seen from the current load terminals is thereby reduced. Alternatively, at high frequencies, the capacitors decouple the high impedance paths of the power delivery network from the load. These capacitors are therefore referred to as decoupling capacitors [147], [148]. Several stages of decoupling capacitors are typically utilized to maintain the output impedance of a power distribution system below a target impedance [107], [149], as described in Section 6.2.3.

6.2.2 Antiresonance

Decoupling capacitors are a powerful technique to reduce the impedance of a power distribution system over a significant range of frequencies. A decoupling capacitor, however, reduces the resonant frequency of a power delivery network, making the system susceptible to resonances. Unlike the classic self-resonance in a series circuit formed by a decoupling capacitor combined with a parasitic resistance and inductance [109], [150] or by an on-chip decoupling capacitor and the parasitic inductance of the package (*i.e.*, chip-package resonance) [151], [152], antiresonance occurs in a circuit formed by two capacitors connected in parallel. At the resonant frequency, the impedance of the series circuit decreases in the vicinity of the resonant frequency, reaching the absolute minimum at the resonant frequency determined by the ESR of the decoupling capacitor. At antiresonance, however, the circuit impedance drastically increases, producing a distinctive peak, as illustrated in Fig. 6.9. This antiresonant peak can result in system failures as the impedance of the power distribution system becomes greater than the maximum tolerable impedance Z_{target}. The antiresonance phenomenon in a system with parallel decoupling capacitors is the subject of this section.

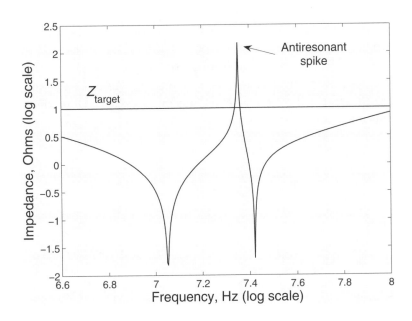

Fig. 6.9. Antiresonance of the output impedance of a power distribution network. Antiresonance results in a distinctive peak, exceeding the target impedance specification.

To achieve a low impedance power distribution system, multiple decoupling capacitors are placed in parallel. The effective impedance of a power distribution system with several identical capacitors placed in parallel is illustrated in Fig. 6.10. Observe that the impedance of the power delivery network is reduced by a factor of two as the number of capacitors is doubled. Also note that the effective drop in the impedance of a power distribution system diminishes rapidly with each additional decoupling capacitor. It is therefore desirable to utilize decoupling capacitors with a sufficiently low ESR in order to minimize the number of capacitors required to satisfy a target impedance specification [107].

A number of decoupling capacitors with different magnitudes is typically used to maintain the impedance of a power delivery system below a target specification over a wide frequency range. Capacitors with different magnitudes connected in parallel, however, result in a sharp antiresonant peak in the system impedance [27]. The antiresonance phenomenon for different capacitive values is illustrated in Fig. 6.11. The antiresonance of parallel decoupling capacitors can be explained as follows. In the frequency range from f_1 to f_2, the impedance of the capacitor C_1 has become inductive whereas the impedance of the capacitor C_2 remains capacitive (see Fig. 6.11). Thus, an LC tank is formed in the frequency range from f_1 to f_2, producing a peak at the resonant frequency located between f_1 and f_2. As a result, the total impedance drastically increases and becomes greater than the target impedance, causing a system to fail.

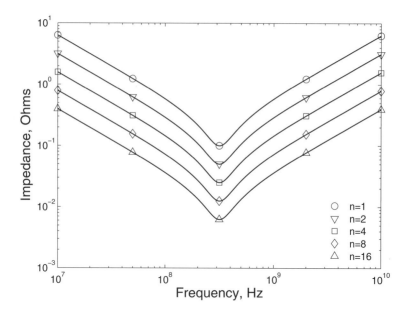

Fig. 6.10. Impedance of a power distribution system with n identical decoupling capacitors connected in parallel. The ESR of each decoupling capacitor is $R = 0.1\,\Omega$, the ESL is $L = 100\,\mathrm{pH}$, and the capacitance is $C = 1\,\mathrm{nF}$. The impedance of a power distribution system is reduced by a factor of two as the number of capacitors is doubled.

The magnitude of the antiresonant spike can be effectively reduced by lowering the parasitic inductance of the decoupling capacitors. For instance, as discussed in [107], the magnitude of the antiresonant spike is significantly reduced if board decoupling capacitors are mounted on low inductance pads. The magnitude of the antiresonant spike is also determined by the ESR of the decoupling capacitor, decreasing with larger parasitic resistance. Large antiresonant spikes are produced when low ESR decoupling capacitors are placed on inductive pads. A high inductance and low resistance result in a parallel LC circuit with a high quality factor Q,

$$Q = \frac{L}{R}. \tag{6.11}$$

In this case, the magnitude of the antiresonant spike is amplified by Q. Decoupling capacitors with a low ESR should therefore always be used on low inductance pads (with a low ESL).

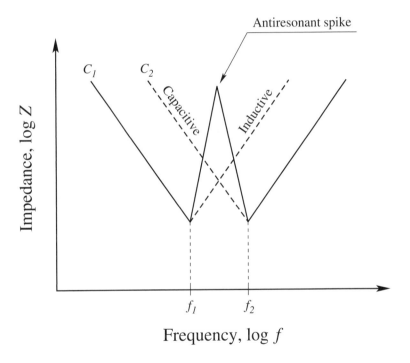

Fig. 6.11. Antiresonance of parallel capacitors, $C_1 > C_2$, $L_1 = L_2$, and $R_1 = R_2$. A parallel LC tank is formed in the frequency range from f_1 to f_2. The total impedance drastically increases in the frequency range from f_1 to f_2 (the solid line), producing an antiresonant spike.

Antiresonance also becomes well pronounced if a large variation exists between the capacitance values. This phenomenon is illustrated in Fig. 6.12. In the case of two capacitors with distinctive nominal values ($C_1 \gg C_2$), a significant gap between two capacitances results in a sharp antiresonant spike with a large magnitude in the frequency range from f_1 to f_2, violating the target specification Z_{target}, as shown in Fig. 6.12(a). If another capacitor with nominal value $C_1 > C_3 > C_2$ is added, the antiresonant spike is canceled by C_3 in the frequency range from f_1 to f_2. As a result, the overall impedance of a power distribution system is maintained below the target specification over a broader frequency range, as shown in Fig. 6.12(b). As described in [153], the high frequency impedance of two parallel decoupling capacitors is only reduced by a factor of two (or 6 dB) as compared to a single capacitor. It is also shown that adding a smaller capacitor in parallel with a large capacitor results in only a small reduction in the high frequency impedance. Antiresonances are effectively managed by utilizing decoupling capacitors with a low ESL and by placing a greater number of decoupling capacitors with progressively decreasing magnitude, shifting the antiresonant spike to the higher frequencies (out of the range of the operating frequencies of the circuit) [154].

6.2.3 Hydraulic analogy of hierarchical placement of decoupling capacitors

As discussed in Section 6.1.2, an ideal decoupling capacitor should provide a high capacity and be able to release and accumulate energy at a sufficiently high rate. Constructing a device with both high energy capacity and high power capability is, however, challenging. It is expensive to satisfy both of these requirements in an ideal decoupling capacitor. Moreover, these requirements are typically contradictory in most practical applications. The physical realization of a large decoupling capacitance requires the use of discrete capacitors with a large nominal capacity and, consequently, a large form factor. The large physical dimensions of the capacitors have two implications. The parasitic series inductance of a physically large capacitor is relatively high due to the increased area of the current loop within the capacitors. Furthermore, due to technology limitations, the large physical size of the capacitors prevents placing the capacitors sufficiently close to the current load. A greater physical separation increases the inductance of the current path from the capacitors to the load. A tradeoff therefore exists between the

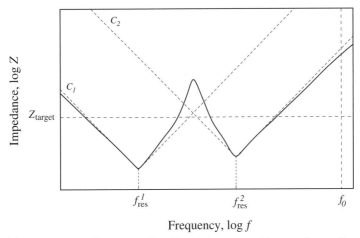

(a) Impedance of a power distribution system with two decoupling capacitors, $C_1 \gg C_2$

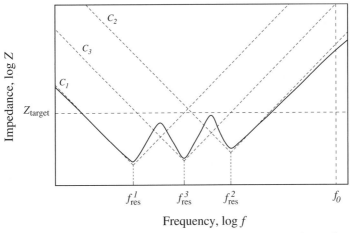

(b) Impedance of a power distribution system with three decoupling capacitors, $C_1 > C_3 > C_2$

Fig. 6.12. Antiresonance of parallel capacitors. (a) A large gap between two capacitances results in a sharp antiresonant spike with a large magnitude in the frequency range from f_1 to f_2, violating the target specification Z_{target}. (b) If another capacitor with magnitude $C_1 > C_3 > C_2$ is added, the antiresonant spike is canceled by C_3 in the frequency range from f_1 to f_2. As a result, the overall impedance of the power distribution system is maintained below the target specification over the desired frequency range.

high capacity and low parasitic inductance of a decoupling capacitor
for an available component technology.

Gate switching times of a few tens of picoseconds are common in
modern high performance ICs, creating high transient currents in the
power distribution system. At high frequencies, only those on-chip decoupling capacitors with a low ESR and a low ESL can effectively
maintain a low impedance power distribution system. Placing a sufficiently large on-chip decoupling capacitor requires a die area many
times greater than the area of a typical circuit. Thus, while technically
feasible, a single-tier decoupling solution is prohibitively expensive. A
large on-chip decoupling capacitor is therefore typically built as a series of small decoupling capacitors connected in parallel. At high frequencies, a large on-chip decoupling capacitor exhibits a distributed
behavior. Only on-chip decoupling capacitors located in the vicinity of
the switching circuit can effectively provide the required charge to the
current load within the proper time. An efficient approach to this problem is to hierarchically place multiple stages of decoupling capacitors,
progressively smaller and closer to the load.

Utilizing hierarchically placed decoupling capacitors produces a low
impedance, high frequency power distribution system realized in a cost
effective way. The capacitors are placed in several stages: on the board,
package, and circuit die. Arranging the decoupling capacitors in several
stages eliminates the need to satisfy both the high capacitance and low
inductance requirements in the same decoupling stage [28].

The hydraulic analogy of the hierarchical placement of decoupling
capacitors is shown in Fig. 6.13. Each decoupling capacitor is represented by a water tank. All of the water tanks are connected to the
main water pipe connected to the consumer (current load). Water tanks
at different stages are connected to the main pipe through the local water pipes, modeling different interconnect levels. The goal of the water
supply system (power delivery network) is to provide uninterrupted
water flow to the consumer at the required rate (switching time). The
amount of water released by each water tank is proportional to the
tank size. The rate at which the water tank is capable of providing water is inversely proportional to the size of the water tank and directly
proportional to the distance from the consumer to the water tank.

A power supply is typically treated as an infinite amount of charge.
Due to large physical dimensions, the power supply cannot be placed
close to the current load (the consumer). The power supply therefore

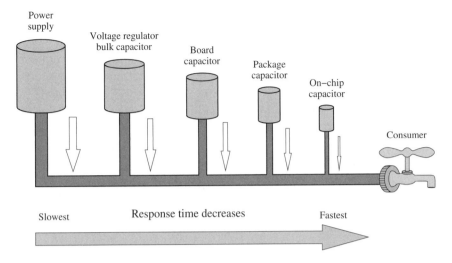

Fig. 6.13. Hydraulic analogy of the hierarchical placement of decoupling capacitors. The decoupling capacitors are represented by the water tanks. The response time is proportional to the size of the capacitor and inversely proportional to the distance from a capacitor to the consumer. The on-chip decoupling capacitor has the shortest response time (located closest to the consumer), but is capable of providing the least amount of charge.

has a long response time. Unlike the power supply, an on-chip decoupling capacitor can be placed sufficiently close to the consumer. The response time of an on-chip decoupling capacitor is significantly shorter as compared to the power supply. An on-chip decoupling capacitor is therefore able to respond to the consumer demand in a much shorter period of time but is capable of providing only a small amount of water (or charge). Allocating decoupling capacitors with progressively decreasing magnitudes and closer to the current load, an uninterrupted flow of charge can be provided to the consumer. In the initial moment, charge is only supplied to the consumer by the on-chip decoupling capacitor. As the on-chip decoupling capacitor is depleted, the package decoupling capacitor is engaged. This process continues until the power supply is activated. Finally, the power supply is turned on and provides the necessary charge with relatively relaxed timing constraints. The voltage regulator, board, package, and on-chip decoupling capacitors therefore serve as intermediate reservoirs of charge, relaxing the timing constraints for the power delivery supply.

A hierarchy of decoupling capacitors is utilized in high performance power distribution systems in order to extend the frequency region of the low impedance characteristics to the maximum operating frequency f_0. The impedance characteristics of a power distribution system with board, package, and on-chip decoupling capacitors (see Fig. 6.6) are illustrated in Fig. 6.14. By utilizing the hierarchical placement of decoupling capacitors, the antiresonant spike is shifted outside the range of operating frequencies (beyond f_0). The overall impedance of a power distribution system is also maintained below the target impedance over the entire frequency range of interest (from DC to f_0).

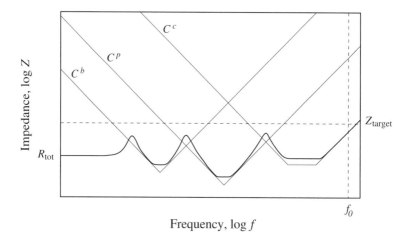

Fig. 6.14. Impedance of a power distribution system with board, package, and on-chip decoupling capacitances. The overall impedance is shown with a black line. The impedance of a power distribution system with three levels of decoupling capacitors is maintained below the target impedance (dashed line) over the frequency range of interest. The impedance characteristics of the decoupling capacitors are shown by the thin solid lines.

Fully compensated system

A special case in the impedance of an RLC circuit formed by a decoupling capacitor and the parasitic inductance of the P/G lines is achieved when the zeros of a tank circuit impedance cancel the poles, making the impedance purely resistive and independent of frequency,

$$R_L = R_C = R_0 = \sqrt{\frac{L}{C}}, \tag{6.12}$$

$$\frac{L}{R_L} = R_C C, \tag{6.13}$$

where R_L and R_C are the parasitic resistance of the P/G lines and the ESR of the decoupling capacitor, respectively. In this case, the impedance of the RLC tank is fully compensated. Equations (6.12) and (6.13) are equivalent to two conditions, *i.e.*, the impedance at the lower frequencies is matched to the impedance at the high frequencies and the time constants of the inductor and capacitor currents are also matched. A constant, purely resistive impedance, characterizing a power distribution system with decoupling capacitors, is achieved across the entire frequency range of interest, if each decoupling stage is fully compensated [108], [155]. The resistance and capacitance of the decoupling capacitors in a fully compensated system are completely determined by the impedance characteristics of the power and ground interconnect and the location of the decoupling capacitors.

The hierarchical placement of decoupling capacitors exploits the tradeoff between the capacity and the parasitic inductance of a capacitor to achieve an economically effective solution. The total decoupling capacitance of a hierarchical scheme $C_{\text{total}} = C^b + C^p + C^c$ is larger than the total decoupling capacitance of a single-tier solution, where C^b, C^p, and C^c are the board, package, and on-chip decoupling capacitances, respectively. The primary advantage of utilizing a hierarchical placement is that the inductive limit is imposed only on the final stage of decoupling capacitors which constitutes a small fraction of the total required decoupling capacitance. The constraints on the physical dimensions and parasitic impedance of the capacitors in the remaining stages are therefore significantly reduced. As a result, cost efficient electrolytic and ceramic capacitors can be used to provide medium size and high capacity decoupling capacitors [28].

6.3 Intrinsic vs intentional on-chip decoupling capacitance

Several types of on-chip capacitances contribute to the overall on-chip decoupling capacitance. The *intrinsic* decoupling capacitance is the inherent capacitance of the transistors and interconnects that exists between the power and ground terminals. The thin gate oxide capacitors placed on-chip to solely provide power decoupling are henceforth referred to as an *intentional* decoupling capacitance. The intrinsic decoupling capacitance is described in Section 6.3.1. The intentional decoupling capacitance is reviewed in Section 6.3.2.

6.3.1 Intrinsic decoupling capacitance

An intrinsic decoupling capacitance (or symbiotic capacitance) is the parasitic capacitance between the power and ground terminals within an on-chip circuit structure. The intrinsic capacitance is comprised of three types of parasitic capacitances [156].

One component of the intrinsic capacitance is the parasitic capacitance of the interconnect lines. Three types of intrinsic interconnect capacitances are illustrated in Fig. 6.15. The first type of interconnect capacitance is the capacitance C_1^i between the signal line and the power/ground line. Capacitance C_2^i is the capacitance between signal lines at different voltage potentials. The third type of intrinsic interconnect capacitance is the capacitance C_3^i between the power and ground lines (see Fig. 6.15).

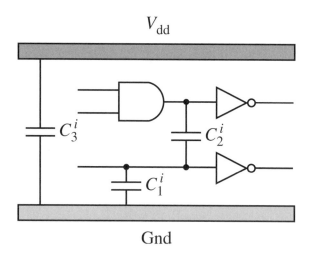

Fig. 6.15. Intrinsic decoupling capacitance of the interconnect lines. C_1^i denotes the capacitance between the signal line and the power/ground line. C_2^i denotes the capacitance between signal lines. C_3^i denotes the capacitance between the power and ground lines.

Parasitic device capacitances, such as the drain junction capacitance and gate-to-source capacitance, also contribute to the overall intrinsic decoupling capacitance where the terminals of the capacitance are connected to power and ground. For example, in the simple inverter circuit depicted in Fig. 6.16, if the input is one (high) and the output is zero (low), the NMOS transistor is turned on, connecting C_p from V_{dd} to G_{nd}, providing a decoupling capacitance to the other switching circuits, as illustrated in Fig. 6.16(a). Alternatively, if the input is zero (low) and the output is one (high), the PMOS transistor is turned on, connecting C_n from G_{nd} to V_{dd}, providing a decoupling capacitance to the other switching circuits, as illustrated in Fig. 6.16(b).

Depending upon the total capacitance $(C_p + C_n)$ and the switching factor SF, the decoupling capacitance from the non-switching circuits is [157]

$$C_{\text{circuit}} = \frac{P}{V_{dd}^2 \, f} \frac{(1 - SF)}{SF}, \tag{6.14}$$

where P is the circuit power, V_{dd} is the power supply voltage, and f is the switching frequency. The time constant for C_{circuit} is determined by $R_{\text{PMOS}} C_n$ or $R_{\text{NMOS}} C_p$ and usually varies in a $0.18\,\mu$m CMOS technology from about 50 ps to 250 ps [157].

The contribution of the transistor and interconnect capacitance to the overall decoupling capacitance is difficult to determine precisely. The transistor terminals as well as the signal lines can be connected either to power or ground, depending upon the internal state of the digital circuit at a particular time. The transistor and interconnect decoupling capacitance therefore depends on the input pattern and the internal state of the circuit. The input vectors that produce the maximum intrinsic decoupling capacitance in a digital circuit are described in [158].

Another source of intrinsic capacitance is the p-n junction capacitance of the diffusion wells. The N-type wells, P-type wells, or wells of both types are implanted into a silicon substrate to provide an appropriate body doping for the PMOS and NMOS transistors. The N-type wells are ohmically connected to the power supply while the P-type wells are connected to ground to provide a proper body bias for the transistors. The N-well capacitor is the reverse-biased p-n junction capacitor between the N-well and p-substrate, as shown in Fig. 6.17. The total on-chip N-well decoupling capacitance C_{nw} is determined by the area, perimeter, and depth of each N-well. Multiplying C_{nw} by the

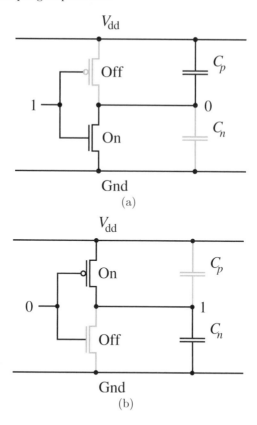

Fig. 6.16. Intrinsic decoupling capacitance of a non-switching circuit. (a) Inverter input is high. (b) Inverter input is low.

series and contact resistance in the N-well and p-substrate, the time constant $(R_p + R_n + R_{\text{contact}})C_{nw}$ for an N-well capacitor is typically in the range of 250 ps to 500 ps in a 0.18 μm CMOS technology [157]. The parasitic capacitance of the wells usually dominates the intrinsic decoupling capacitance of ICs fabricated in an epitaxial CMOS process [159], [160]. The overall intrinsic on-chip decoupling capacitance consists of several components and is

$$C_{\text{intrinsic}} = C_{\text{inter}} + C_{pn} + C_{\text{well}} + C_{\text{load}} + C_{gs} + C_{gb}, \quad (6.15)$$

where C_{inter} is the interconnect capacitance, C_{pn} is the p-n junction capacitance, C_{well} is the capacitance of the well, C_{load} is the load capacitance, C_{gs} is the gate-to-source (drain) capacitance, and C_{gb} is the gate-to-body capacitance.

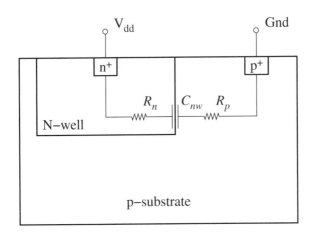

Fig. 6.17. N-well junction intrinsic decoupling capacitance. The capacitor C_{nw} is formed by the reverse-biased p-n junction between the N-well and the p-substrate.

Silicon-on-insulator (SOI) CMOS circuits lack diffusion wells and therefore do not contribute to the intrinsic on-chip decoupling capacitance. A reliable estimate of the contribution of the interconnect and transistors to the on-chip decoupling capacitance is thus particularly important in SOI circuits. Several techniques for estimating the intrinsic decoupling capacitance are presented in [111], [161]. The overall intrinsic decoupling capacitance of an IC can also be determined experimentally. In [162], the signal response of a power distribution system versus frequency is measured with a vector network analyzer. An RLC model of the system is constructed to match the observed response. The magnitude of the total on-chip decoupling capacitance is determined from the frequency of the resonant peaks in the response of the power system. Alternatively, the total on-chip decoupling capacitance can be experimentally determined from the package-chip resonance, as described in [152]. The intentional decoupling capacitance placed on-chip during the design process is known within the margins of the process variations. Subtracting the intentional capacitance from the measured overall capacitance yields an estimate of the on-chip intrinsic capacitance.

6.3.2 Intentional decoupling capacitance

Intentional decoupling capacitance is often added to a circuit during the design process to increase the overall on-chip decoupling capacitance to a satisfactory level. The intentional decoupling capacitance is typically realized as a gate capacitance in large MOS transistors placed on-chip specifically for this purpose. In systems with mixed memory and logic, however, the intentional capacitance can also be realized as a trench capacitance [163], [164].

Banks of MOS decoupling capacitors are typically placed among the on-chip circuit blocks, as shown in Fig. 6.18. The space between the circuit blocks is often referred to as "white" space, as this area is primarily used for global routing and does not contain any active devices. Unless noted otherwise, the term "on-chip decoupling capacitance" commonly

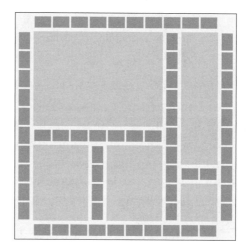

Fig. 6.18. Banks of on-chip decoupling capacitors (the dark gray rectangles) placed among circuit blocks (the light gray rectangles).

refers to the intentional decoupling capacitance. Using more than 20% of the overall die area for intentional on-chip decoupling capacitance is common in modern high speed integrated circuits [165], [166].

A MOS capacitor uses the thin oxide layer between the N-well and polysilicon gate to provide the additional decoupling capacitance

needed to mitigate the power noise, as shown in Fig. 6.19. An optional
fuse (or control gate) is typically provided to disconnect the thin oxide
capacitor from the rest of the circuit in the undesirable situation of
a short circuit due to process defects. As the size and shape of MOS
capacitors vary, the $R_n C_{ox}$ time constant typically ranges from 40 ps
to 200 ps in a 0.18 μm CMOS process. Depending upon the switching
speed of the circuit, typical on-chip MOS decoupling capacitors are
effective for RC time constants below 200 ps [157].

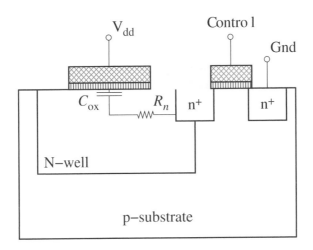

Fig. 6.19. Thin oxide MOS decoupling capacitor.

A MOS capacitor is formed by the gate electrode on one side of the
oxide layer and the source-drain inversion channel under the gate on the
other side of the oxide layer. The resistance of the channel dominates
the ESR of the MOS capacitor. Due to the resistance of the transistor
channel, the MOS capacitor is modeled as a distributed RC circuit,
as shown in Fig. 6.20. The impedance of the distributed RC structure
shown in Fig. 6.20 is frequency dependent, $Z(\omega) = R(\omega) + \frac{1}{j\omega C(\omega)}$. Both
the resistance $R(\omega)$ and capacitance $C(\omega)$ decrease with frequency. The
low frequency resistance of the MOS capacitor is approximately one
twelfth of the source-drain resistance of the MOS transistor in the linear
region [167]. The low frequency capacitance is the entire gate-to-channel
capacitance of the transistor. At high frequencies, the gate-to-channel
capacitance midway between the drain and source is shielded from the

capacitor terminals by the resistance of the channel, decreasing the
effective capacitance of the MOS capacitor. The higher the channel
resistance per transistor width, the lower the frequency at which the
capacitor efficiency begins to decrease. Capacitors with a long channel
(with a relatively high channel resistance) are therefore less effective
at high frequencies as compared to short-channel capacitors. A higher
series resistance of the on-chip MOS decoupling capacitor, however, is
beneficial in damping the resonance of a die-package RLC tank circuit [167].

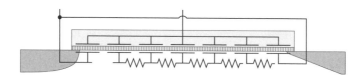

Fig. 6.20. Equivalent RC model of a MOS decoupling capacitor.

Long channel transistors, however, are more area efficient. In transistors with a minimum length channel, the source and body contacts
dominate the transistor area, while the MOS capacitor stack occupies
a relatively small fraction of the total area. For longer channels, the
area of the MOS capacitor increases while the area overhead of the
source/drain contacts remain constant, increasing the capacitance per
total area [28]. A tradeoff therefore exists between the area efficiency
and the ESR of the MOS decoupling capacitor. Transistors with a channel length twelve times greater than the minimum length are a good
compromise [167]. In this case, the RC time constant is smaller than
the switching time of the logic gates, which typically are composed of
transistors with a minimum channel length, while the source and drain
contacts occupy a relatively small fraction of the total area.

6.4 Types of on-chip decoupling capacitors

Multiple on-chip capacitors are utilized in ICs to satisfy various design
requirements. Four types of widely utilized on-chip decoupling capacitors are the subject of this section. Polysilicon-insulator-polysilicon

(PIP) capacitors are presented in Section 6.4.1. Three types of MOS decoupling capacitors, accumulation, depletion, and inversion, are described in Section 6.4.2. Metal-insulator-metal (MIM) decoupling capacitors are reviewed in Section 6.4.3. In Section 6.4.4, lateral flux decoupling capacitors are described. The design and performance characteristics of the different on-chip decoupling capacitors are compared in Section 6.4.5.

6.4.1 Polysilicon-insulator-polysilicon (PIP) capacitors

Both junction and MOS capacitors use diffusion for the lower electrodes. The junction isolating the diffused electrode exhibits substantial parasitic capacitance, limiting the voltage applied across the capacitor. These limitations are circumvented in PIP capacitors, which employ two polysilicon electrodes in combination with either an oxide or an oxide-nitride-oxide (ONO) dielectric [168], as illustrated in Fig. 6.21. Since typical CMOS and BiCMOS processes incorporate multiple polysilicon layers, PIP capacitors do not require any additional masking steps. The gate polysilicon can serve as the lower electrode of the PIP capacitor, while the resistor polysilicon (doped with a suitable implant) can form the upper electrode. The upper electrode is typically doped with either an N-type source/drain (NSD) or P-type source/drain (PSD) implant. The implant resulting in the lowest sheet resistance is preferable, since heavier doping reduces the ESR and minimizes voltage modulation due to polysilicon depletion [168].

PIP capacitors require additional process steps. Even if both of the electrodes consist of existing depositions, the capacitor dielectric is unique to this structure and consequently requires a process extension. The simplest way to form this dielectric is to eliminate the interlevel oxide (ILO) deposition that normally separates the two polysilicon layers and add a thin oxide layer on the lower polysilicon electrode. With this technique, a capacitor can be built between the two polysilicon layers as long as the second polysilicon layer is not used as an interconnection.

Silicon dioxide has a relatively low permittivity. A higher permittivity, and therefore a higher capacitance per unit area, is achieved using a stacked ONO dielectric (see Fig. 6.21(b)). Observe from Fig. 6.21 that the PIP capacitors normally reside over the field oxide. The oxide steps should not intersect the structure, since those steps cause surface irregularities in the lower capacitor electrode, resulting in localized thinning of the dielectric, thereby concentrating the electric field. As a result of

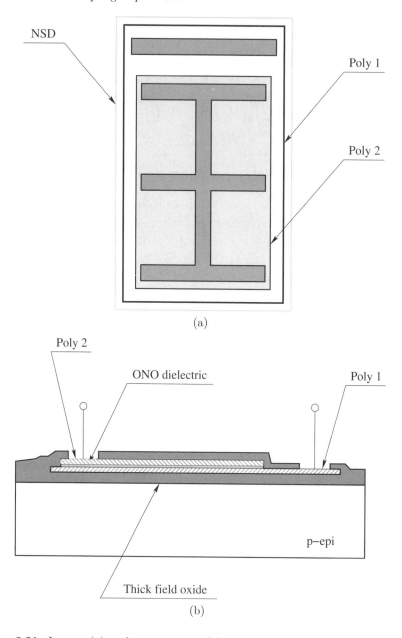

Fig. 6.21. Layout (a) and cross section (b) of a PIP oxide-nitride-oxide (ONO) capacitor. The entire capacitor is enclosed in an N-type source/drain region, reducing the sheet resistance of the polysilicon layer.

the intersection, the breakdown voltage of the capacitor can be severely compromised.

Selecting the dielectric material in a PIP capacitor, several additional issues should be considered. Composite dielectrics experience hysteresis effects at high frequencies (above 10 MHz) due to the incomplete redistribution of static charge along the oxide-nitride interface. Pure oxide dielectrics are used for PIP capacitors to achieve a relatively constant capacitance over a wide frequency range. Oxide dielectrics, however, typically have a lower capacitance per unit area. Low capacitance dielectrics are also useful for improving matching among the small capacitors.

Voltage modulation of the PIP capacitors is relatively small, as long as both electrodes are heavily doped. A PIP capacitor typically exhibits a voltage modulation of 150 ppm/volt [168]. The temperature coefficient of a PIP capacitor also depends on voltage modulation effects and is typically less than 250 ppm/°C [169].

6.4.2 MOS capacitors

A MOS capacitor consists of a metal-oxide-semiconductor structure, as illustrated in Fig. 6.22. A top metal contact is referred to as the gate, serving as one plate of the capacitor. In digital CMOS ICs, the gate is often fabricated as a heavily doped n^+-polysilicon layer, behaving as a metal. A second metal layer forms an ohmic contact to the back of the semiconductor and is called the bulk contact. The semiconductor layer serves as the other plate of the capacitor. The bulk resistivity is typically 1 to $10 \, \Omega \cdot \text{cm}$ (with a doping of $10^{15} \, \text{cm}^{-3}$).

The capacitance of a MOS capacitor depends upon the voltage applied to the gate with respect to the body. The dependence of the capacitance upon the voltage across a MOS capacitor (a capacitance versus voltage (CV) diagram) is plotted in Fig. 6.23. Depending upon the gate-to-body potential V_{gb}, three regions of operation are distinguished in the CV diagram of a MOS capacitor. In the accumulation mode, mobile carriers of the same type as the body (holes for an NMOS capacitor with a p-substrate) accumulate at the surface. In the depletion mode, the surface is devoid of any mobile carriers, leaving only a space charge (depletion layer). In the inversion mode, mobile carriers of the opposite type of the body (electrons for an NMOS capacitor with a p-substrate) aggregate at the surface, inverting the conductivity type. These three regimes are roughly separated by the two voltages

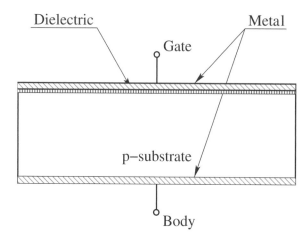

Fig. 6.22. The structure of an n-type MOS capacitor.

(see Fig. 6.23). A flat band voltage V_{fb} separates the accumulation regime from the depletion regime. The threshold voltage V_t demarcates the depletion regime from the inversion regime. Based on the mode of operation, three types of MOS decoupling capacitors exist and are described in the following three subsections.

Accumulation

In MOS capacitors operating in accumulation, the applied gate voltage is lower than the flat band voltage ($V_{gb} < V_{fb}$) and induces negative charge on the metal gate and positive charge in the semiconductor. The hole concentration at the surface is therefore above the bulk value, leading to surface accumulation. The charge distribution in a MOS capacitor operating in accumulation is shown in Fig. 6.24. The flat band voltage is the voltage at which there is no charge on the plates of the capacitor (there is no electric field across the dielectric). The flat band voltage depends upon the doping of the semiconductor and any residual charge existing at the interface between the semiconductor and the insulator. In the accumulation mode, the charge per unit area Q_n at the semiconductor/oxide interface is a linear function of the applied voltage V_{gb}. The oxide capacitance per unit area C_{ox} is determined by the slope of Q_n, as illustrated in Fig. 6.25. The capacitance of a MOS capacitor operating in accumulation achieves the maximum value and is

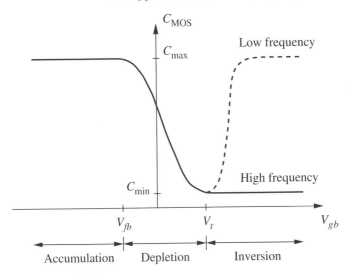

Fig. 6.23. Capacitance versus gate voltage (CV) diagram of an n-type MOS capacitor. The flat band voltage V_{fb} separates the accumulation region from the depletion region. The threshold voltage V_t separates the depletion region from the inversion region.

$$C_{\text{MOS}_{\text{accum}}} = C_{\max} = A\,C_{\text{ox}} = A\,\frac{\epsilon_{\text{ox}}}{t_{\text{ox}}}, \qquad (6.16)$$

where A is the area of the gate electrode, ϵ_{ox} is the permittivity of the oxide, and t_{ox} is the oxide thickness.

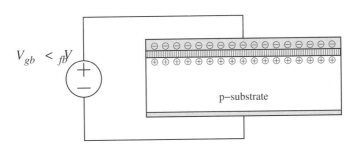

Fig. 6.24. Charge distribution in an NMOS capacitor operating in accumulation ($V_{gb} < V_{fb}$).

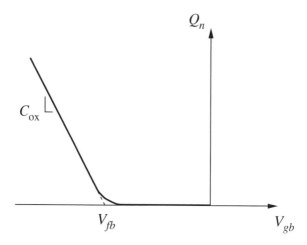

Fig. 6.25. Accumulation charge density as a function of the applied gate voltage. The capacitance per unit area C_{ox} is determined by the slope of the line.

Depletion

In MOS capacitors operating in depletion, the applied gate voltage is brought above the flat band voltage and below the threshold voltage $(V_{fb} < V_{gb} < V_t)$. A positive charge is therefore induced at the interface between metal gate and the oxide. A negative charge is induced at the oxide/semiconductor interface. This scenario is accomplished by pushing all of the mobile positive carriers (holes) away, exposing the fixed negative charge from the donors. Hence, the surface of the semiconductor is depleted of mobile carriers, leaving behind a negative space charge. The charge distribution in the MOS capacitor operating in depletion is illustrated in Fig. 6.26.

The resulting space charge behaves like a capacitor with an effective capacitance per unit area C_d. The effective capacitance C_d depends upon the gate voltage V_{gb} and is

$$C_d(V_{gb}) = \frac{\epsilon_{\text{Si}}}{x_d(V_{gb})}, \tag{6.17}$$

where ϵ_{Si} is the permittivity of the silicon and x_d is the thickness of the depletion layer (space charge). Observe from Fig. 6.26 that the oxide capacitance per unit area C_{ox} and depletion capacitance per unit area C_d are connected in series. The capacitance of a MOS structure in the depletion region is therefore

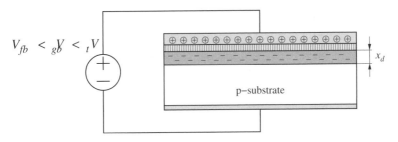

$$V_{fb} < V_{gb} < V_t$$

Fig. 6.26. Charge distribution in an NMOS capacitor operating in depletion ($V_{fb} < V_{gb} < V_t$). Under this bias condition, all of the mobile positive carriers (holes) are pushed away, depleting the surface of the semiconductor, resulting in a negative space charge with thickness x_d.

$$C_{\mathrm{MOS_{deplet}}} = A \frac{C_{\mathrm{ox}} C_d}{C_{\mathrm{ox}} + C_d}. \qquad (6.18)$$

Note that the thickness of the silicon depletion layer becomes wider as the gate voltage is increased, since more holes are pushed away, exposing more fixed negative ionized dopants, leading to a thicker space charge layer. As a result, the capacitance of the depleted silicon decreases, reducing the overall MOS capacitance.

Inversion

In MOS capacitors operating in inversion, the applied gate voltage is further increased above the threshold voltage ($V_t < V_{gb}$). The conduction type of the semiconductor surface is inverted (from p-type to n-type). The threshold voltage is referred to as the voltage at which the conductivity type of the surface layer changes from p-type to n-type (in the case of an NMOS capacitor). This phenomenon is explained as follows. As the gate voltage is increased beyond the threshold voltage, holes are pushed away from the Si/SiO$_2$ interface, exposing the negative charge. Note that the density of holes decreases exponentially from the surface into the bulk. The number of holes decreases as the applied voltage increases. The number of electrons at the surface therefore increases with applied gate voltage and becomes the dominant type of carrier, inverting the surface conductivity. The charge distribution of a MOS capacitor operating in inversion is depicted in Fig. 6.27.

Note that the depletion layer thickness reaches a maximum in the inversion region. The total voltage drop across the semiconductor also

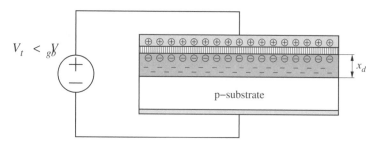

Fig. 6.27. Charge distribution of an NMOS capacitor operating in inversion ($V_t < V_{gb}$). Under this bias condition, a negative charge is accumulated at the semi-conductor surface, inverting the conductivity of the semiconductor surface (from p-type to n-type).

reaches the maximum value. Further increasing the gate voltage, the applied voltage drops primarily across the oxide layer. If the gate voltage approaches the threshold voltage, the depleted layer capacitance per unit area C_d^{min} reaches a minimum [170]. In this case, the overall MOS capacitance reaches the minimum value and is

$$C_{\mathrm{MOS_{inv}}} = C_{\mathrm{MOS}}^{\mathrm{min}} = A \frac{C_{\mathrm{ox}} \, C_d^{\mathrm{min}}}{C_{\mathrm{ox}} + C_d^{\mathrm{min}}}, \qquad (6.19)$$

where

$$C_d^{\mathrm{min}} = \frac{\epsilon_{\mathrm{Si}}}{x_d^{\mathrm{max}}}. \qquad (6.20)$$

Note that at low frequencies (quasi-static conditions), the generation rate of holes (electrons) in the depleted silicon surface layer is sufficiently high. Electrons are therefore swept to the Si/SiO$_2$ interface, forming a sheet charge with a thin layer of electrons. The inversion layer capacitance under quasi-static conditions therefore reaches the maximum value. At high frequencies, however, the generation rate is not sufficiently high, prohibiting the formation of the electron charge at the Si/SiO$_2$ interface. In this case, the thickness of the silicon depletion layer reaches the maximum. Hence, the inversion layer capacitance reaches the minimum.

A MOS transistor operated as a capacitor has a substantial ESR, most of which is associated with the lower electrode. This parasitic resistance can be reduced by using a fairly short channel length ($25\,\mu$m or less) [168]. If the source and drain diffusions are omitted, the backgate contact is typically placed entirely around the gate.

A layout and cross section of a MOS capacitor formed in a BiCMOS process are illustrated in Fig. 6.28. Since the N-type source/drain layer follows the gate oxide growth and polysilicon deposition, the lower plate should consist of some other diffusion (typically deep-n^+). Deep-n^+ has a higher sheet resistance than the N-type source/drain layer (typically $100\,\Omega/\square$), resulting in a substantial parasitic resistance of the lower plate. The heavily concentrated n-type doping thickens the gate oxide by 10% to 30% through dopant-enhanced oxidation, resulting in higher working voltages but a lower capacitance per unit area. The deep-n^+ is often placed inside the N-well to reduce the parasitic capacitance to the substrate. The N-well can be omitted, however, if the larger parasitic capacitance and lower breakdown voltage of the deep-n^+/p-epi junction can be tolerated.

Regardless of how a MOS capacitor is constructed, the two capacitor electrodes are never entirely interchangeable. The lower plate always consists of a diffusion with substantial parasitic junction capacitance. This junction capacitance is eliminated by connecting the lower plate of the capacitor to the substrate potential. The upper plate of the MOS capacitor consists of a deposited electrode with a relatively small parasitic capacitance. The lower plate of a MOS capacitor should therefore be connected to the driven node (with the lower impedance). Swapping the two electrodes of a MOS capacitor can load a high impedance node with a high parasitic impedance, compromising circuit performance.

The major benefit of MOS capacitors is the natural compatibility with CMOS technology. MOS capacitors also provide a high capacitance density [171], providing a cost effective on-chip decoupling capacitance. MOS capacitors result in relatively high matching: the gate oxide capacitance is typically controlled within 5% error [169]. MOS capacitors, however, are non-linear devices that exhibit strong voltage dependence (more than 100 ppm/volt [172]) due to the variation of both the dielectric constant and the depletion region thickness within each plate. The performance of the MOS capacitors is limited at high frequencies due to the large diffusion-to-substrate parasitic capacitance. As technology scales, the leakage currents of MOS capacitors also increase substantially, increasing the total power dissipation. High leakage current is the primary issue with MOS capacitors.

A MOS on-chip capacitance is typically realized as accumulation and inversion capacitors. Note that capacitors operating in accumulation are more linear than capacitors operating in inversion [173]. The

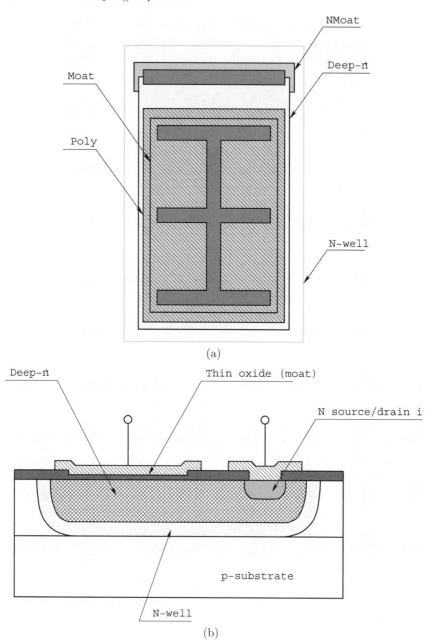

Fig. 6.28. Layout (a) and cross section (b) of a deep-n$^+$ MOS capacitor constructed in a BiCMOS process.

MOS capacitance operating in accumulation is almost independent of frequency. Moreover, MOS decoupling capacitors operating in accumulation result in an approximately 15 X reduction in leakage current as compared to MOS decoupling capacitors operating in inversion [174]. MOS decoupling capacitors operating in accumulation should therefore be the primary form of MOS decoupling capacitors in modern high performance ICs.

6.4.3 Metal-insulator-metal (MIM) capacitors

A MIM capacitor consists of two metal layers (plates) separated by a deposited dielectric layer. A cross section of a MIM capacitor is shown in Fig. 6.29. A thick oxide layer is typically deposited on the substrate, reducing the parasitic capacitance to the substrate. The parasitic substrate capacitance is also lowered by utilizing the top metal layers as plates of a MIM capacitor. For instance, in comb MIM capacitors [175], the parasitic capacitance to the substrate is less than 2% of the total capacitance.

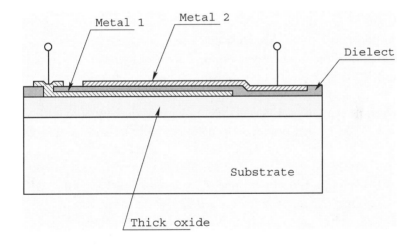

Fig. 6.29. Cross section of a MIM capacitor. A thick oxide (SiO_2) layer is typically deposited on the substrate to reduce the parasitic capacitance to the substrate.

Historically, MIM capacitors have been widely used in RF and mixed-signal ICs due to the low leakage, high linearity, low process variations (high accuracy), and low temperature variations [176], [177], [178] of MIM capacitors. Conventional circuits utilize SiO_2 as a dielectric deposited between two metal layers. Large MIM capacitors therefore require significant circuit area, prohibiting the use of MIM capacitors as decoupling capacitors in high complexity ICs. The capacitance density can be increased by reducing the dielectric thickness and employing high-k dielectrics. Reducing the dielectric thickness, however, results in a substantial increase in leakage current which is highly undesirable.

MIM capacitors with a capacitance density comparable to MOS capacitors (8 to $10\,fF/\mu m^2$) have been fabricated using Al_2O_3 and $AlTiO_x$ dielectrics [179], $AlTaO_x$ [180], and HfO_2 dielectric using atomic layer deposition (ALD) [181]. A higher capacitance density ($13\,fF/\mu m^2$) is achieved using laminate ALD HfO_2–Al_2O_3 dielectrics [182], [183]. Laminate dielectrics also result in higher voltage linearity and reliability. Recently, MIM capacitors with a capacitance density approximately two times greater than the capacitance density of MOS capacitors have been fabricated [184]. A capacitance density of $17\,fF/\mu m^2$ is achieved using a Nb_2O_5 dielectric with HfO_2–Al_2O_3 barriers.

Unlike MOS capacitors, MIM capacitors require high temperatures for thin film deposition. Integrating MIM capacitors into a standard low temperature ($\leq 400\,^\circ C$) back-end high complexity digital process is therefore a challenging problem [185]. This problem can be overcome by utilizing MIM capacitors with plasma enhanced chemical vapor deposition (PECVD) nitride dielectrics [186], [187]. Previously, MIM capacitors were unavailable in CMOS technology with copper metallization. Recently, MIM capacitors have been successfully integrated into CMOS and BiCMOS technologies with a copper dual damascene metallization process [188], [189], [190]. In [191], a high density MIM capacitor with a low ESR using a plug-in copper plate is described, making MIM capacitors highly efficient for use as a decoupling capacitor.

MIM capacitors are widely utilized in RF and mixed-signal ICs due to low voltage coefficients, good capacitor matching, precision control of capacitor values, small parasitic capacitance, high reliability, and low defect densities [192]. MIM capacitors also exhibit high linearity over a wide frequency range. Additionally, a high capacitance density with lower leakage currents has recently been achieved, making MIM capacitors the best candidate for decoupling power and ground lines in modern high performance, high complexity ICs. For instance, for a

MIM capacitor with a dielectric thickness $t_{ox} = 1\,\text{nm}$, a capacitance density of $34.5\,\text{fF}/\mu\text{m}^2$ has been achieved [193].

6.4.4 Lateral flux capacitors

The total capacitance per unit area can be increased by using more than one pair of interconnect layers. Current technologies offer up to ten metal layers, increasing the capacitance nine times through the use of a sandwich structure. The capacitance is further increased by exploiting the lateral flux between the adjacent metal lines within a specific interconnect layer. In scaled technologies, the adjacent metal spacing (on the same level) shrinks faster than the spacing between the metal layers (on different layers), resulting in substantial lateral coupling.

A simplified structure of an interdigitated capacitor exploiting lateral flux is shown in Fig. 6.30. The two terminals of the capacitor are shown in light grey and dark grey. Note that the two plates built in the same metal layer alternate to better exploit the lateral flux. Ordinary vertical flux can also be exploited by arranging the segments of a different metal layer in a complementary pattern [194], as illustrated in Fig. 6.31. Note that a higher capacitance density is achieved by using a lateral flux together with a vertical flux (parallel plate structure).

An important advantage of using a lateral flux capacitor is reducing the bottom plate parasitic capacitance as compared to an ordinary parallel plate structure. This reduction is due to two reasons. First, the higher density of the lateral flux capacitor results in a smaller area for a specific value of total capacitance. Second, some of the field lines originating from one of the bottom plates terminate on the adjacent plate rather than the substrate, further reducing the bottom plate capacitance, as shown in Fig. 6.32. Such phenomenon is referred to as flux stealing. Thus, some portion of the bottom plate parasitic capacitance is converted into a useful plate-to-plate capacitance. Three types of enhanced lateral flux capacitors with a higher capacitance density are described in the following three subsections.

Fractal capacitors

Since the lateral capacitance is dependent upon the perimeter of the structure, the maximum capacitance can be obtained with those geometries that maximize the total perimeter. Fractals are therefore good

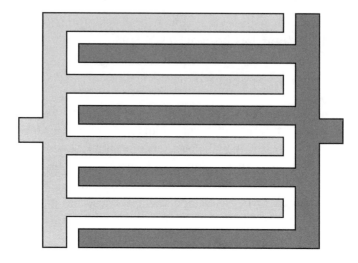

Fig. 6.30. A simplified structure of an interdigitated lateral flux capacitor (top view). Two terminals of the capacitor are shown in light grey and dark grey.

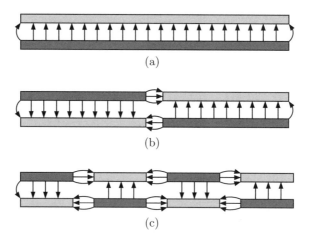

Fig. 6.31. Vertical flux versus lateral flux. (a) A standard parallel plate structure, (b) divided by two cross-connected metal layers, and (c) divided by four cross-connected metal layers.

candidates for use in lateral flux capacitors. A fractal is a structure that encloses a finite area with an infinite perimeter [195]. Although lithography limitations prevent fabrication of a real fractal, quasi-fractal geometries with feature sizes limited by lithography have been successfully fabricated in fractal capacitors [196]. It has been demonstrated that in certain cases, the effective capacitance of fractal capacitors can be increased by more than ten times.

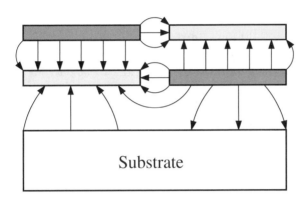

Fig. 6.32. Reduction of the bottom plate parasitic capacitance through flux stealing. Shades of grey denote the two terminals of the capacitor.

The final shape of a fractal can be tailored to almost any form. The flexibility arises from the characteristic that a wide variety of geometries exists, determined by the fractal initiator and generator [195]. It is also possible to use different fractal generators during each step. Fractal capacitors of any desired form can therefore be constructed due to the flexibility in the shape of the layout. Note that the capacitance per unit area of a fractal capacitor depends upon the fractal dimensions. Fractals with large dimensions should therefore be used to improve the layout density [196].

In addition to the capacitance density, the quality factor Q is important in RF and mixed-signal applications. In fractal capacitors, the degradation in quality factor is minimal, since the fractal structure naturally limits the length of the thin metal sections to a few micrometers, maintaining a reasonably small ESR. Hence, smaller dimension fractals should be used to achieve a low ESR. Alternatively, a tradeoff exists between the capacitance density and the ESR in fractal capacitors.

Existing technologies typically provide tighter control over the lateral spacing of the metal layers as compared to the vertical thickness of the oxide layers (both from wafer to wafer and across the same wafer). Lateral flux capacitors shift the burden of matching from the oxide thickness to the lithography. The matching characteristics are therefore greatly improved in lateral flux capacitors. Furthermore, the pseudorandom nature of the lateral flux capacitors compensate for the effects of nonuniformity in the etching process.

Comparing fractal and conventional interdigitated capacitors, note the inherent parasitic inductance of an interdigitated capacitor. Most fractal geometries randomize the direction of the current flow, reducing the ESL. In an interdigitated capacitor, however, the current flows in the same direction for all of the parallel lines. Also in fractal structures, the electric field concentrates around the sharp edges, increasing the effective capacitance density (about 15%) [196]. Nevertheless, due to simplicity, interdigitated capacitors are widely used in ICs.

Woven capacitors

A woven structure is also utilized to achieve high capacitance density. A woven capacitor is depicted in Fig. 6.33. Two orthogonal metal layers are used to construct the plates of the capacitor. Vias connect the metal lines of a specific capacitor plate at the overlap sites. Note that in a woven structure, the current in the adjacent lines flows in the opposite direction. The woven capacitor has therefore much less inherent parasitic inductance as compared to an interdigitated capacitor [122], [197]. In addition, the ESR of a woven capacitor contributed by vias is smaller than the ESR of an interdigitated capacitor. A woven capacitor, however, results in a smaller capacitance density as compared to an interdigitated capacitor with the same metal pitch due to the smaller vertical capacitance.

Vertical parallel plate (VPP) capacitors

Another way to utilize a number of metal layers in modern CMOS technologies is to construct conductive vertical plates out of vias in combination with the interconnect metal. Such a capacitor is referred to as a vertical parallel plate (VPP) capacitor [198]. A VPP capacitor

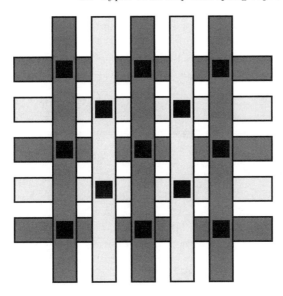

Fig. 6.33. Woven capacitor. The two terminals of the capacitor are shown in light grey and dark grey. The vias are illustrated by the black colored squares.

consists of metal slabs connected vertically using multiple vias between the vertical plates. This structure fully exploits lateral scaling trends as compared to fractal structures [197].

6.4.5 Comparison of on-chip decoupling capacitors

On-chip decoupling capacitors can be implemented in ICs in a number of ways. The primary characteristics of four common types of on-chip decoupling capacitors, discussed in Sections 6.4.1 – 6.4.4, are listed in Table 6.1. Note that typical MIM capacitors provide a lower capacitance density $(1\,\mathrm{fF}/\mu\mathrm{m}^2 - 10\,\mathrm{fF}/\mu\mathrm{m}^2)$ than MOS capacitors. Recently, a higher capacitance density $(13\,\mathrm{fF}/\mu\mathrm{m}^2)$ of MIM capacitors has been achieved using laminate ALD HfO_2–Al_2O_3 dielectrics [182], [183]. A capacitance density of $34.5\,\mathrm{fF}/\mu\mathrm{m}^2$ has been reported in [193] for a MIM capacitor with a dielectric thickness of 1 nm.

Note that the quality factor of the MOS and lateral flux capacitors is limited by the channel resistance and the resistance of the multiple vias. Decoupling capacitors with a low quality factor produce wider antiresonant spikes with a significantly reduced magnitude [199]. It is therefore highly desirable to limit the quality factor of the on-chip

Table 6.1. Four common types of on-chip decoupling capacitors in a 90 nm CMOS technology

Feature	PIP capacitor	MOS capacitor	MIM capacitor	Lateral flux capacitor
Capacitance density (fF/μm^2)	$1-5$	$10-20$	$1-30$	$10-20$
Bottom plate capacitance (%)	$5-10$	$20-30$	$2-5$	$1-5$
Linearity (ppm/volt)	$50-150$	$300-500$	$10-50$	$50-100$
Quality factor	$5-15$	$1-10$	$50-150$	$10-50$
Parasitic resistance (mΩ)	$500-2000$	$1000-10000$	$50-250$	$100-500$
Leakage current (A/cm^2)	$10^{-10}-10^{-9}$	$10^{-2}-10^{-1}$	$10^{-9}-10^{-8}$	$10^{-10}-10^{-9}$
Temperature dependence (ppm/$^\circ$C)	$150-250$	$300-500$	$50-100$	$50-100$
Process complexity	Extra steps	Standard	Standard	Standard

decoupling capacitors. Note that in the case of a low ESR (high quality factor), an additional series resistance should be provided, lowering the magnitude of the antiresonant spike. This additional resistance, however, is limited by the target impedance of the power distribution system [26].

The parasitic resistance is another important characteristic of on-chip decoupling capacitors. The parasitic resistance characterizes the efficiency of a decoupling capacitor. Alternatively, both the amount of charge released by the decoupling capacitor and the rate with which the charge is restored on the decoupling capacitor are primarily determined by the parasitic resistance [200]. The parasitic resistance of PIP capacitors is mainly determined by the resistive polysilicon layer. MIM capacitors exhibit the lowest parasitic resistance due to the highly conductive metal layers used as the plates of the capacitor. The increased parasitic resistance of the lateral flux capacitors is due to the multiple resistive vias, connecting metal plates at different layers [197]. In MOS capacitors, both the channel resistance and the resistance of the metal plates contribute to the parasitic resistance. The performance of MOS capacitors is therefore limited by the high parasitic resistance.

Observe from Table 6.1 that MOS capacitors result in prohibitively large leakage currents. As technology scales, the leakage power is expected to become the major component of the total power dissipation. Thick oxide MOS decoupling capacitors are often used to reduce the leakage power. Thick oxide capacitors, however, require a larger die area for the same capacity as a thinner oxide capacitance. Note that the leakage current in MOS capacitors increases exponentially with temperature, further exacerbating the problem of heat removal. Also note that leakage current is reduced in MIM capacitors as compared to MOS capacitors by about seven orders of magnitude. The leakage current of MIM capacitors is also fairly temperature independent, increasing twofold as the temperature rises from 25 °C to 125 °C [188].

Note that PIP capacitors typically require additional process steps, adding extra cost. From the information listed in Table 6.1, MIM capacitors and stacked lateral flux capacitors (fractal, VPP, and woven) are the best candidates for decoupling the power and ground lines in modern high performance, high complexity ICs.

6.5 On-chip switching voltage regulator

The efficiency of on-chip decoupling capacitors can be enhanced by an on-chip switching voltage regulator [201]. The decoupling capacitors reduce the impedance of the power distribution system by serving as an energy source when the power voltage decreases, as discussed in previous sections. The smaller the power voltage variation, the smaller the energy transferred from a decoupling capacitor to the load.

Consider a group of N decoupling capacitors C placed on a die. Where connected in parallel between the power and ground network, the capacitors behave as a single capacitor NC. As the power supply level decreases from the nominal level V_{dd} to a target minimum $V_{\mathrm{dd}} - \delta V$, the non-switching decoupling capacitors release only a small fraction k of the stored charge into the network,

$$\frac{\delta Q}{Q_0} = \frac{NCV_{\mathrm{dd}} - NC(V_{\mathrm{dd}} - \delta V)}{NCV_{\mathrm{dd}}} = \frac{\delta V}{V_{\mathrm{dd}}} \equiv k. \qquad (6.21)$$

A correspondingly small fraction of the total energy E stored in the capacitors is transfered to the load,

$$\frac{\delta E}{E_0} = \frac{V_{dd}^2 - (V_{dd} - \delta V)^2}{V_{dd}^2} \approx \frac{2\delta V}{V_{dd}} = 2k. \tag{6.22}$$

Switching the on-chip capacitors can increase the charge (and energy) transferred from the capacitors to the load as the power voltage decreases below the nominal voltage level [201]. Rather than a fixed connection in parallel as in the traditional non-switching case, the connection of capacitors to the power and ground networks can be changed from parallel to series using switches, as shown in Fig. 6.34 for the case of two capacitors. When the rate of variation in the power supply voltage is relatively small, the capacitors are connected in parallel, as shown in Fig. 6.34(a), and charged to V_{dd}. When the instantaneous power supply variation exceeds a certain threshold, the capacitors are reconnected in series, as shown in Fig. 6.34(b), transforming the circuit into a capacitor of C/N capacity carrying a charge of CV_{dd}. In this configuration, the circuit can release

$$\delta Q_{sw} = CV_{dd} \left(1 - \frac{1 - k}{N} \right) \tag{6.23}$$

amount of charge before the drop in the voltage supply level exceeds the noise margin $\delta V = kV_{dd}$. This amount of charge is greater than the charge released in the non-switching case, as determined by (6.21), if $k < \frac{1}{N+1}$. The effective charge storage capacity of the on-chip decoupling capacitors is thereby enhanced. The area of the on-chip capacitors required to lower the peak resonant impedance of the power network to a satisfactory level is decreased.

This technique is employed in the UltraSPARC III microprocessor, as described by Ang, Salem, and Taylor [201]. In addition to the 176 nF of non-switched on-chip decoupling capacitance, 134 nF of switched on-chip capacitance is placed on the die. The switched capacitance occupies 20 mm^2 of die area and is distributed in the form of 99 switching regulator blocks throughout the die to maintain a uniform power supply voltage. The switching circuitry is designed to minimize the short-circuit current when the capacitor is switching. Feedback loop control circuitry ensures stable behavior of the switching capacitors. The switching circuitry occupies 0.4 mm^2, a small fraction of the overall regulator area. The regulator blocks are connected directly to the global

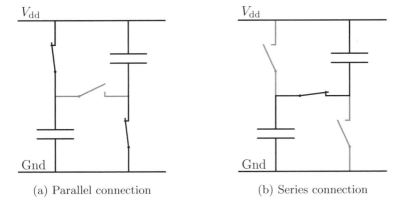

Fig. 6.34. Switching decoupling capacitors from a parallel to a series connection.

power distribution grid. In terms of the frequency domain characteristics, the switching regulator lowers the magnitude of the die-package resonance impedance. The switched decoupling capacitors decrease the on-chip power noise by roughly a factor of two and increase the operating frequency of the circuit by approximately 20%.

6.6 Summary

A brief overview of decoupling capacitors has been presented in this chapter. The primary characteristics of decoupling capacitors can be summarized as follows:

- A decoupling capacitor serves as an intermediate and temporary storage of charge and energy located between the power supply and current load, which is electrically closer to the switching circuit

- To be effective, a decoupling capacitor should have a high capacity to store a sufficient amount of energy and be able to release and accumulate energy at a sufficient rate

- In order to ensure correct and reliable operation of an IC, the impedance of the power distribution system should be maintained below the target impedance in the frequency range from DC to the maximum operating frequency

- The high frequency impedance is effectively reduced by placing decoupling capacitors across the power and ground interconnects, permitting the current to bypass the inductive interconnect

- A decoupling capacitor has an inherent parasitic resistance and inductance and therefore can only be effective within a certain frequency range

- Several stages of decoupling capacitors are typically utilized to maintain the output impedance of a power distribution system below a target impedance

- Antiresonances are effectively managed by utilizing decoupling capacitors with low ESL and by placing a large number of decoupling capacitors with progressively decreasing magnitude, shifting the antiresonant spike to a higher frequency

- MIM capacitors and stacked lateral flux capacitors (fractal, VPP, and woven) are preferable candidates for decoupling power and ground lines in modern high speed, high complexity ICs

7

On-chip Power Distribution Networks

The impedance characteristics of a power distribution system are analyzed in the previous chapter based on a one-dimensional circuit model. While useful for understanding the principles of the overall operation of a power distribution system, a one-dimensional model is not useful in describing the distribution of power and ground across a circuit die. The size of an integrated circuit is usually considerably greater than the wavelength of the signals in the power distribution network. Furthermore, the power consumption of on-chip circuitry (and, consequently, the current drawn from the power distribution network) varies across the die area. The voltage across the on-chip power and ground distribution networks is therefore non-uniform. It is therefore necessary to consider the two-dimensional structure of the on-chip power distribution network to ensure that target performance characteristics of a power distribution system are satisfied. The on-chip power distribution network should also be considered in the context of a die-package system as the properties of the die-package interface significantly affect the constraints imposed on the electrical characteristics of the on-chip power distribution network.

The objectives of this chapter is to describe the structure of an on-chip power distribution network as well as review related tradeoffs. Various structural styles of on-chip power distribution networks are described in Section 7.1. The influence of the electrical characteristics of the die-package interface on the on-chip power and ground distribution is analyzed in Section 7.2. The influence of the on-chip power distribution network on the integrity of the on-chip signals is discussed in Section 7.3. The chapter concludes with a summary.

7.1 Styles of on-chip power distribution networks

Several topological structures are typically used in the design of on-chip power distribution networks. The power network structures range from completely irregular, essentially *ad hoc*, structures, as in routed power distribution networks, to highly regular and uniform structures, as in gridded power networks and power planes. These topologies and other basic types of power distribution networks are described in Section 7.1.1. Design approaches to improve the impedance characteristics of on-chip power distribution networks are presented in Section 7.1.2. The evolution of on-chip power distribution networks in the family of Alpha microprocessors is presented in Section 7.1.3.

7.1.1 Basic structure of on-chip power distribution networks

Several structural types of on-chip power distribution networks are described in this section. Different parts of an on-chip network can be of different types, forming a hybrid network.

Routed networks

In routed power distribution networks, the local circuit blocks are connected with dedicated routed power trunks to the power I/O pads along the periphery of the die [202], as shown in Fig. 7.1. A power mesh is typically used to distribute the power and ground within a circuit block. The primary advantage of routed networks is the efficient use of interconnect resources, favoring this design approach in circuits with limited interconnect resources. The principal drawback of this topology is the relatively low redundancy of the power network. All of the current supplied to any circuit block is delivered through only a few power trunks. The failure of a single segment in a power distribution network jeopardizes the integrity of the power supply voltage levels in several circuit blocks and, consequently, the correct operation of the entire circuit. Routed power distribution networks are predominantly used in low power, low cost integrated circuits with limited interconnect resources.

Mesh networks

Improved robustness and reliability are offered by power and ground mesh networks. In mesh structured power distribution networks, parallel power and ground lines in the upper metal layers span an entire circuit or a specific circuit block. These lines are relatively thick and

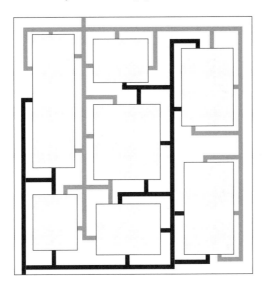

Fig. 7.1. Routed power and ground distribution networks. The on-chip circuit blocks are connected to the I/O power terminals with dedicated power (black) and ground (gray) trunks. The structure of the power distribution networks within the individual circuit blocks is not shown.

wide, globally distributing the power current. The lines are interconnected by relatively short orthogonal straps in the lower metal layer, forming an irregular mesh, as shown in Fig. 7.2. The power and ground lines in the lower metal layer distribute current in the direction orthogonal to the upper metal layer lines and facilitate the connection of on-chip circuits to the global power distribution network. Mesh networks are used to distribute power in relatively low power circuits with limited interconnect resources [203]. It is often the case in semiconductor processes with only three or four metal layers that a regular power distribution grid in the upper two metal layers cannot be utilized due to an insufficient amount of metal resources and an ensuing large number of routing conflicts. Mesh networks are also used to distribute power within individual circuit blocks.

Grid structured networks

Grid structured power distribution networks, shown in Fig. 7.3, are commonly used in high complexity, high performance integrated circuits. Each layer of a power distribution grid consists of many equidistantly spaced lines of equal width. The direction of the power and ground lines within each layer is orthogonal to the direction of the

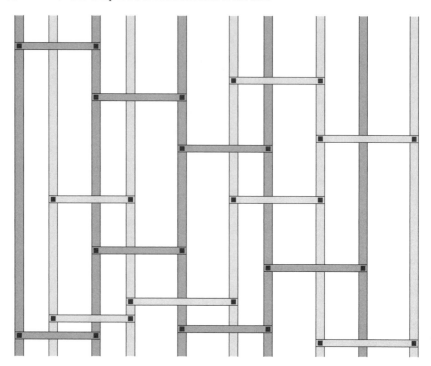

Fig. 7.2. A mesh structured power distribution network. Power (dark gray) and ground (light gray) lines in the vertically routed metal layer span an entire die or circuit block. These lines are connected by short straps of horizontally routed metal to form a mesh. The lines and straps are connected by vias (the dark squares).

lines in the adjacent layers. The power and ground lines are typically interdigitated within each layer. Each power and ground line is connected by vias to other power and ground lines, respectively, in the adjacent layers at the overlap sites. In a typical integrated circuit, the lower the metal layer, the smaller the width and pitch of the lines. The coarse pitch of the upper metal layer improves the utilization of the metal resources, conforming to the pitch of the I/O pads of the package, while the fine pitch of the lower grid layers brings the power and ground supplies in close proximity to each on-chip circuit, facilitating the connection of these circuits to power and ground.

Power distribution grids are significantly more robust than routed distribution networks. Multiple redundant current paths exist between the power terminals of each load circuit and the power supply pads. Due to this property, the power supply integrity is less sensitive to changes in the power current requirements of the individual circuit blocks. The

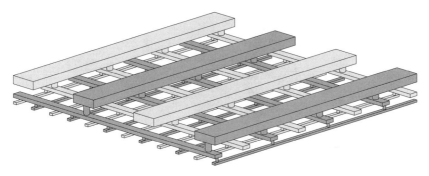

Fig. 7.3. A multi-layer power distribution grid. The ground lines are light gray, the power lines are dark gray. The pitch, width, and thickness of the lines are smaller in the lower grid layers than in the upper layers.

failure of any single segment of the grid is not critical to delivering power to any circuit block. An additional advantage of power distribution grids is the enhanced integrity of the on-chip data signals due to the capacitive and inductive shielding properties of the power and ground lines. These advantages of power distribution grids, however, are achieved at the cost of a significant share of on-chip interconnect resources. It is not uncommon to use from 20% to 40% of the metal resources to build a high density power distribution grid in modern high performance microprocessors [24], [25], [202].

Power and ground planes

Dedicated power and ground planes, shown in Fig. 7.4, have also been used in the design of power distribution networks [204]. In this scheme, an entire metal layer is used to distribute power current across a die, as shown in Fig. 7.4. The signal lines above and below the power/-ground plane are connected with vias through the holes in the plane. Power planes also provide a close current return path for the surrounding signal lines, reducing the inductance of the signal lines and therefore the signal-to-signal coupling. This advantage, however, diminishes as the interconnect aspect ratios are gradually increased with technology scaling. While power planes provide a low impedance path for the power current and are highly robust, the interconnect overhead is typically prohibitively large, as entire metal layers are unavailable for signal routing.

Cascaded power/ground rings

A novel topology for on-chip power distribution networks, called a

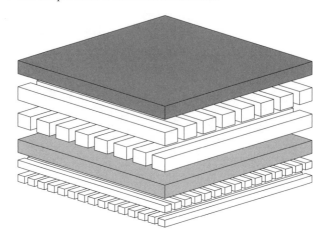

Fig. 7.4. On-chip power distribution scheme using power and ground planes. Two entire metal layers are dedicated to the distribution of power (dark gray layer) and ground (light gray layer).

"cascaded power/ground ring," has been proposed by Lao and Krusius [205] for integrated circuits with peripheral I/O. This approach is schematically illustrated in Fig. 7.5. The power and ground lines are routed from the power supply pads at the periphery of the die toward the die center. The power and ground lines at the periphery of the die, where the power current density is the greatest, are placed on the thick topmost metal layers with the highest current capacity. As the power current decreases toward the die center, the power and ground interconnect is gradually transfered to the thinner lower metal layers.

Hybrid-structured networks

Note that the boundaries between these network topologies are not well defined. A routed network with a large number of links looks quite similar to a meshed network, which, in turn, resembles a grid structure if the number of "strapping" links is large. The terms "mesh" and "grid" are often used interchangeably in the literature.

Furthermore, the structure of a power distribution network in a complex circuit often comprises a variety of styles. For example, the global power distribution can be performed through a routed network, while a meshed network is used for the local power distribution. Or the global power distribution network is structured as a regular grid, while the local power network is structured as a mesh network within one circuit block and a routed network within another circuit block. The common

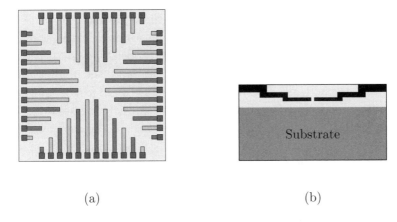

Fig. 7.5. Power distribution network structured as a cascaded power/ground ring; (a) cross-sectional view and (b) top view.

style of a *global* power distribution network has, however, evolved from a routed network to a global power grid, as the power requirements of integrated circuits have gradually increased with technology scaling.

7.1.2 Improving the impedance characteristics of on-chip power distribution networks

The on-chip power and ground interconnect carries current from the I/O pads to the on-chip capacitors and from the on-chip capacitors to the switching circuits, acting as a load to the power distribution network. The flow of the power current through the on-chip power distribution network produces a power supply noise proportional to the network impedance, $Z = R + j\omega L$. The primary design objective is to ensure that the resistance and inductance of a power distribution system is sufficiently small so as to satisfy a target noise margin.

Several techniques have been employed to reduce the parasitic impedance of on-chip power distribution networks. The larger width and smaller pitch of the power and ground lines increase the metal area of the power distribution network, decreasing the network resistance. The resistance is effectively lowered by increasing the area of the power lines in the upper metal layers since these layers have a low sheet resistance. It is not uncommon to allocate over half of the topmost metal layer for global power and ground distribution [24], [165].

The inductance of on-chip power and ground lines has traditionally been neglected because the overall inductance of a power distribution network has been dominated by the parasitic inductance of the package pins, planes, vias, and bond wires. This situation is changing due to the increasing switching speed of integrated circuits [206], [207], the lower inductance of advanced flip chip packaging, and the higher on-chip decoupling capacitance which terminates the high frequency current paths. The requirement of achieving a low inductive impedance is in conflict with the requirement of a low resistance, as the use of wide lines to lower the resistance of a global power distribution network increases the network inductance. Replacing a few wide power and ground lines with multiple narrow interdigitated power and ground lines, as shown in Fig. 7.6, reduces the self inductance of the supply network [208], [209] but increases the resistance. The tradeoffs among the area, resistance, and inductance of on-chip power distribution grids are explored in greater detail in Chapter 11.

Fig. 7.6. Replacing wide power and ground lines (left) with multiple narrow interdigitated lines (right) reduces the inductance and characteristic impedance of the power distribution network.

7.1.3 Evolution of power distribution networks in Alpha microprocessors

The evolution of on-chip power distribution networks in high speed, high complexity integrated circuits is well illustrated by several generations of Digital Equipment Corporation Alpha microprocessors, as described by Gronowski, Bowhill, and Preston [165]. Supporting the reliable and efficient distribution of rising power currents have required adapting the structure of these on-chip power distribution networks, utilizing a greater amount of on-chip metal resources.

Alpha 21064

The Alpha 21064, the first microprocessor in the Alpha family, is implemented in a 0.7 μm CMOS process in 1992. The Alpha 21064 consumes 30 watts of power at 3.3 volts, resulting in nine amperes of average power current. Distributing this current across a 16.8 mm × 13.9 mm die would not have been possible in an existing two metal layer 0.75 μm CMOS process. An additional thick metal layer was added to the process, which is used primarily for the power and clock distribution networks. The power and ground lines in the third metal layer are alternated. All of the power lines are interconnected at one edge of the die with a perpendicular line in metal level three, and all of the ground lines are interconnected at the opposite edge of the die, as shown in Fig. 7.7. The resulting comb-like power and ground global distribution networks are interdigitated. The parallel lines of the global power and ground distribution networks are strapped with the power and ground lines of the second metal layer, forming a mesh-structured power distribution network.

Fig. 7.7. Global power distribution network in Alpha 21064 microprocessor.

Alpha 21164

The second generation microprocessor in the series, the Alpha 21164, appeared in 1995 and was manufactured in a 0.5 μm CMOS process. The Alpha 21164 nearly doubled the power current requirements to 15 amperes, dissipating 50 watts from a 3.3 volt power supply. The type of power distribution network used in the previous generation could not support these requirements. An additional forth metal layer was therefore added to a new 0.5 μm CMOS process. The power and ground lines in the forth metal layer are routed orthogonally to the lines in the third layer, forming a two layer global power distribution grid.

Alpha 21264

The Alpha 21264, the third generation of Alpha microprocessors, was introduced in 1998 in a 0.35 µm CMOS process. The Alpha 21264 consumes 72 watts from a 2.2 volt power supply, requiring 33 amperes of average power current distributed with reduced power noise margins. Utilization of conditional clocking techniques to reduce the power dissipation of the circuit increased the cycle-to-cycle variation in the power current to 25 amperes, exacerbating the overall power distribution problem [165]. The two layer global power distribution grid used in the 21164 could not provide the necessary power integrity characteristics in an integrated circuit with peripheral I/O. Therefore, two thick metal layers were added to the four layer process to allow the exclusive use of two metal layers as power and ground planes. More recent implementations of the Alpha 21264 microprocessor in newer process technologies utilize flip-chip packaging with a high density area array of I/O contacts. This approach obviates the use of on-chip metal planes for distributing power [210].

Power distribution grids are the design style of choice in most modern high performance integrated circuits [165], [204], [211], [212]. The focus of the material presented herein is therefore on on-chip power distribution grids.

7.2 Die-package interface

At high frequencies, the impedance of a power distribution system is determined by the impedance characteristics of the on-chip and package power distribution networks, as discussed in Chapter 5. On-chip decoupling is essential to maintain a low impedance power distribution network at the highest signal frequencies of interest, as discussed in Section 5.5. The required on-chip decoupling capacitance is determined by the frequency where the package decoupling capacitors become inefficient. This frequency, in turn, is determined by the inductive impedance of the current path between the package capacitance and the integrated circuit. The required minimum on-chip decoupling capacitance is proportional to this inductance, as expressed by (5.20) (or by the analogous constraints presented in Section 5.6). Minimizing this inductance achieves the target impedance characteristics of a power distribution network with the smallest on-chip decoupling capacitance.

Achieving a low impedance connection between the package capacitors and an integrated circuit is, however, difficult. Delivering power is only one of the multiple functions of an IC package, which also include connecting the I/O signals to the outside world, maintaining an acceptable thermal environment, providing mechanical support, and protecting the circuit from the environment. These package functions all compete for physical resources within a small volume in the immediate vicinity of the die. Complex tradeoffs among these package design goals are made in practice, often preventing the realization of a resonance-free die-to-package interface.

Wire-bond packaging

Maintaining a low impedance die-package interface is particularly challenging in wire-bonded integrated circuits. The self inductance of a wire bond connection typically ranges from 4 nH to 6 nH. The number of bond wires is limited by the perimeter of the die and the pitch of the wire bond connections. The total number of power and ground connections typically does not exceed several hundred. It is therefore difficult to decrease the inductance of the package capacitor connection to the die significantly below one to two nanohenrys. In wire-bond packages, the inductance of the current loop terminated by the package capacitors is not significantly smaller than the inductance of the current loop terminated by the board decoupling capacitors. Under these conditions, the package capacitors do not significantly improve the impedance characteristics of the power distribution system and therefore no appreciable gain in circuit speed is achieved [105], [213]. Decoupling a high inductive impedance at gigahertz frequencies typically requires an impractical amount of on-chip decoupling capacitance (and therefore die area), limiting the operational frequency of a wire bonded circuit.

Providing a low impedance connection between the off-chip decoupling capacitors and a wire-bonded integrated circuit requires special components and packaging solutions. For example, a so-called "closely attached capacitor" has been demonstrated to be effective for this purpose in wire-bonded circuits [214]. A thin flat capacitor is placed on the active side of an integrated circuit. The dimensions of the capacitor are slightly smaller than the die dimensions. The bonding pads of an integrated circuit and the edge of the capacitor are in close proximity, permitting a connection with a short bond wire, as illustrated in Fig. 7.8. The die-to-capacitor wires are several times shorter than the wire connecting the die to the package. The impedance between the circuit and the off-chip decoupling capacitance is therefore significantly decreased.

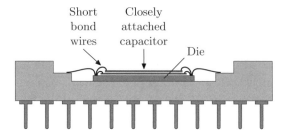

Fig. 7.8. Closely attached capacitor. Stacking a thin flat capacitor on top of a circuit die allows connecting the capacitor and the circuit with bond wires of much shorter length as compared to bond wires connecting the circuit to the package.

This wiring technique reduces the power switching noise in CMOS circuits by at least two to three times, as demonstrated by Hashemi *et al.* [214]. For example, attaching a 12 nF capacitor to a 32-bit microprocessor decreases the internal logic power supply noise from 900 mV to 150 mV, a sixfold improvement (the microprocessor is packaged in a 169-pin pin grid array package and operates at 20 MHz with a 5 volt power supply). Similarly, in a 3.3 volt operated bus interface circuit, a 12 nF closely attached capacitor lowers the internal power noise from 215 mV to 100 mV. A closely attached capacitor is used in the Alpha 21264 microprocessor, as the on-chip decoupling capacitance is insufficient to maintain adequate power integrity [165], [215].

Flip-chip packaging

The electrical characteristics of the die-package interface are significantly improved in flip-chip packages. Flip-chip bonding refers to attaching a die to a package with an array of solder balls (or bumps) typically 50 μm to 150 μm in diameter. In cost sensitive circuits, the ball connections, similar to wire-bond connections, can be restricted to the periphery of the die in order to reduce the interconnect complexity of the package. In high complexity, high speed integrated circuits, however, an area array flip-chip technology is typically used where solder ball connections are distributed across (almost) the entire area of the die. The inductance of a solder ball connection, typically from 0.1 nH to 0.5 nH, is much smaller than the 4 nH to 10 nH typical for a bond wire [103], [132], [216]. Area array flip-chip bonding also provides a larger number of die to package connections as compared to wire bonding. Modern high performance microprocessors have thousands of flip-chip contacts dedicated to the power distribution network [25], [217], [218], [219]. A larger number of lower inductance power and ground connectors significantly decreases the overall inductance of the die to package connection.

Also important in integrated circuits with peripheral I/O, the power current is distributed on-chip across a significant distance: from the die edge to the die center, as shown in Fig. 7.9. In integrated circuits with a flip-chip area array of I/O bumps, the power current is distributed on-chip over a distance comparable to the size of the power pad pitch, as shown in Fig. 7.10. This distance is significantly smaller than half the die size. Flip-chip packaging with high density I/O, therefore, significantly reduces the effective resistance and inductance of the on-chip power distribution network, mitigating the resistive [220], [221] and inductive voltage drops.

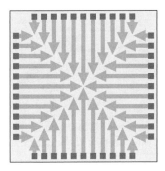

Fig. 7.9. Flow of power current in an integrated circuit with peripheral I/O. The power current is distributed on-chip across a significant distance: from the edge of the die to the die center.

Flip-chip packaging with high density I/O therefore decreases the area requirements of the on-chip power distribution network, improving the overall performance of a circuit [210], [222], [223], [224]. The dependence of the on-chip power voltage drop on the flip-chip I/O pad density and related power interconnect requirements are discussed in greater detail in Chapter 12.

Another advantage of area array flip-chip packaging is the possibility of placing the package decoupling capacitors in close physical proximity to the die, significantly enhancing the efficacy of the capacitors. A bank

of low parasitic inductance capacitors can be placed on the underside
of the package immediately below the die, as shown in Fig. 7.11. The
separation between the capacitors and the on-chip circuitry is reduced
to 1 mm to 2 mm, minimizing the area of the current loop and associated inductance. The parasitic inductance of a package decoupling
capacitor with the package vias and solder bumps connecting the capacitor to the die can be reduced well below 1 nH [213], enhancing the
capacitor efficiency at high frequencies. Placing a package decoupling
capacitor immediately below the circuit region with the greatest power
current requirements further improves the efficiency of the decoupling
capacitors of the package [225].

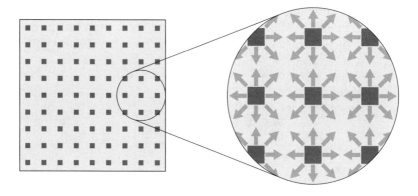

Fig. 7.10. Flow of power current in an integrated circuit with flip-chip I/O. The
power current is distributed on-chip over a distance comparable to the pitch of the
I/O pads.

Overall, flip-chip packaging significantly decreases the impedance
between the integrated circuit and the package decoupling capacitors,
relaxing the constraints on the resistance of the on-chip power distribution network and on-chip decoupling capacitors. Flip-chip packaging
therefore can significantly improve the power supply integrity while
reducing the die area.

Future packaging solutions

The increasing levels of current consumed by CMOS integrated
circuits as well as the high switching speeds require power distribution systems with a lower impedance over a wider frequency range.
An increasingly lower inductance between the integrated circuit and
the package decoupling capacitance is essential to maintain a lower

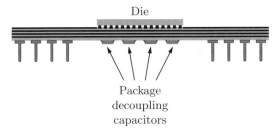

Fig. 7.11. Flip-chip pin grid array package. The package decoupling capacitors are mounted on the bottom side of the package immediately below the die mounted on the top side. In this configuration, the package capacitors are in physical proximity to the die, minimizing the impedance between the capacitors and the circuit.

impedance at higher frequencies. Providing a low impedance die-to-package connection remains a challenging task [226]. Future generations of packaging solutions, such as chip-scale and bumpless build-up layer (BBUL) packaging, are addressing this problem with higher density die-to-package contacts, a smaller separation between the power and ground planes, and a lower package height [114], [227], [228].

The electrical characteristics of a package have become one of the primary factors that limit the performance of an integrated circuit [229], [230]. The package design is now crucial in satisfying both the speed and overall cost targets of high performance integrated circuits. Achieving these goals will require explicit co-design of the on-chip global interconnect and the package interconnect networks [112], [229].

7.3 Other considerations

Dependence of the power supply integrity on the impedance characteristics of the on-chip power distribution network has been discussed previously. The on-chip power distribution network also significantly affects the integrity of the on-chip data and clock signals. The integrity of the on-chip signals depends upon the structure of the power distribution network through two primary mechanisms: inductive interaction among the power and signal interconnect and substrate coupling. These two phenomena are briefly discussed in this section.

Dependence of on-chip signal integrity on the structure of the power distribution network

The structure of the power distribution grid is one of the primary factors determining the integrity of on-chip signals. The power and ground lines shield adjacent signal lines from capacitive crosstalk. Sensitive signals are typically routed adjacent to the power and ground lines. Co-designing the power and signal interconnect has become an important consideration in the design of high speed integrated circuits [231], [232], [233].

Power and ground networks provide a low impedance path for the signal return currents. The structure of the on-chip power distribution network is therefore a primary factor that determines the inductive properties of on-chip signal lines, such as the self and mutual inductance. Modeling the inductive properties of the power distribution grid is necessary for accurately analyzing high frequency phenomena, such as return current distribution, signal overshoot, and signal delay variations, as demonstrated by on-chip interconnect structures using full-wave partial element equivalent circuit (PEEC) models [234], [235]. These conclusions are also supported by the analysis of commercial microprocessors. Inadequate design of the local power distribution network can lead to significant inductive coupling of the signals, resulting in circuit failure [212].

Interaction between the substrate and the power distribution network

In high complexity digital integrated circuits, the ground distribution network is typically connected to the substrate to provide an appropriate body bias for the NMOS transistors. The substrate provides additional current paths in parallel to the ground distribution network, affecting the current distribution in the network, as illustrated in Fig. 7.12. This effect is significant in most digital CMOS processes which utilize a low resistivity substrate to prevent device latch-up [236]. A methodology for analyzing power distribution networks together with the silicon substrate requires a complete model, which includes both the power distribution system and the substrate, as described by Panda, Sundareswaran, and Blaauw [159]. The substrate significantly reduces the voltage drop in the ground distribution network (assuming an N-well process) by serving as an additional parallel path for the ground current to flow, as demonstrated by an analysis of three Motorola processor circuits [159]. The placement of substrate contacts also affects the on-chip power supply noise and the substrate noise [159], [237].

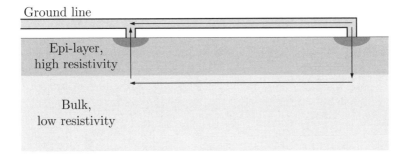

Fig. 7.12. Interaction of the substrate and power distribution network. The low resistivity bulk substrate provides additional current paths between the points where the ground network is connected to the substrate. These current paths are connected in parallel to the ground distribution network.

7.4 Summary

The structure of the on-chip power distribution network and related design considerations are described in this chapter. The primary conclusions are summarized as follows.

- On-chip power distribution grids are the preferred design style in high speed, high complexity digital integrated circuits

- Constraints placed on the impedance characteristics of the on-chip power distribution networks are greatly affected by the electrical properties of the package

- The high frequency impedance characteristics of a power distribution system are significantly enhanced in packages with an area array of low inductance I/O contacts

8

Computer-Aided Design and Analysis

The process of computer-aided design and analysis of on-chip power distribution networks is discussed in this chapter. The necessity for designing and analyzing the integrity of the power supply arises at various stages of the integrated circuit design process as well as during the verification phase. The design and analysis of power distribution networks, however, poses unique challenges and requires different approaches as compared to the design and analysis of logic circuits.

The requirement for analyzing on-chip power distribution networks arises throughout the design process, from the onset of circuit specification to the final verification phase, as discussed in Section 8.1. The primary tasks and difficulties in analyzing the power supply vary at different phases of the design process. At the initial and intermediate design phases, the specification of the power distribution network is incomplete. The primary goal of the power supply analysis process is to guide the general design of the on-chip power distribution network based on information characterizing the power current requirements of the on-chip circuits. The information characterizing the power current requirements is limited, giving rise to the principal difficulty of the analysis process: producing efficient design guidance based on data of limited accuracy. The character of the analysis process gradually changes toward the final phases of the design process. The design of both the power distribution network and the on-chip logic circuits becomes more detailed, making a more accurate analysis possible. The principal goal of the analysis process shifts to verifying the design and identifying those locations where the target specifications are not satisfied. The dramatically increased complexity of the analysis process is the primary difficulty, requiring utilization of specialized computational methods.

The chapter is organized as follows. A typical flow of the power distribution network design process is described in Section 8.1. An approach for reducing the analysis of power distribution system to a linear problem is presented in Section 8.2. The process of constructing circuit models that characterize a power distribution system is discussed in Section 8.3. Techniques for characterizing the power current requirements of the on-chip circuits are described in Section 8.4. Numerical techniques used in the analysis of power distribution networks are briefly described in Section 8.5. Three strategies for allocating on-chip decoupling capacitors are described in Section 8.6. The chapter is summarized in Section 8.7.

8.1 Design flow for on-chip power distribution networks

In high performance circuits, the high level design of the global power distribution network typically begins before the physical design of the circuit blocks. This approach ensures preferential allocation of sufficient metal resources, simplifying the design process. The principal decisions on the structure of the power distribution network are therefore made when little is known about the specific power requirements of the on-chip circuits. The early design of the power grid is therefore based on conservative design tradeoffs and is gradually refined in the subsequent phases of the design process.

The design flow for a power distribution grid is shown in Fig. 8.1. As the circuit design becomes better specified, a more accurate characterization of the power requirements is possible and, consequently, the design of the power distribution network becomes more precise. The design process can be roughly divided into three phases: preliminary pre-floorplan design, floorplan-based refinement, and layout-based verification [161], [211]. These phases are described in the rest of this section.

Preliminary pre-floorplan design

In the initial pre-floorplan phase, little is known about the power current requirements of the circuit. Preliminary estimates of the power current consumed by a circuit are typically made by scaling the power consumption of previously designed circuits considering the target die

POWER GRID DESIGN FLOW

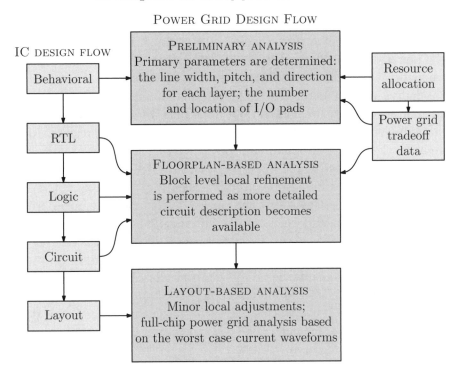

Fig. 8.1. Design flow for on-chip power distribution networks.

area, operating frequency, power supply voltage, and other circuit characteristics.

At this phase of the design process, the power distribution grid is often laid out as a regular periodic structure. A preliminary DC analysis of the IR drops within a power distribution grid is performed, assuming that the power current requirements are uniform across the die. The average power current requirements used in a DC analysis are increased by three to seven times to estimate the maximum power current. The power distribution network is assumed to be uniformly loaded with constant current sources. Basic parameters of the power distribution grid, such as the width and pitch of the power lines in each metal layer and the location of the power/ground pads, are determined based on a preliminary DC analysis of the network. This initial structure largely determines the tradeoff between robustness and the amount of metal resources used by an on-chip power distribution grid.

Floorplan-based refinement

Once the floorplan of a circuit is determined, the initial design of the power distribution grid is refined to better match the local current capacity of the power distribution grid to the power requirements of the individual circuit blocks [202]. The maximum and average power current of each circuit block is determined based on the function of an individual block (*e.g.*, the memory, floating point unit, register file), area, block architecture, and the specific circuit style (*e.g.*, static, dynamic, pass transistor [238], *etc.*). The current distribution is assumed uniform within the individual circuit blocks.

Block specific estimates of the power current provide an approximation of the non-uniform power requirements across the circuit die. The structure of the power distribution grid is tailored according to a DC analysis of a non-uniform power current distribution. Many of the primary problems in the design of power distribution networks are identified at this phase. Moderate computational requirements permit iterative application of a static analysis of the network. Large scale deficiencies in the coverage and capacity of the power distribution network are detected and repaired.

As the structure of the circuits blocks becomes better specified, the local power consumption of an integrated circuit can be characterized with more detail and accuracy. After the logic structure of the circuits is determined, the accuracy of the current requirements are enhanced based on the number of gates and clocking requirements of the circuit blocks. Gate level simulations provide a per cycle estimate of the DC power current for a chosen set of input vectors [211]. Cycle-to-cycle variations of the average power current provide an approximation of the temporal variations of the power current, permitting a preliminary dynamic AC analysis of the power distribution system. The accuracy of the dynamic analysis can be improved if more detailed current waveforms are obtained through gate level simulations. The worst case current waveform of each type of gate and circuit macro is precharacterized. The current waveforms of the constituent gates are arranged according to the timing information obtained in the simulations and are combined into an effective power current waveform for an entire circuit block. As the circuit structure and operating characteristics become better specified, the structure of the power distribution grid within each of the circuit blocks is refined to provide sufficient reliability and integrity of the on-chip power supply while minimizing the required routing resources.

As the precise placement of the circuit gates is not known in the pre-layout phase, the spatial resolution of the floorplan-based models is relatively coarse. The die area is divided into a grid of $N \times M$ cells. The power and ground distribution networks within each cell are reduced to a simplified macromodel. These macromodels form a coarse RC/RLC grid model of the on-chip power distribution network, as shown in Fig. 8.2. The power current of the circuits located in each cell is combined and modeled by a current source connected to the appropriate node of the macromodel. The number of cells in each dimension of the circuit typically varies from several cells up to a hundred cells, depending on the size of the circuit and the accuracy of the power consumption estimate. The computational requirements of the analysis increase with the specificity of the circuit description. The total number of nodes, however, remains relatively small, permitting an analysis with conventional nonlinear circuit simulation tools such as SPICE.

Layout-based verification

When the physical design of a circuit is largely completed, a detailed analysis of the power distribution network is performed to verify that the target power supply noise margins are satisfied at the power/-ground terminals of each on-chip circuit. A detailed analysis is first performed at the level of the individual circuit blocks. Those areas where the noise margins are violated are identified during this analysis phase. The current capacity of the power distribution grid is locally increased in these areas by widening the existing power lines, adding lines, and placing additional on-chip decoupling capacitance. The detailed verification process is repeated on the modified circuit. The iterative process of analysis and modification is continued until the design targets are satisfied. Finally, the verification process is performed for the entire circuit.

An analysis of an entire integrated circuit is necessary to verify the design of a power distribution network. Analyzing the integrity of the power supply at the circuit block level is insufficient as neighboring blocks affect the flow of the current through the power grid. For example, the design of a power grid within a circuit block drawing a relatively low power current (*e.g.*, a memory block) may appear satisfactory at a block-level analysis. However, it is likely to fail if the block is placed in

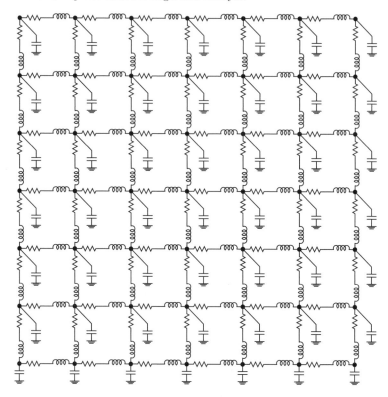

Fig. 8.2. An *RLC* model of an on-chip power distribution network [157].

close proximity to a block drawing high power current. The high power blocks can increase the current flowing through the power network of adjacent low power units [161]. It is therefore necessary to verify the design of the power distribution network at the entire circuit level.

The principal difficulty in verifying an entire power distribution network in a high complexity integrated circuit is the sheer magnitude of the problem. The on-chip power network of a modern high complexity integrated circuit often comprises tens of millions of interconnect line segments and circuit nodes forming a multi-layer power distribution grid, as described in Section 7.1. The circuits loading the power distribution network also consist of tens or hundreds of millions of interconnects and transistors. A transistor level circuit simulation of an entire circuit is therefore infeasible due to prohibitive memory and CPU time requirements. Final analysis and verification is therefore one of the most challenging tasks in the design of on-chip high complexity power distribution networks. The remainder of this chapter is largely focused on techniques and methodologies to manage the complexity of the analysis and verification process of power distribution networks.

This methodology is successful if the noise margin violations are local and can be corrected with available metal resources. However, if the necessary changes in the power grid require significant changes in the routing of the critical signal lines, the timing and noise performance characteristics of these critical signals can be significantly impaired. The laborious task of signal routing and timing verification of a circuit is repeated, drastically decreasing design productivity and increasing the time to market. This difficulty of making significant changes in the structure of the power distribution grid at late phases of the design process is the primary reason for using a highly conservative approach in the design of an on-chip power distribution network. Worst case scenarios are assumed throughout the design process. The resulting power distribution network is therefore typically overdesigned, significantly increasing the area of power distribution networks in modern interconnect-limited integrated circuits.

8.2 Linear analysis of power distribution networks

The process of analysis consists of building a circuit model of the power and ground networks including the circuits loading the networks. This step is followed by a numerical analysis of the resulting model. The problem is inherently nonlinear as the digital circuits loading the power distribution grid exhibit highly nonlinear behavior. The current drawn by the load circuits from the power distribution network varies nonlinearly with the voltage across the power terminals of the load. Analyzing a network with tens of millions of nodes is infeasible using a nonlinear circuit simulator such as SPICE due to the enormous computational and memory requirements. To permit the use of efficient numerical analysis techniques, the nonlinear part of the problem is separated from the linear part [161], [202], [211], as illustrated in Fig. 8.3. The current drawn from the power distribution network by nonlinear on-chip circuits is characterized assuming a nominal power and ground supply voltage. The load circuits are replaced by time dependent current sources emulating the original power current characteristics. The resulting network consists of power distribution conductors, decoupling capacitors, and time dependent current sources. This network is linear, permitting the use of efficient numerical techniques.

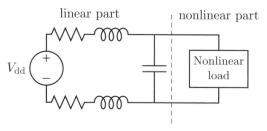

(a) The original problem of analyzing a linear power distribution network with a nonlinear load

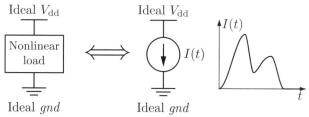

(b) The current requirements of the nonlinear load are characterized under an ideal supply voltage

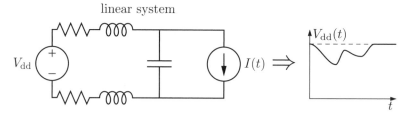

(c) The nonlinear load is replaced with an AC current source. The resulting system can be analyzed with linear methods.

Fig. 8.3. Approximation for analyzing a power distribution network by replacing a nonlinear load with a time-dependent current source.

Partitioning the problem into a power current characterization part and a linear system analysis part ignores the negative feedback between the power current and the power supply noise. The current flowing through the power and ground networks causes the power supply levels to deviate from the nominal voltage. In turn, the reduced voltage between the power and ground networks decreases the current drawn by the power load. This typical approach to the power noise analysis process is therefore conservative, overestimating the magnitude of the power supply noise. The relative decrease in the power current due to the reduced power supply voltage is comparable to the relative decrease in the power-to-ground voltage, typically maintained below 10%. The accuracy of these conservative estimates of the power noise is acceptable for most applications. To achieve greater accuracy, the analysis can be performed iteratively. The power current requirements are recharacterized at each iteration based upon the power supply voltage obtained in the previous step. Each iteration yields a more accurate approximation of the power supply voltage across the power distribution network.

The process of power distribution network analysis therefore proceeds in three phases: model construction, load current characterization, and numerical analysis. These tasks are described in greater detail in the following sections.

8.3 Modeling power distribution networks

It is essential that the on-chip power distribution network is considered in the context of the entire power distribution system, including the package and board power distribution networks. As discussed in Chapter 5, the package and board power distribution networks determine the impedance characteristics of the overall power distribution system at low and intermediate frequencies. It is therefore important to analyze the entire power distribution system, including the package and board power distribution networks and the decoupling capacitors, in order to obtain an accurate analysis of the on-chip power supply noise [157].

The complexity of a model depends upon the objectives of the analysis. Models for a DC analysis performed at the preliminary and floorplan-based design phases need to capture only the resistive characteristics of the interconnect structures. The inductive and capacitive circuit characteristics are unimportant in a DC analysis, greatly simplifying the model. As discussed in the previous section, the spatial resolution of these models is typically limited, further simplifying the process of model characterization.

The reactive impedances of the system are, however, essential for an accurate AC analysis of a power distribution system. The capacitance of the board, package, and on-chip decoupling capacitors as well as the inductive properties of the network should be characterized with high accuracy.

The analysis and verification step towards the end of the design process requires highly detailed models that capture the smallest features of a power distribution system. These models are typically constructed through a back annotation process. The complexity of the board and package power distribution networks is relatively moderate, with the number of conductors ranging from hundreds to thousands. The moderate complexity supports the use of relatively sophisticated analysis tools, such as two- and three-dimensional quasi-static electromagnetic field analyzers [105], [111], [157], [239]. Characterizing the on-chip power distribution network is the most difficult part of the modeling process. The on-chip power distribution network comprises tens of million of nodes and interconnect elements. This level of complexity necessitates the use of highly efficient algorithms to extract the parasitic impedances of the on-chip circuit structures.

Resistance of the on-chip power distribution network

The resistance of on-chip interconnect can be efficiently characterized either with simple resistance formulas based on the sheet resistance of a metal layer [211], [239] or using well developed shape-based extraction algorithms [240], [241]. The temperature dependence of the interconnect resistance should also be included in the model. If R_{25} is the nominal metal resistance at a room temperature of 25°C, the metal resistance at the operating temperature of the circuit T_{op} is $R_{25}(1 + k_T(T_{op} - 25))$, where k_T is the temperature coefficient of the metal resistance. For a temperature coefficient of copper doped aluminum metalization of $0.003\,°C^{-1}$ and an operating temperature of 85°C, the temperature induced per cent increase in the resistance is 18%, a significant change. Furthermore, the interconnect resistance increases over the circuit lifetime due to electromigration induced defects in the metal structure. This increase in resistance is typically considered in the design process by increasing the nominal metal resistance by a coefficient K_{em}, typically ranging from 10% to 20% [239]. The overall resistance of the on-chip metal R_{eff} can therefore be characterized as [157]

$$R_{\text{eff}} = R_{25}\left(1 + (T_{\text{op}} - 25)\right)\left(1 + K_{\text{em}}\right). \tag{8.1}$$

Characterization of the on-chip decoupling capacitance

Characterizing the capacitive impedances within the power distribution system is more difficult as compared to resistance characterization. The intrinsic capacitance of the power and ground lines is dominated by other sources of the decoupling capacitance, *i.e.*, the intrinsic circuit capacitance, well capacitance, and intentional capacitance, as discussed in Section 6.3. The capacitance of the power and ground lines can therefore be neglected in this analysis. The intentional and well diffusion decoupling capacitances can be readily characterized by shape-based extraction methods. The intrinsic decoupling capacitance of the on-chip circuits depends upon the state of the digital circuits, making this capacitance difficult to characterize.

The intrinsic circuit decoupling capacitance can be estimated based on the power consumption of the circuit, as described by Chen and Ling [239] and by Larsson [151]. Assuming that the total power P_0 is dominated by the dynamic switching power $P_{\text{switching}}$,

$$P_0 \approx P_{\text{switching}} = \alpha C_{\text{total}} f_{\text{clk}} V_{\text{dd}}^2, \tag{8.2}$$

where α is the switching factor of the circuit, C_{total} is the total intrinsic capacitance of the circuit, f_{clk} is the clock frequency, and V_{dd} is the supply voltage. The total capacitance C_{total} can therefore be determined from an estimate of the total circuit power: $C_{\text{total}} = \frac{P_0}{\alpha f V_{\text{dd}}^2}$. The fraction of the total capacitance being switched, *i.e.*, αC_{total} on average, is the load capacitance of the circuit. The rest of the total capacitance, $(1 - \alpha)C_{\text{total}}$, is quiescent and effectively serves as a decoupling capacitance. The intrinsic decoupling capacitance of the circuit is, therefore,

$$C_{\text{decap}}^{\text{ckt}} \approx \frac{P_0}{f V_{\text{dd}}^2} \frac{1 - \alpha}{\alpha}. \tag{8.3}$$

An estimate of the intrinsic decoupling capacitance represented by (8.3) is, however, strongly dependent on the switching factor α. The switching factor varies significantly depending upon the specific switching pattern and circuit type. The switching factor is therefore difficult to determine with sufficient accuracy in complex digital circuits.

Alternatively, the decoupling capacitance of quiescent circuits can be characterized by simulating a small number of representative circuit blocks, as described by Panda et al. [111]. A complete circuit model of each selected circuit block, including the parasitic impedances of the interconnect, is constructed through a back annotation process. The input terminals of a circuit block are randomly set to either the high or low state. The power terminals of the circuit are biased with the power supply voltage V_{dd}. A sinusoidal AC voltage of relatively small amplitude (5% to 15% of V_{dd}) is added to the power terminals of the circuit, modeling the power supply noise, as shown in Fig. 8.4(a). The current flowing through the power terminals is obtained and the small signal impedance of the circuit block *as seen from the power terminals* is determined for the specific frequency of the AC excitation. A series RC model is subsequently constructed, such that the model impedance approximates the impedance of the original circuit block, as shown in Fig. 8.4(b). The model capacitance is scaled by a factor $(1 - \alpha)$ to account for the switching of the circuit capacitance α which does not participate in the decoupling process. The resulting model is an equivalent circuit of the decoupling capacitance of the quiescent circuits, including the decoupling capacitance of both the transistors and interconnect structures. This estimate of the decoupling analysis is significantly less sensitive to the value of α, as compared to (8.3).

The elements R_{eff} and C_{eff} of the equivalent model depend on the state of the digital circuit and the frequency of the applied AC excitation. Nevertheless, these model parameters typically vary little with the input pattern and the excitation frequency in the range of 0.2 to 2 times the clock frequency [111]. For example, the model parameters exhibit less than a 3% variation over all of the input states of an example circuit block consisting of 240 transistors with ten primary inputs [111]. The decoupling characteristics of a larger circuit block are extrapolated from the characteristics of one or several of the precharacterized blocks, depending upon the circuit structure of the larger block. This technique allows for variations in the intrinsic decoupling capacitance for different circuit types.

In many circuits, however, the circuit decoupling capacitance is dominated by the well diffusion capacitance and the intentional capacitance [159]. The overall accuracy of the power supply noise analysis in these circuits is only moderately degraded by the inaccuracies in characterizing the circuit decoupling capacitance.

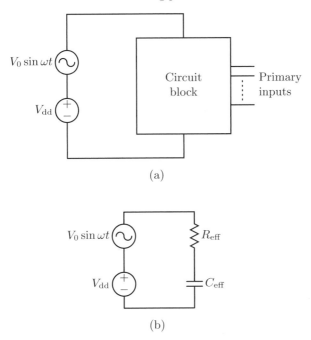

(a)

(b)

Fig. 8.4. Characterization of the intrinsic decoupling capacitance of the quiescent circuits; (a) circuit model to characterize the capacitance, (b) an equivalent circuit model of the intrinsic decoupling capacitance [111].

Inductance of the on-chip power distribution network

The inductance of the on-chip power and ground lines has historically been neglected [111]. The relatively high resistance R of the on-chip interconnect has dominated the inductive impedance ωL, suppressing the inductive behavior, such as signal reflections, oscillations, and overshoots. As the switching time of the on-chip circuits decreases with technology scaling, the spectral content of the on-chip signals has extended to higher frequencies, making on-chip inductive effects more pronounced. The significance of the on-chip inductance has been demonstrated by Chen and Schuster in an investigation of the sensitivity of the power supply noise to various electrical characteristics of the power distribution system [242]. Assuming the package leads provide an ideal nominal voltage of 2.5 volts, an RLC analysis of the on-chip power grid predicts a minimum on-chip voltage V_{dd} of 2.307 volts, 0.193 volts below the nominal level. If the inductance of the on-chip power grid is neglected, the analysis predicts a minimum on-chip power voltage V_{dd} of 2.396 volts, underestimating the on-chip power noise by 50% as compared to a more complete RLC model. Including the package model in the analysis further reduces the on-chip power supply to 2.199 volts. Modeling the inductive properties of the on-chip power interconnect is therefore necessary to ensure an accurate analysis.

Incorporating the inductive properties of on-chip interconnect into the model of a power distribution network poses two challenges. First, existing techniques for characterizing the inductive properties of complex interconnect structures are computationally intensive, greatly reducing the efficiency of the back annotation process. This issue is discussed further below. Second, including inductance in the model precludes the use of highly efficient techniques for numerically analyzing complex power distribution networks, as discussed in Section 8.5.

The inductive properties of power and ground interconnect lines are difficult to characterize. Characterizing the inductance by a conventional method, $i.e.$, determining the loop inductance of the on-chip circuits based on the shape and size of the current loops is difficult as the current path consists of multiple conductors and the path of the current flow is, generally, not known a $priori$. The inductive properties of regular on-chip power distribution grids can be estimated based on an electromagnetic analysis of the grid structure [157], [239]. Alternatively, the inductive properties can be extracted in the form of a partial inductance matrix. While extracting the partial inductance matrix of an entire circuit is computationally efficient, this matrix is highly dense. The density of a complete partial inductance matrix drastically degrades the efficiency of the subsequent numerical analysis of the circuit model, as the computational efficiency of the most effective numerical methods is conditioned on the sparsity of the matrices characterizing the system. Techniques for sparsifying partial inductance matrices are an active area of research and several techniques has been proposed [63], [64], [65], [66]. The computational efficiency of these techniques is currently insufficient to make the analysis of multimillion conductor systems practical. A method to characterize the inductance of on-chip power distribution grids is described in Chapter 9.

Exploiting symmetry to reduce model complexity

The magnitude of the current flowing through the power and ground distribution networks is the same. The power and ground networks have the same electrical requirements and the structures of these networks are often (close to) symmetric, particularly at the initial and intermediate phases of the design process. This symmetry can be exploited to reduce the complexity of the power distribution network model by half [111], as illustrated in Fig. 8.5. The model reduction is achieved by circuit "folding." The original symmetric circuit, as shown in Fig. 8.5(a), is transformed into an equivalent circuit, where the sources, loads, and decoupling capacitors are replaced with equivalent symmetric networks, as shown in Fig. 8.5(b). The nodes on the axis of symmetry of the circuit (shown with a dashed line in Fig. 8.5(b)) are equipotential. It is convenient to use the potential of these nodes as a reference potential. These nodes are therefore referred to as a *virtual ground*. The original circuit is transformed into two independent circuits, as shown in Fig. 8.5(c). The independent circuits are symmetric; consequently, an analysis of only one circuit is necessary. The currents and voltages in one circuit have an opposite polarity as compared to the currents and voltages in the symmetric circuit. Where the impedances of the power and ground distribution networks are symmetric, the voltages in the power and ground networks (with reference to the virtual ground) are also symmetric. That is, wherever the power voltage is decreased by δV, the ground voltage is increased by δV (thereby decreasing the power rail-to-rail voltage by $2\delta V$).

8.4 Characterizing the power current requirements of on-chip circuits

Accurate characterization of the power current requirements is an integral part of the power distribution analysis process, as discussed in Section 8.2. A brief overview of the methods for power current characterization is presented in this section. As the structure of an integrated circuit becomes specified in greater detail, the power current characteristics can be specified with greater accuracy. The complexity of the power current characterization process dramatically increases with the complexity of the circuit description.

Preliminary evaluation of power current requirements

Early estimates of the power current are static. The temporal variation of the power current cannot be assessed in this phase as only the high level structure of the circuit has been developed. Static approaches are based on estimates of the average load currents drawn from the power network [161], [211], permitting static estimates of the IR

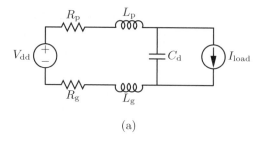

(a)

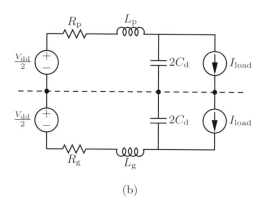

(b)

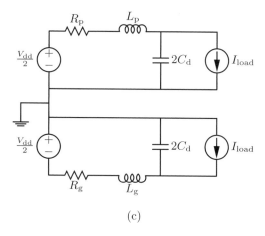

(c)

Fig. 8.5. Exploiting the symmetry of the power and ground distribution networks to reduce the model complexity by a factor of two. (a) The power and ground current paths in the original model are symmetric. (b) A virtual ground is introduced between the power and ground networks in the equivalent circuit as shown by the dashed line. (c) The resulting circuit model contains two independent symmetric circuits.

drops and electromigration reliability. Estimating average power current consumption is equivalent to estimating average circuit power, as the power supply voltage is maintained approximately constant. At the onset of the design process, the average power current per circuit area is estimated based on the function of a particular circuit block, circuit style used to implement a circuit block, and a scaling analysis of previously designed similar circuits. These estimates can be augmented by estimates of average circuit power based on a circuit description at the behavioral, register transfer, or microarchitectural levels [243], [244].

Gate level estimates of the power current requirements

Estimates of the power current requirements are refined once the logic structure of the circuit is determined. The primary difficulty in determining the power requirements of a CMOS circuit with sufficient accuracy is the dependency of the power current on the circuit input pattern [245]. The time of switching and the magnitude of the power current of a particular gate are determined by the temporal and functional relationships with other gates. The switching patterns that produce the greatest variation of the power supply voltage from a nominal specification (*i.e.*, the greatest power supply noise) are referred to as the worst case switching patterns. These worst case switching patterns are difficult to identify.

The worst case power current of small circuit structures, such as individual logic gates and circuit macrocells, are relatively easy to determine as the number of possible switching patterns is small and the patterns can be readily evaluated. The number of possible switching patterns increases exponentially with the number of inputs and internal state variables. The worst case switching patterns of relatively large circuit blocks comprising thousands of circuit gates and macrocells cannot be determined from an exhaustive analysis. Assuming that all of the gates draw the worst case power current at the same time is overly conservative. Incorporating the logical dependencies among the logic gates, however, greatly increases the complexity of the analysis. A tradeoff therefore exists between the accuracy and efficiency of the power current model.

Estimates of the average power current are typically based on a probabilistic or statistical analysis of the average switching activity and the output load (*i.e.*, the switched capacitance) of the gates [245]. Simple estimates of the average load currents can be obtained by determining the saturation current of each gate in a block and scaling this

current to account for the quiescent state of the majority of the gates at
any particular time. More accurate estimates of the average current I_{avg}
can be obtained through gate level simulations by evaluating the average switching activity P_s and capacitive load C_L of the gates. The av-
erage power current of the circuit is evaluated as $I_{\mathrm{avg}} = \frac{1}{2} P_s f_{\mathrm{clk}} C_L V_{\mathrm{dd}}$,
where f_{clk} is the clock frequency of the circuit. Several methods have
been developed to determine the upper bound of the power current
consumed by the circuit and the associated bound on the power supply
noise in an input pattern independent manner [246], [247], [248], [249].

8.5 Numerical methods for analyzing power distribution networks

The circuit model of a power distribution network is combined with
time dependent current sources emulating the worst case load to form
a linear model of a power distribution network. The linear model is
described by a system of linear differential equations. The system of
differential equations is reduced to a system of linear equations, which
can be numerically analyzed using a number of efficient linear system
solution methods [250]. These linear solution methods are classified
into direct and iterative methods [251]. The direct methods rely on
factoring the coefficient matrix that characterizes the linear system.
Once the matrix decomposition is performed, the system solution at
each simulation time step is obtained by forward and backward substitution. Alternatively, iterative methods can be used to obtain the
solution through a series of successive approximations. Assuming sufficient memory capacity to store the factorization matrices, the use of
direct methods is preferable in analyses requiring a large number of
time steps, as the solution at each step is obtained through an efficient
substitution procedure. Iterative methods are more efficient in solving
large systems with limited memory resources.

Numerical techniques exploiting special properties of the system are
commonly employed to enhance the efficiency of the analysis process.
The coefficient matrix of a linear system describing a power distribution
network is highly sparse, with non-zero elements typically constituting
only a 10^{-6} to 10^{-8} fraction of the total number of elements [202]. Furthermore, in a modified nodal analysis approach, the matrix is symmetric and, for an RC model of a power distribution network, positive

definite [202]. Of the direct methods, Cholesky factorization is particularly well suited, requiring moderate memory resources to store the
factorization data. Of the iterative methods, the conjugate gradient
method is more memory efficient for denser and larger systems [202].
Several techniques to further enhance the efficiency of analyzing power
distribution networks are described in the remainder of this section.

Model partitioning in RC and RLC parts

These numerical methods are modified if the mutual inductance
among the interconnect segments are considered, as described by Panda
et al. [111]. The matrix describing the system is no longer guaranteed
to be positive definite, preventing the use of efficient methods based on
this property and forcing the use of more general (and computationally expensive) methods. Including the mutual inductance elements is
virtually always necessary to accurately describe the electrical properties of the power distribution networks of a printed circuit board and
an integrated circuit package. An RC-only model is often adequate
for describing the on-chip power distribution network. In many cases,
therefore, only a relatively small part of the overall model (describing
the package and board power interconnect) contains inductive elements.
The computational complexity of the problem can be significantly reduced in these cases [111], as illustrated in Fig. 8.6. A comprehensive
model of a power distribution system is partitioned into an RLC part
containing all of the inductive elements (at the package and board
level) and an RC-only part (the on-chip network). The RC-only part
contains the vast majority of elements comprising the overall power
distribution system. The complexity of the RC part of the system can
be reduced by exploiting efficient techniques based on solving a symmetric positive definite system of equations. The RC part of the model
is replaced with the equivalent admittance at the ports of the interface
with the RLC part. The resulting system is significantly smaller than
the original system and can be solved with general solution methods.

An approach to analyzing power distribution networks composed of
RLC segments (with no mutual inductance terms) has been proposed
by Zheng and Tenhunen [252], [253]. An enhanced matrix formulation
of the power distribution network problem is proposed and numerical
techniques to solve this formulation are described. As demonstrated on
sample networks consisting of several hundred segments, this analysis
approach is three to four hundred times faster as compared to SPICE
simulations, while maintaining an accuracy within 5% of SPICE.

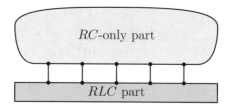

(a) A circuit model of a power distribution system can be partitioned into a relatively small *RLC* part and an *RC*-only part containing the vast majority of the circuit elements.

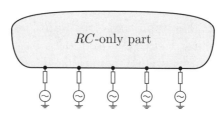

(b) An equivalent admittance macromodel of the *RC*-only part is constructed.

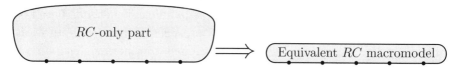

(c) The *RC* part is replaced with a reduced model and the system is analyzed using robust numerical methods. The voltages and equivalent admittances at the ports of the *RC*-only part are determined.

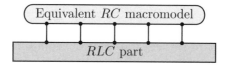

(d) The *RC*-only part is analyzed using efficient numerical methods. The *RLC* part of the model is replaced with equivalent circuits at the appropriate ports, as has been determined in the previous step.

Fig. 8.6. Reducing the computational complexity of the analysis process by separating the analysis of the *RLC* and *RC*-only parts of a power distribution system.

Improving the initial condition accuracy of the AC analysis

The efficiency of the transient analysis can be enhanced by accurate estimates of the steady state condition, *i.e.*, the currents passing through the network inductors and the voltages across the network capacitors. The steady state condition is not known before the analysis. The analysis starts with a rough estimate of the initial conditions, for example, with the voltages and currents determined in a DC analysis. In the beginning of the AC analysis, the initial excitation conditions are maintained and the system is allowed to relax to the AC steady state. After the steady state is reached, the switching pattern of interest can be applied to the circuit inputs, permitting a transient analysis to be initiated. No useful information is produced as the system settles to the AC steady state. In systems with a low damping factor, the time required to reach the steady state can be a substantial portion of the overall time span of the simulation, significantly increasing the computational overhead of the analysis [111].

An accurate estimate of the initial conditions is therefore desirable. This estimate can be efficiently obtained as follows [111]. A simplified circuit model of the power distribution network is constructed. Elements of the simplified model are determined based on the elements of the original network and the worst case voltage drop obtained by a DC analysis of the original network. The simplified circuit is simulated and the steady state inductor currents and capacitor voltages are determined. These currents and voltages are used as steady state values in the transient analysis of the original network. Using this technique to analyze the power distribution network of a 300 MHz PowerPC microprocessor, the initial conditions are estimated with an accuracy of 6.5% as compared to a 62% accuracy based on a DC analysis. The greater accuracy of the initial conditions shortens by a factor of three the time required to determine the AC steady state.

Global-local hierarchical analysis

A hierarchical approach to the electrical analysis of an on-chip power distribution network reduces the CPU time and memory requirements as compared to a flat (non-hierarchical) analysis, as described by Zhao *et al.* [254]. A power network is partitioned into a global grid and many local grids, as depicted in Fig. 8.7. A macromodel is built for every local partition. A macromodel is a linear multi-port network characterized by the same relationship between the port currents and voltages as the original local partition. The power network is simulated with each local partition substituted by the respective macromodel. The problem size is thereby reduced from the total number of nodes in the original power distribution network to the number of nodes in the global partition plus the total sum of the local partition ports. Subsequently, to determine the voltage at the nodes of the local partitions, each local partition can be independently analyzed with the port currents determined during the analysis of the global grid with macromodels. The efficiency of the methodology therefore depends upon judicial partitioning, the computational cost of constructing a macromodel, and the complexity of the macromodel.

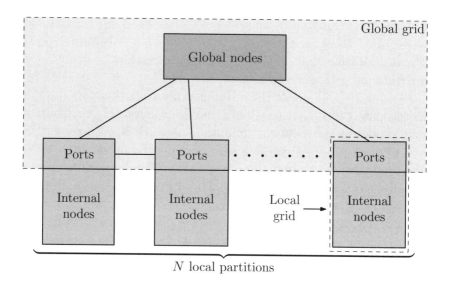

Fig. 8.7. A hierarchical model of a power distribution network. In a global analysis, the local grids are represented by multi-port linear macromodels.

The greatest reduction in the complexity of the analysis is achieved if the system is partitioned into subnetworks with the number of internal nodes much larger than the square of the respective number of ports [254]. The macromodel matrices tend to have a higher density than the matrix representation of the original power distribution network. The higher density can limit the choice of suitable numerical solution methods and, therefore, the efficiency of the analysis. Sparsification of the macromodels can be performed to address this problem [254].

The performance gains of the proposed hierarchical method over a conventional flat analysis have been assessed for power distribution networks of several industrial DSP and microprocessor circuits [254]. The memory requirements are reduced severalfold. The size of the linear system describing the hierarchical system is approximately ten times smaller then the size of the system in a flat (non-hierarchical) analysis. The memory requirement is reduced by ten to twenty times. The one-time overhead of the analysis setup (partitioning, macromodel generation, sparsification, *etc.*, in the case of the hierarchical methodology) is reduced by a factor of two to five. The run time of the subsequent time steps, however, is greater as compared to a flat analysis. The difference in the runtime decreases with the size of the system, becoming relatively small in networks with ten million nodes or more.

Since each local partition of the network is solved independently, a hierarchical analysis has two other desirable properties [254]. First, the hierarchical analysis is easily amenable to parallel computation, permitting additional speedup in the subsequent time steps. Being performed in parallel, hierarchical analysis is two to five times faster than a flat analysis. Second, the mutual independence of the macromodels renders the hierarchical analysis flexible. Local changes in the circuit structure necessitate regeneration of only a single macromodel. The rest of the setup can be reused, permitting an efficient incremental analysis. Alternatively, if a detailed power analysis of a specific block is only of interest, the local solution of other partitions can be omitted, accelerating the analysis while preserving the effect of these partitions on the partition of interest.

Ad hoc analysis techniques

No assumptions regarding the topological structure of the on-chip power distribution network are made for the numerical analysis techniques described in the previous section. These techniques can therefore be applied to a network of general topology. A number of *ad hoc* techniques have also been developed. These techniques are either tailored to a specific network topology or exploit specific properties of the network. These techniques are briefly discussed below.

Multi-grid analysis

The efficient analysis of power distribution grids can be performed through the use of multi-grid methods, as described by Nassif and

Kozhaya [255], [256]. Power distribution grids are spatially and temporally well behaved (*i.e.*, smooth and damped) systems. General purpose robust techniques are unnecessary to achieve an accurate solution of such systems. A system of linear equations describing such well behaved systems is analogous to a finite element discretization of a two-dimensional parabolic partial differential equation [256]. Efficient numerical methods developed for parabolic partial differential equations can therefore be exploited to analyze power distribution grids. The multi-grid method is most commonly used for solving parabolic partial differential equations [257]. Using a fixed time step requires only a single inversion of a large and sparse matrix during the numerical analysis process.

Hierarchical analysis of networks with mesh-tree topology

The analysis and optimization of power distribution networks structured as a global mesh feeding local trees can be performed by a specially formulated hierarchical method, as proposed by Su, Gala, and Sapatnekar [258]. The process of hierarchical analysis proceeds in three stages. First, each tree is replaced with an equivalent circuit model obtained from the passive reduced-order interconnect macromodeling algorithm (PRIMA) [259]. The system is solved to determine all of the nodal voltages in step two. Each tree is analyzed independently based on the voltage at the root of the tree obtained in step two. The method produces results within 10% of SPICE with a greater than ten fold speedup.

Efficient analysis of RL trees

A worst case IR and ΔI noise analysis is efficiently performed in power and ground distribution networks structured as RL trees originating from a single I/O pad, as described by Zhao, Roy, and Koh [260]. The worst case power current requirements of each circuit attached to a power distribution tree are approximated by a trapezoidal waveform. The intrinsic and intentional decoupling capacitances are neglected in the analysis, allowing the power voltage to be efficiently calculated at each node of the tree.

A method for the frequency domain analysis of the noise in RL power distribution trees has also been developed [261]. A frequency domain noise spectrum is computed by analyzing the effective output impedance of the power distribution network at each current source and the spatial correlation among the trees. A time domain noise waveform is obtained by applying an inverse Fast Fourier transform to the frequency domain spectrum. This approach is more than two orders of magnitude faster than HSPICE simulations, while maintaining an accuracy within 10% as compared to circuit simulation.

8.6 Allocation of on-chip decoupling capacitors

The allocation of on-chip decoupling capacitors is commonly performed iteratively. Each iteration of the allocation process consists of two steps, as shown in Fig. 8.8. In the power noise analysis phase, the magnitude of the power supply noise is determined throughout the circuit. The size and placement of the decoupling capacitors are then modified during the allocation phase based on the results of the noise analysis. This process continues until all of the target power noise constraints are satisfied. Occasionally, the power noise constraints cannot be satisfied for a specific circuit. In this case, the area dedicated to the on-chip decoupling capacitors should be increased. In some cases, large functional blocks should be partitioned, permitting the allocation of decoupling capacitors around the smaller circuit blocks.

Although a sufficiently large amount of on-chip decoupling capacitance distributed across an IC will ensure adequate power supply integrity, the on-chip decoupling capacitors consume considerable die area and leak significant amounts of current. Interconnect limited circuits typically contain a certain amount of white space (area not occupied by the circuit) where intentional decoupling capacitors can be placed without increasing the overall die size. After this area is utilized, accommodating additional decoupling capacitors increases the overall circuit area. The amount of intentional decoupling capacitance should therefore be minimized. A strategy guiding the capacitance allocation process is therefore required to achieve target specifications with fewer iterations while utilizing the minimum amount of on-chip decoupling capacitance.

Different allocation strategies are the focus of this section. A charge-based allocation methodology is presented in Section 8.6.1. An allocation strategy based on an excessive noise amplitude is described in Section 8.6.2. An allocation strategy based on excessive charge is discussed in Section 8.6.3.

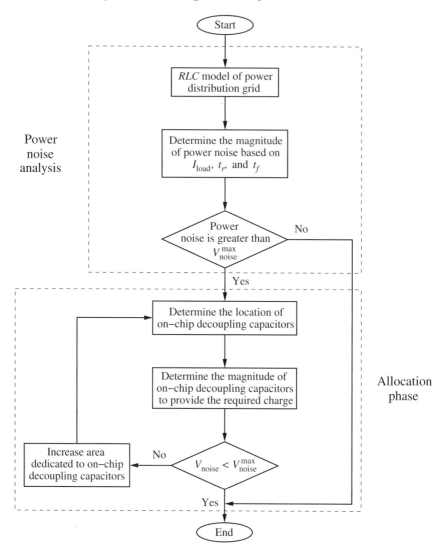

Fig. 8.8. Flow chart for allocating on-chip decoupling capacitors.

8.6.1 Charge-based allocation methodology

One of the first approaches is based on the average power current drawn by a circuit block [262]. The decoupling capacitance C_i^{dec} at node i is selected to be sufficiently large so as to supply an average power current I_i^{avg} drawn at node i for a duration of a single clock period, *i.e.*, to release charge $\delta Q_i = \frac{I_i^{\text{avg}}}{f_{\text{clk}}}$ as the power voltage level varies by a noise margin δV_{dd},

$$C_i^{\text{dec}} = \frac{\delta Q_i}{\delta V_{\text{dd}}} = \frac{I_i^{\text{avg}}}{f_{\text{clk}} \, \delta V_{\text{dd}}}, \qquad (8.4)$$

where f_{clk} is the clock frequency.

The rationale behind the approach represented by (8.4) is that the power current during a clock period is provided by the on-chip decoupling capacitors. This allocation methodology is based on two assumptions. First, at frequencies higher than the clock frequency, the on-chip decoupling capacitors are effectively disconnected from the package and board power delivery networks (*i.e.*, at these frequencies, the impedance of the current path to the off-chip decoupling capacitors is much greater than the impedance of the on-chip decoupling capacitors). Second, the on-chip decoupling capacitors are fully recharged to the nominal power supply voltage before the next clock cycle begins.

Both of these assumptions cannot be simultaneously satisfied with high accuracy. The required on-chip decoupling capacitance as determined by (8.4) is neither sufficient nor necessary to limit the power supply fluctuations within the target margin δV_{dd}. If the impedance of the package-to-die interface is sufficiently low, a significant share of the power current during a single clock period is provided by the decoupling capacitors of the package, overestimating the required on-chip decoupling capacitance as determined by (8.4). Conversely, if the impedance of the package-to-die interface is relatively high, the time required to recharge the on-chip decoupling capacitors is greater than the clock period, making the requirement represented by (8.4) insufficient. This inconsistency is largely responsible for the unrealistic dependence of the decoupling capacitance as determined by (8.4) on the circuit frequency, *i.e.*, the required decoupling capacitance decreases with frequency. Certain assumptions concerning the impedance characteristics of the power distribution network of the package and package-die interface should therefore be considered to accurately estimate the required on-chip decoupling capacitance.

The efficacy of the charge-based allocation strategy has been evaluated on the Pentium II and Alpha 21264 microprocessors using micro-architectural estimation of the average current drawn by a circuit block [263], [264], [265]. The characteristics of the power distribution network based on (8.4) are simulated and compared in both the frequency and time domains to three other cases: no decoupling capacitance is added, decoupling capacitors are placed at the center of each functional unit, and a uniform distribution of the decoupling capacitors. The AC current requirements of the microprocessor functional units are estimated based on the average power current obtained with architectural simulations. The charge-based allocation strategy has been demonstrated to result in the lowest impedance power distribution system in the frequency domain and the smallest peak-to-peak magnitude of the power noise in the time domain.

8.6.2 Allocation strategy based on the excessive noise amplitude

More aggressive capacitance budgeting is proposed in [266], [267] to amend the allocation strategy described by (8.4). In this modified scheme, the circuit is first analyzed without an intentional on-chip decoupling capacitance and the worst case power noise inside each circuit block is determined. No decoupling capacitance is allocated to those blocks where the power noise target specifications have already been achieved. Alternatively, the intrinsic decoupling capacitance of these circuit blocks is sufficient. In those circuit blocks where the maximum power noise V_{noise} exceeds the target margin δV_{dd}, the amount of decoupling capacitance is

$$C_{dec} = \frac{V_{noise} - \delta V_{dd}}{V_{noise}} \frac{\delta Q}{\delta V_{dd}}, \qquad (8.5)$$

where δQ is the charge drawn from the power distribution system by the current load during a single clock period.

The rationale behind (8.5) is that in order to reduce the power noise from V_{noise} to δV_{dd} (*i.e.*, by a factor of $\frac{V_{noise}}{\delta V_{dd}}$), the capacitance C_{dec} should supply a $1 - \frac{\delta V_{dd}}{V_{noise}}$ share of the total current. Consequently, the same share of charge as the power voltage is decreased by δV_{dd}, making $C_{dec} \, \delta V_{dd} = \frac{V_{noise} - \delta V_{dd}}{V_{noise}} \delta Q$. Adding a decoupling capacitance to only those circuit blocks with a noise margin violation, the allocation strategy based on the excessive noise amplitude implicitly considers the decoupling effect of the on-chip intrinsic decoupling capacitance and the off-chip decoupling capacitors [28].

The efficacy of a capacitance allocation methodology based on (8.5) has been tested on five MCNC benchmark circuits [268]. For a 0.25 μm CMOS technology, the proposed methodology requires, on average, 28%

lower overall decoupling capacitance as compared to the more conservative allocation methodology based on (8.4) [262]. A noise aware floorplanning methodology based on this allocation strategy has also been developed [268]. The noise aware floorplanning methodology results, on average, a 20% lower peak power noise and a 12% smaller decoupling capacitance as compared to a post-floorplanning approach. The smaller required decoupling capacitance occupies less area and produces, on average, a 1.2% smaller die size.

8.6.3 Allocation strategy based on excessive charge

The allocation strategy presented in Section 8.6.2 can be further refined. Note that (8.5) uses only the excess of the power voltage over the noise margin as a metric of the severity of the noise margin violation. This metric does not consider the duration of the voltage disturbance. Longer variations of the power supply voltage have a greater impact on signal timing and integrity. A time integral of the excess of the signal variation above the noise margin is proposed in [269], [270] as a more accurate metric characterizing the severity of the noise margin violation. According to this approach, a metric of the ground supply quality at node j is

$$M_j = \int_0^T \max\left[\left(V_j^{\text{gnd}}(t) - \delta V\right), 0\right] dt, \tag{8.6}$$

or, assuming a single peak noise violates the noise margin between times t_1 and t_2,

$$M_j = \int_{t_1}^{t_2} \left(V_j^{\text{gnd}}(t) - \delta V\right) dt, \tag{8.7}$$

where $V_j^{\text{gnd}}(t)$ is the ground voltage at node j of the power distribution grid.

Worst case switching patterns are used to calculate (8.6) and (8.7). This metric is illustrated in Fig. 8.9. The value of the integral in (8.7) equals the area of the shaded region. Note that if the variation of the ground voltage does not exceed the noise margin at any point in time, the metric M_j is zero. The overall power supply quality M is calculated by summing the quality metrics of the individual nodes,

$$M = \sum_j M_j. \tag{8.8}$$

This metric becomes zero when the power noise margins are satisfied at all times throughout the circuit.

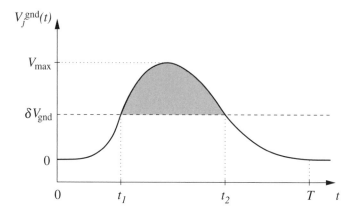

Fig. 8.9. Variation of ground supply voltage with time. The integral of the excess of the ground voltage deviation over the noise margin δV_{gnd} (the shaded area) is used as a quality metric to guide the process of allocating the decoupling capacitors.

Application of (8.6) and (8.8) to the decoupling capacitance allocation process requires a more complex procedure as compared to (8.4) and (8.5). Note that utilizing (8.6) requires detailed knowledge of the power voltage waveform V_j^{gnd} at each node of the power distribution grid rather than just the peak magnitude of the deviation from the nominal power supply voltage. Computationally expensive techniques are therefore necessary to obtain the power voltage waveform. Furthermore, the metric of power supply quality as expressed in (8.8) does not explicitly determine the distribution of the decoupling capacitance. A multi-variable optimization is required to determine the distribution of the decoupling capacitors that minimizes (8.8). The integral formulation expressed by (8.6) is, fortunately, amenable to efficient optimization algorithms. The primary motivation for the original integral formulation of the excessive charge metric is, in fact, to facilitate incorporating these noise effects into the circuit optimization process.

The efficacy of the allocation strategy represented by (8.8) in application-specific ICs has been demonstrated in [271], [272]. The distribution of the decoupling capacitance in standard-cell circuit blocks has been analyzed. The total decoupling capacitance within the circuit is determined by the empty space between the standard cells within the rows of cells. The total budgeted decoupling capacitance (the amount of empty space) remains constant. As compared to a uniform distribution of the decoupling capacitance across the circuit area, the proposed methodology results in a significant reduction in the number of circuit nodes exhibiting noise margin violations and a significant reduction in the maximum power supply noise.

8.7 Summary

The process of designing and analyzing on-chip power distribution networks has been presented in this chapter. The primary conclusions of the chapter are summarized as follows.

- The design of on-chip power distribution networks typically begins prior to the physical design of the on-chip circuitry and is gradually refined as the structure of the on-chip circuits is developed

- The primary difficulty in the early stages of the design process is accurately assessing the on-chip power current requirements

- The primary challenge shifts to the efficient analysis of the on-chip power distribution network, once the circuit structure is specified in sufficient detail

- The complexity of analyzing an entire power distribution network loaded by millions of nonlinear transistors is well beyond the capacity of nonlinear circuit simulators

- Approximating nonlinear loads by time-varying current sources and thereby rendering the problem amenable to the methods of linear analysis is a common approach to manage the complexity of the power distribution network analysis process

- Several techniques have been developed to enhance the efficiency of the numerical analysis process

- The local impedance characteristics of an on-chip power distribution network depend upon the distribution of the decoupling capacitors

- Existing capacitance allocation methodologies place large decoupling capacitances near those on-chip circuits with the greater power requirements

- The time integral of the excess of the signal variation above the noise margin is a useful metric for characterizing the severity of a noise margin violation

Inductive Properties of On-Chip Power Distribution Grids

The inductive properties of power distribution grids are investigated in this chapter. As discussed in Section 1.3, the inductance of power grids is an important factor in determining the impedance characteristics of a power distribution network operating at high frequencies.

The chapter is organized as follows. In Section 9.1, the inductive characteristics of a current loop formed by a power transmission circuit are established. The analysis approach is outlined in Section 9.2. The three types of grid structures considered in this study and analysis setup are described in Section 9.3. The dependence of the grid inductance characteristics on the line width is discussed in Section 9.4. The differences in the inductive properties among the three types of grids are reviewed in Section 9.5. The dependence of the grid inductance on the grid dimensions is described in Section 9.6. The chapter concludes with a summary.

9.1 Power transmission circuit

Consider a power transmission circuit as shown in Fig. 9.1(a). The circuit consists of the forward current (power) path and the return current (ground) path forming a transmission current loop between the power supply at one end of the loop and a power consuming circuit at the other end. In a simple case, the forward and return paths each consist of a single conductor. In general, a "path" refers to a multi-conductor structure carrying current in a specific direction (to or from the load), which is the case in power distribution grids. The circuit dimensions are assumed to be sufficiently small for a lumped circuit approximation to be valid. The inductance of both terminating devices is assumed negligible as compared to the inductance of the transmission line. The inductive characteristics of the current loop are, therefore, determined by the inductive properties of the forward and return current paths.

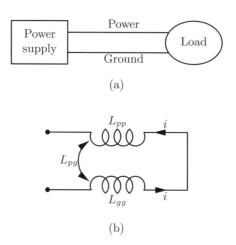

(a)

(b)

Fig. 9.1. A simple power transmission circuit; (a) block diagram, (b) equivalent inductive circuit.

The power transmission loop consists of forward and return current paths. The equivalent inductive circuit is depicted in Fig. 9.1(b). The partial inductance matrix for this circuit is

$$L_{ij} = \begin{bmatrix} L_{pp} & -L_{pg} \\ -L_{gp} & L_{gg} \end{bmatrix}, \tag{9.1}$$

where L_{pp} and L_{gg} are the partial self inductances of the forward and return current paths, respectively, and L_{pg} is the absolute value of the partial mutual inductance between the paths. The loop inductance of the power transmission loop is

$$L_{\text{loop}} = L_{pp} + L_{gg} - 2L_{pg}. \tag{9.2}$$

The mutual coupling L_{pg} between the power and ground paths reduces the loop inductance. This behavior can be formulated more generally: *the greater the mutual coupling between antiparallel (flowing in opposite directions) currents, the smaller the loop inductance of a circuit.* The effect is particularly significant when the mutual inductance is comparable to the self inductance of the current paths. This is the case when the line separation is comparable to the dimensions of the line cross section.

The effect of coupling on the net inductance is reversed when currents of the two inductive coupled paths flow in the same direction. For example, in order to reduce the partial inductance L_{11} of line segment 1, another line segment 2 with partial self inductance L_{22} and coupling L_{12} is placed in parallel with segment 1. A schematic of the equivalent inductance is shown in Fig. 9.2. The resulting partial inductance of the current path is

$$L_{1\|2} = \frac{L_{11}L_{22} - L_{12}^2}{L_{11} + L_{22} - 2L_{12}}. \tag{9.3}$$

For the limiting case of no coupling, this expression simplifies to $\frac{L_{11}L_{22}}{L_{11}+L_{22}}$. In the opposite case of full coupling of segment 1, $L_{11} = L_{12}$ ($L_{11} \leq L_{22}$) and the total inductance becomes L_{11}. For two identical parallel elements, (9.3) simplifies to $L_\| = (L_{\text{self}} + L_{\text{mutual}})/2$. In general, *the greater the mutual coupling between parallel (flowing in the same directions) currents, the greater the loop inductance of a circuit.* To present this concept in an on-chip perspective, consider a $1000\,\mu$m long line with a $1\,\mu$m $\times\,3\,\mu$m cross section, and a partial self inductance of $1.342\,$nH (at $1\,$GHz). Adding another identical line in parallel with the first line with a $17\,\mu$m separation ($20\,\mu$m line pitch) results in a mutual line coupling of $0.725\,$nH and a net inductance of $1.033\,$nH ($\sim 55\%$ higher than $L_{\text{self}}/2 = 0.671\,$nH).

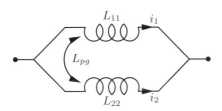

Fig. 9.2. Two parallel coupled inductors.

Inductive coupling among the conductors of the same circuit can, therefore, either increase or decrease the total inductance of the circuit. To minimize the circuit inductance, coupling of conductors carrying current in the same direction can be reduced by increasing the distance between the conductors. Coupling of conductors carrying current in opposite directions should be increased by physically placing the conductors closer to each other.

This optimization naturally occurs in grid structured power distribution networks with alternating power and ground lines. Three types of power distribution grid are analyzed to demonstrate this effect, as described in Section 9.3.

9.2 Simulation setup

The inductance extraction program FastHenry [67] is used to explore the inductive properties of grid structures. FastHenry efficiently calculates the frequency dependent self and mutual impedances, $R(\omega) + \omega L(\omega)$, in complex three-dimensional interconnect structures. A magnetoquasistatic approximation is assumed, meaning the distributed capacitance of the line and any related displacement currents associated with the capacitance are ignored. The accelerated solution algorithm employed in the program provides approximately a 1% worst case accuracy as compared with the direct solution of the system of linear equations characterizing the system [67].

A conductivity of $58\,\mathrm{S}/\mu\mathrm{m} \simeq (1.72\,\mu\Omega \cdot \mathrm{cm})^{-1}$ is assumed for the interconnect material. The inductive portion of the impedance is relatively insensitive to the interconnect resistivity in the range of $1.7\,\mu\Omega \cdot \mathrm{cm}$ to $2.5\,\mu\Omega \cdot \mathrm{cm}$ (typical for advanced processes with copper interconnect [273], [274], [275]). A conductivity of $40\,\mathrm{S}/\mu\mathrm{m}$ $(2.5\,\mu\Omega \cdot \mathrm{cm})^{-1}$ yields an inductance that is less than 4% larger than the inductance obtained for a conductivity of $58\,\mathrm{S}/\mu\mathrm{m}$.

A line thickness of $1\,\mu\mathrm{m}$ is assumed for the interconnect structures. In the analysis, the lines are split into multiple filaments to account for skin and proximity effects, as discussed in Section 2.2. The number of filaments is chosen to be sufficiently large to achieve a 1% accuracy of the computed values. Simulations are performed at three frequencies, $1\,\mathrm{GHz}$, $10\,\mathrm{GHz}$, and $100\,\mathrm{GHz}$. Typical simulation run times for the structures are under one minute on a Sun Blade 100 workstation.

9.3 Grid types

To assess the dependence of the inductive properties on the power and ground lines, the coupling characteristics of three types of power/-ground grid structures have been analyzed. In the grids of the first type, called *non-interdigitated grids*, the power lines fill one half of the

grid and the ground lines fill the other half of the grid, as shown in
Fig. 9.3(a). In *interdigitated grids*, the power and ground lines are alternated and equidistantly spaced, as shown in Fig. 9.3(b). The grids
of the third type are a variation of the interdigitated grids. Similar
to interdigitated grids, the power and ground lines are alternated, but
rather than placed equidistantly, the lines are placed in equidistantly
spaced pairs of adjacent power and ground lines, as shown in Fig. 9.3(c).
These grids are called *paired grids*. Interdigitated and paired grids are
grids with alternating power and ground lines.

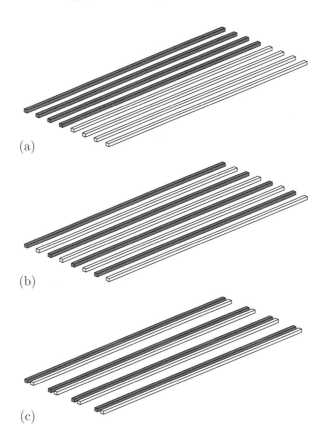

(a)

(b)

(c)

Fig. 9.3. Power/ground grid structures under investigation; (a) a non-interdigitated
grid, (b) a grid with the power lines interdigitated with the ground lines, (c) a paired
grid, the power and ground lines are in close pairs. The power lines are grey colored,
the ground lines are white colored.

The number of power lines matches the number of ground lines in all of the grid structures. The number of power/ground line pairs is varied from one to ten. The grid lines are assumed to be 1 mm long and are placed on a 20 μm pitch. The specific line length is unimportant since at these high length to line pitch ratios the inductance scales nearly linearly with the line length, as discussed in Section 9.6.

An analysis of these structures has been performed for two line cross sections, 1 μm × 1 μm and 1 μm × 3 μm. For each of these structures, the following characteristics have been determined: the partial self inductance of the power (forward current) and ground (return current) paths L_{pp} and L_{gg}, respectively, the power to ground path coupling L_{pg}, and the loop inductance L_{loop}. When determining the loop inductance, all of the ground lines at one end of the grid are short circuited to form a ground terminal, all of the power lines at the same end of the grid are short circuited to form a power terminal, and all of the lines at the other end of the grid are short circuited to complete the current loop. This configuration assumes that the current loop is completed on-chip. This assumption is valid for high frequency signals which are effectively terminated through the on-chip decoupling capacitance which provides a low impedance termination as compared to the inductive leads of the package. The on-chip inductance affects the signal integrity of the high frequency signals. If the current loop is completed on-chip, the current in the power lines is always antiparallel to the current in the ground lines.

The loop inductance of the three types of grid structures operating at 1 GHz is displayed in Fig. 9.4 as a function of the number of lines in the grid. The partial self and mutual inductance of the power and ground current paths is shown in Fig. 9.5 for grid structures with 1 μm × 1 μm cross section lines and in Fig. 9.6 for grids with 1 μm × 3 μm cross section lines. The inductance data for 1 GHz and 100 GHz are summarized in Table 9.1. The data depicted in Figs. 9.4, 9.5, and 9.6 are discussed in the following three sections.

The signal lines surrounding the power grids are omitted from the analysis. This omission is justified by the following considerations. First, current returning through signal lines rather than the power/-ground network causes crosstalk noise on the lines. To minimize this undesirable effect, the circuits are designed in such a way that the majority of the return current flows through the power and ground lines. Second, the signal lines provide additional paths for the return current and only decrease the inductance of the power distribution network. The value of the grid inductance obtained in the absence of signal lines can therefore be considered as an upper bound.

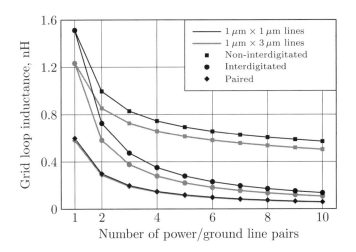

Fig. 9.4. Loop inductance of the power/ground grids as a function of the number of power/ground line pairs (at a 1 GHz signal frequency).

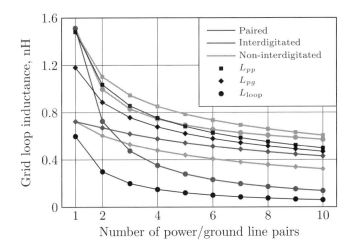

Fig. 9.5. Loop and partial inductance of the power/ground grids with $1\,\mu\text{m} \times 1\,\mu\text{m}$ cross section lines (at a 1 GHz signal frequency).

Table 9.1. Inductive characteristics of power/ground grids with a 1000 μm length and a 40 μm line pair pitch operating at 1 GHz and 100 GHz

# of P/G pairs	Cross section (μm)	L_{pp}, L_{gg} (nH)			L_{pg} (nH)			L_{loop} (nH)		
		N/int	Int	Paired	N/int	Int	Paired	N/int	Int	Paired
					1 GHz					
1	1 × 1	1.481	1.481	1.481	0.724	0.724	1.181	1.513	1.513	0.599
1	1 × 3	1.342	1.342	1.342	0.725	0.725	1.053	1.235	1.235	0.579
2	1 × 1	1.102	1.035	1.035	0.604	0.671	0.886	0.996	0.726	0.299
2	1 × 3	1.031	0.963	0.966	0.604	0.672	0.822	0.853	0.582	0.288
3	1 × 1	0.945	0.856	0.857	0.530	0.619	0.757	0.829	0.474	0.199
3	1 × 3	0.894	0.807	0.810	0.531	0.618	0.714	0.726	0.377	0.192
4	1 × 1	0.851	0.753	0.753	0.478	0.577	0.678	0.745	0.351	0.149
4	1 × 3	0.809	0.714	0.717	0.479	0.575	0.645	0.658	0.279	0.144
5	1 × 1	0.785	0.682	0.682	0.439	0.542	0.622	0.693	0.279	0.120
5	1 × 3	0.748	0.649	0.652	0.440	0.539	0.594	0.614	0.221	0.115
6	1 × 1	0.735	0.628	0.629	0.407	0.513	0.579	0.656	0.231	0.100
6	1 × 3	0.700	0.600	0.602	0.409	0.509	0.555	0.582	0.183	0.096
7	1 × 1	0.694	0.586	0.587	0.380	0.488	0.544	0.628	0.197	0.085
7	1 × 3	0.661	0.561	0.563	0.383	0.483	0.522	0.557	0.156	0.082
8	1 × 1	0.660	0.552	0.552	0.357	0.466	0.515	0.606	0.172	0.075
8	1 × 3	0.629	0.529	0.531	0.361	0.461	0.495	0.536	0.136	0.072
9	1 × 1	0.631	0.523	0.523	0.338	0.446	0.490	0.588	0.152	0.066
9	1 × 3	0.601	0.502	0.504	0.342	0.441	0.472	0.518	0.120	0.064
10	1 × 1	0.606	0.497	0.498	0.321	0.429	0.468	0.571	0.137	0.060
10	1 × 3	0.577	0.478	0.480	0.326	0.424	0.451	0.503	0.108	0.057
					100 GHz					
1	1 × 1	1.468	1.468	1.457	0.724	0.724	1.181	1.486	1.486	0.551
1	1 × 3	1.315	1.315	1.291	0.725	0.725	1.062	1.180	1.180	0.457
2	1 × 1	1.088	1.022	1.023	0.604	0.670	0.886	0.968	0.703	0.275
2	1 × 3	1.010	0.945	0.940	0.605	0.670	0.826	0.810	0.548	0.228
3	1 × 1	0.928	0.845	0.848	0.533	0.616	0.756	0.789	0.458	0.183
3	1 × 3	0.873	0.793	0.792	0.534	0.615	0.716	0.678	0.355	0.152
4	1 × 1	0.830	0.741	0.745	0.483	0.571	0.676	0.695	0.340	0.138
4	1 × 3	0.788	0.702	0.702	0.484	0.570	0.645	0.607	0.263	0.114
5	1 × 1	0.762	0.671	0.674	0.444	0.536	0.619	0.634	0.270	0.110
5	1 × 3	0.727	0.639	0.640	0.446	0.535	0.594	0.560	0.208	0.091
6	1 × 1	0.710	0.618	0.621	0.414	0.506	0.575	0.591	0.224	0.092
6	1 × 3	0.680	0.591	0.592	0.416	0.505	0.554	0.527	0.173	0.076
7	1 × 1	0.668	0.576	0.579	0.389	0.480	0.540	0.559	0.191	0.079
7	1 × 3	0.641	0.553	0.554	0.391	0.479	0.522	0.501	0.147	0.065
8	1 × 1	0.633	0.542	0.545	0.367	0.458	0.510	0.533	0.167	0.069
8	1 × 3	0.609	0.522	0.523	0.369	0.457	0.494	0.480	0.128	0.057
9	1 × 1	0.604	0.513	0.516	0.348	0.439	0.485	0.511	0.148	0.061
9	1 × 3	0.582	0.495	0.496	0.351	0.438	0.471	0.463	0.114	0.050
10	1 × 1	0.578	0.488	0.491	0.332	0.422	0.463	0.493	0.133	0.055
10	1 × 3	0.558	0.472	0.473	0.334	0.421	0.450	0.448	0.102	0.045

N/int — non-interdigitated grids, Int — interdigitated grids,
Paired — paired grids

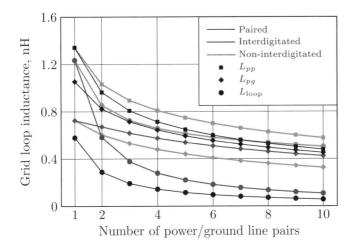

Fig. 9.6. Loop and partial inductance of the power/ground grids with $1\,\mu\text{m} \times 3\,\mu\text{m}$ cross section lines (at a 1 GHz signal frequency).

9.4 Inductance versus line width

The loop inductance of the grid depends relatively weakly on the line width. Grids with $1\,\mu\text{m} \times 3\,\mu\text{m}$ cross section lines have a lower loop inductance than grids with $1\,\mu\text{m} \times 1\,\mu\text{m}$ cross section lines. This decrease in inductance is dependent upon the grid type, as shown in Fig. 9.4. The largest decrease, approximately 21%, is observed in interdigitated grids. In non-interdigitated grids, the inductance decreases by approximately 12%. In paired grids, the decrease in inductance is limited to 3% to 4%.

This behavior can be explained in terms of the partial inductance, L_{pp} and L_{pg} [69], [70]. Due to the symmetry of the power and ground paths, $L_{gg} = L_{pp}$, the relation of the loop inductance to the partial inductance (9.2) simplifies to

$$L_{\text{loop}} = L_{pp} + L_{gg} - 2L_{pg} = 2(L_{pp} - L_{pg}). \tag{9.4}$$

According to (9.4), L_{loop} increases with larger L_{pp} and decreases with larger L_{pg}. That is, decreasing the self inductance of the forward and return current paths forming the current loop decreases the loop inductance, while increasing the inductive coupling of the two paths decreases the loop inductance. The net change in the loop inductance depends, therefore, on the relative effect of increasing the line width on L_{pp} and L_{pg} in the structures of interest.

The self inductance of a single line is a weak function of the line cross-sectional dimensions, see (3.1) [41]. This behavior is also true for the complex multi-conductor structures under investigation. Comparison of the data shown in Fig. 9.5 with the data shown in Fig. 9.6 demonstrates that changing the line cross section from $1\,\mu$m $\times\,1\,\mu$m to $1\,\mu$m $\times\,3\,\mu$m decreases L_{pp} by 4% to 6% in all of the structures under consideration.

The dependence of inductive coupling L_{pg} on the line width, however, depends on the grid type. In non-interdigitated and interdigitated grids, the line spacing is much larger than the line width and the coupling L_{pg} changes insignificantly with the line width. Therefore, the loop inductance L_{loop} in non-interdigitated and interdigitated grids decreases with line width primarily due to the decrease in the self inductance of the forward and return current paths, L_{pp} and L_{gg}.

In paired grids, the line width is comparable to the line-to-line separation and the dependence of L_{pg} on the line width is non-negligible: L_{pg} decreases by 4% to 6%, as quantified by comparison of the data shown in Fig. 9.5 with the data shown in Fig. 9.6. In paired grids, therefore, the grid inductance decreases more slowly with line width as compared with interdigitated and non-interdigitated grids, because a reduction in the self inductance of the current paths L_{pp} is significantly offset by a decrease in the inductive coupling of the paths L_{pg}.

9.5 Dependence of inductance on grid type

The grid inductance varies with the configuration of the grid. With the same number of power/ground lines, grids with alternating power and ground lines exhibit a lower inductance than non-interdigitated grids; this behavior is discussed in Section 9.5.1. The inductance of the paired grids is lower than the inductance of the interdigitated grids; this topic is discussed in Section 9.5.2.

9.5.1 Non-interdigitated versus interdigitated grids

The difference in inductance between non-interdigitated and interdigitated grids increases with the number of lines, reaching an approximately 4.2 difference for ten power/ground line pairs for the case of a $1\,\mu$m $\times\,1\,\mu$m cross section line, as shown in Fig. 9.4 (~ 4.7 difference for the case of a $1\,\mu$m $\times\,3\,\mu$m cross section line). This difference is due to two factors.

First, in non-interdigitated grids the lines carrying current in the same direction (forming the forward or return current paths) are spread over half the width of the grid, while in interdigitated (and paired) grids both the forward and return paths are spread over the entire width of the grid. The smaller the separation between the lines, the greater the mutual inductive coupling between the lines and the partial self inductance of the forward and return paths, L_{pp} and L_{gg}, as described in Section 9.1. This trend is confirmed by the data shown in Figs. 9.5 and 9.6, where interdigitated grids have a lower L_{pp} as compared to non-interdigitated grids.

Second, each line in the interdigitated structures is surrounded with lines carrying current in the opposite direction, creating strong coupling between the forward and return currents and increasing the partial mutual inductance L_{pg}. Alternatively, in the non-interdigitated arrays (see Fig. 9.3(a)), all of the lines (except for the two lines in the middle of the array) are surrounded with lines carrying current in the same direction. The power-to-ground inductive coupling L_{pg} is therefore lower in non-interdigitated grids, as shown in Figs. 9.5 and 9.6.

In summary, the interdigitated grids exhibit a lower partial self inductance L_{pp} and a higher partial mutual inductance L_{pg} as compared to non-interdigitated grids [69], [70]. The interdigitated grids, therefore, have a lower loop inductance L_{loop} as described by (9.4).

9.5.2 Paired versus interdigitated grids

The loop inductance of paired grids is 2.3 times lower than the inductance of the interdigitated grids for the case of a $1\,\mu\mathrm{m} \times 1\,\mu\mathrm{m}$ cross section line and is 1.9 times lower for the case of a $1\,\mu\mathrm{m} \times 3\,\mu\mathrm{m}$ cross section line, as shown in Fig. 9.4. The reason for this difference is described as follows in terms of the partial inductance. The structure of the forward (and return) current path in a paired grid is identical to the structure of the forward path in an interdigitated grid (only the relative position of the forward and return current paths differs). The partial self inductance L_{pp} is therefore the same in paired and interdigitated grids; the two corresponding curves completely overlap in Figs. 9.5 and 9.6. The values of L_{pp} and L_{gg} for the two types of grids are equal within the accuracy of the analysis. In contrast, due to the immediate proximity of the forward and return current lines in the paired grids, the mutual coupling L_{pg} is higher as compared to the interdigitated grids, as shown in Figs. 9.5 and 9.6. Therefore, the difference in loop inductance between paired and interdigitated grids is due to the difference in the mutual inductance [69], [70].

Note that although the inductance of a power distribution network (*i.e.*, the inductance of the power-ground current loop) in the case of paired power grids is lower as compared to interdigitated grids, the signal self inductance (*i.e.*, the inductance of the signal to the power/-ground loop) as well as the inductive coupling of the signal lines is higher. The separation of the power/ground line pairs in paired grids is double the separation of the power and ground lines in interdigitated grids. The current loops formed between the signal lines and the power and ground lines are therefore larger in the case of paired grids. Interdigitated grids also provide enhanced capacitive shielding for the signal lines, as each power/ground line has the same number of signal neighbors as a power/ground line pair. Thus, a tradeoff exists between power integrity and signal integrity in the design of high speed power distribution networks. In many circuits, signal integrity is of primary concern, making interdigitated grids the preferred choice.

9.6 Dependence of Inductance on grid dimensions

The variation of grid inductance with grid dimensions, such as the grid length and width (assuming that the width and pitch of the grid lines is maintained constant) is considered in this section. The dependence of the grid loop inductance on the number of lines in the grid (*i.e.*, the grid width) is discussed in Section 9.6.1. The dependence of the grid loop inductance on the length of the grid is discussed in Section 9.6.2. The concept of sheet inductance is described in Section 9.6.3. A technique for efficiently and accurately calculating the grid inductance is outlined in Section 9.6.4.

9.6.1 Dependence of inductance on grid width

Apart from a lower loop inductance, paired and interdigitated grids have an additional desirable property as compared to non-interdigitated grids. The loop inductance of paired and interdigitated grids depends inversely linearly with the number of lines, as shown in Fig. 9.4. That is, for example, the inductance of a grid with ten power/ground line pairs is half of the inductance of a grid with five power/ground line pairs, all other factors being the same. For paired grids, this inversely linear dependence is exact (*i.e.*, any deviation is well within the accuracy of the inductance extracted by FastHenry). For interdigitated grids, the inversely linear dependence is exact within the extraction accuracy at high numbers of power/ground line pairs. As the number of line pairs is reduced to two or three, the accuracy deteriorates to 5% to 8% due to the "fringe" effect. The electrical environment of the lines at the edges of the grid, where a line has only one neighbor, is significantly different from the environment within the grid, where a line has two neighbors. The fringe effect is insignificant in paired grids because the electrical environment of a line is dominated by the pair neighbor, which is physically much closer as compared to other lines in the paired grid.

As discussed in Section 9.1, the inductance of conductors connected in parallel decreases slower than inversely linearly with the number of conductors if inductive coupling of the parallel conductors is present. The inversely linear decrease of inductance with the number of lines may seem to contradict the existence of significant inductive coupling among the lines in a grid. This effect can be explained by (9.4). While the partial self inductance of the power and ground paths L_{pp} and L_{gg} indeed decreases slowly with the number of lines, so does the power to ground coupling L_{pg}, as shown in Figs. 9.5 and 9.6. The nonlinear behavior of L_{pp} and the nonlinear behavior of L_{pg} effectively cancel each other [see (9.4)], resulting in a loop inductance with an approximately inversely linear dependence on the number of lines [68], [70].

From a circuit analysis point of view this behavior can be explained as follows. Consider a paired grid. The coupling of a distant line to a power line in any power/ground pair is nearly the same as the coupling of the same distant line to a ground line in the same pair due to the close proximity of the power and ground lines within each power/-ground pair. The coupling to the power line counteracts the coupling to the ground line. As a result, the two effects cancel. Applying the same argument in the opposite direction, the effect of coupling a specific line to a power line is canceled by the line coupling to the ground line immediately adjacent to the power line. Similar reasoning is applicable to interdigitated grids, however, due to the equidistant spacing between the lines, the degree of coupling cancellation is lower for those lines at the periphery of the grid. The lower degree of coupling cancellation is the cause of the aforementioned "fringe" effect.

9.6.2 Dependence of inductance on grid length

The length of the grid structures described in Section 9.3 is varied to characterize the variation in the grid inductance with grid length. The results are shown in Fig. 9.7. The grid inductance varies virtually

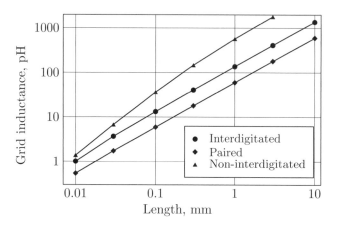

Fig. 9.7. Grid inductance versus grid length.

linearly over a wide range of grid length. This behavior is due to the cancellation of long distance inductive coupling, as described in Chapter 3. The linear dependence of inductance on length is analogous to the dependence of inductance on the number of lines [68]. Similar to the variation in the loop inductance, illustrated in Fig. 3.4, the variation in the grid inductance deviates from the linear behavior at grid lengths comparable to the separation between the forward and return current paths. In non-interdigitated grids, the effective separation between the forward and return currents is greatest. The range of the linear variation of the inductance with grid length is therefore limited as compared to paired and interdigitated grids, as shown in Fig. 9.7.

9.6.3 Sheet inductance of power grids

As discussed in this section, the inductance of grids with alternating power and ground lines is linearly dependent upon the grid length and number of lines. Furthermore, the grid inductance of interdigitated grids is relatively constant with frequency, as described in Chapter 10.

These properties of the grid inductance greatly simplify the procedure for evaluating the inductance of power distribution grids, permitting the efficient assessment of design tradeoffs.

The resistance of the grid increases linearly with grid length and decreases inversely linearly with grid width (*i.e.*, the number of parallel lines). Therefore, the resistive properties of the grid can be conveniently described as a dimension independent grid sheet resistance $R_\square$, similar to the sheet resistance of an interconnect layer. The linear dependence of the grid inductance on the grid dimensions is similar to that of the grid resistance. As with resistance, it is convenient to express the inductance of a power grid as a dimension independent *grid sheet inductance* $L_\square$ (*i.e.*, Henrys per square), rather than to characterize the grid inductance for a particular grid with specific dimensions. Thus,

$$L_\square = L_{\mathrm{grid}} \cdot \frac{PN}{l}, \tag{9.5}$$

where l is the grid length, P is the line (or line pair) pitch, and N is the number of lines (line pairs). This approach is analogous to the *plane* sheet inductance of two parallel power and ground planes (*e.g.*, in a PCB stack), which depends only on the separation between the planes, not on the specific dimensions of the planes. Similarly, the grid sheet inductance reflects the overall structural characteristics of the grid (*i.e.*, the line width and pitch) and is independent of the dimensions of a specific structure (*i.e.*, the grid length and the number of lines in the grid). The sheet inductance is used as a dimension independent measure of the grid inductance in the discussion of power grid tradeoffs in Chapter 11.

9.6.4 Efficient computation of grid inductance

The linear dependence of inductance on the grid length and width (*i.e.*, the number of lines) has a convenient implication. The inductance of a large paired or interdigitated grid can be extrapolated with good accuracy from the inductance of a grid consisting of only a few power/-ground pairs.

For an accurate extrapolation of the interdigitated grids, the line width and pitch of the original and extrapolated grids should be maintained the same. For example, for a grid consisting of $2N$ lines (*i.e.*, N power-ground line pairs) of length l, width W, and pitch P, the inductance L_{2N} can be estimated as

$$L_{2N} \approx \frac{L_2}{N}, \tag{9.6}$$

where L_2 is the inductance of a loop formed by two lines with the same dimensions and pitch. The inductance of a two-line loop L_2 can be calculated using (3.4). The power grid is considered to consist of uncoupled two-line loops. The accuracy of the approximation represented by (9.6) is about 10% for practical line geometries. Alternatively, the grid inductance L_{2N} can be approximated as

$$L_{2N} \approx L_4 \frac{2}{N}, \tag{9.7}$$

where L_4 is the inductance of a power grid consisting of four lines of the original dimensions and pitch. A grid consisting of four lines can be considered as two coupled two-line loops connected in parallel. The loop inductance of a four-line grid can be efficiently calculated using analytic expressions (9.3) and (3.4). For practical geometries, a four-line approximation [see (9.7)] offers an accuracy within 5% as compared to the two-line approximation [see (9.6)], as the coupling to non-neighbor lines is partially considered.

In paired grids, the effective inductive coupling among power/-ground pairs is negligible and (9.6) is practically exact. The effective width of the current loop in paired grids is primarily determined by the separation between the power and ground lines within a pair. The spatial separation between pairs has (almost) no effect on the grid loop inductance. Expression (3.4) should be used with caution in estimating the inductance of adjacent power and ground lines. Expression (3.4) is accurate only for low frequencies and moderate W/T ratios; (3.4) does not consider proximity effects in paired grids at high frequencies.

Alternatively, the grid sheet inductance $L_\square$ can be determined based on (9.6) and (9.7). For example, using (3.4) to determine the loop inductance of a line pair L_{pair}, the grid sheet inductance of a paired grid becomes

$$L_\square = L_{pair}\frac{P}{l} \approx 0.4P \left(\ln \frac{S}{H + W} + \frac{3}{2} \right) \frac{\mu H}{\square}, \tag{9.8}$$

where P is the line pair pitch, and S is the separation between the line centers in the pair. The inductance of grids with the same line width and pitch is

$$L_{\mathrm{grid}} = L_\square \cdot \frac{l}{PN}. \tag{9.9}$$

To summarize, the inductance of regular grids with alternating power and ground lines can be accurately estimated with analytic expressions.

9.7 Summary

The inductive properties of single layer regularly structured grids have been characterized. The primary results are summarized as follows.

- The inductance of grids with alternating power and ground lines varies linearly with grid length and width

- The inductance of grids with alternating power and ground lines varies relatively little with the cross-sectional dimensions of the lines and the signal frequency

- The inductance of grids with alternating power and ground lines can be conveniently expressed in a dimension independent form described as the sheet inductance

- The grid inductance can be analytically calculated based on the grid dimensions and the cross-sectional dimensions of the lines

Variation of Grid Inductance with Frequency

The variation of inductance with frequency in high performance power distribution grids is discussed in this chapter. As discussed in Chapter 5, the on-chip inductance affects the integrity of the power supply in high speed circuits. The frequency of the currents flowing through the power distribution networks in high speed ICs varies from quasi-DC low frequencies to tens of gigahertz. Thus, understanding the variation of the power grid inductance with frequency is important in order to built a robust and efficient power delivery system.

The chapter is organized as follows. A procedure for analyzing the inductance as a function of frequency is described in Section 10.1. The variation of the power grid inductance with frequency is discussed in Section 10.2. The chapter concludes with a summary.

10.1 Analysis approach

The variation of the grid inductance with frequency is investigated for the three types of power/ground grids: non-interdigitated, interdigitated, and paired. These types of grid structures are described in Section 9.3. The grid structures are depicted in Fig. 10.1.

The analysis is analogous to the procedure described in Section 9.3. The inductance extraction program FastHenry [67] is used to explore the inductive properties of these interconnect structures. A conductivity of $58\,\mathrm{S}/\mu\mathrm{m} \simeq (1.72\,\mu\Omega \cdot \mathrm{cm})^{-1}$ is assumed for the interconnect material.

The inductance of grids with alternating power and ground lines, i.e., interdigitated and paired grids, behaves similarly to the grid resistance. With the width and pitch of the lines fixed, the inductance of

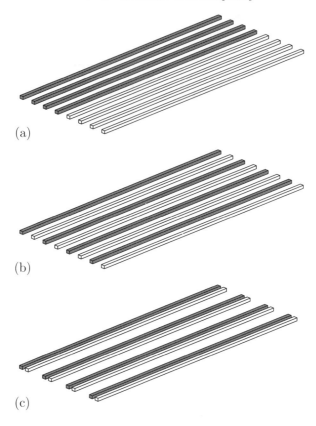

Fig. 10.1. Power/ground grid structures under investigation; (a) a non-interdigitated grid, (b) an interdigitated grid, the power lines are interdigitated with the ground lines, (c) a paired grid, the power and ground lines are in close pairs. The power lines are grey colored, the ground lines are white colored.

these grid types increases linearly with the grid length and decreases inversely linearly with the number of lines, as discussed in Chapter 9. Consequently, the inductive and resistive properties of interdigitated and paired grids with a specific line width and pitch can be conveniently expressed in terms of the sheet inductance $L_\square$, henrys per square, and the sheet resistance $R_\square$, ohms per square [276]. As with the sheet resistance, the sheet inductance is convenient since it is independent of a specific length and width of the grid; this quantity depends only on the pitch, width, and thickness of the grid lines. The impedance properties of interdigitated and paired grids can therefore be studied on structures with a limited number of lines. These results are readily scaled to larger structures, as described in Section 9.6.

The grid structures consist of ten lines, five power lines and five ground lines. The power and ground lines carry current in opposite directions, such that a grid forms a complete current loop. The grid lines are assumed to be 1 mm long and are placed on a 10 μm pitch (20 μm line pair pitch in paired grids). The specific grid length and the number of lines is not significant. As discussed in Section 9.6 and Chapter 3, the inductance scales linearly with the grid length and the number of lines, provided the line length to line separation ratio is high and the number of lines exceeds eight to ten. An analysis of the aforementioned grid structures has been performed for line widths W of 1 μm, 3 μm, and 5 μm. The line thickness is 1 μm. The line separation within power-ground pairs in paired grids is 1 μm.

10.2 Discussion of inductance variation

The variation of grid inductance with frequency is presented and discussed in this section. Simple circuit models are discussed in Section 10.2.1 to provide insight into the variation of inductance with frequency. Based on this intuitive perspective, the data are analyzed and compared in Section 10.2.2.

10.2.1 Circuit models

As discussed in Section 2.2, there are two primary mechanisms that produce a significant decrease in the on-chip interconnect inductance with frequency, the proximity effect and multi-path current redistribution. The phenomenon underlying these mechanisms is, however, the same. Where several parallel paths with significantly different electrical properties are available for current flow, the current is distributed among the paths so as to minimize the total impedance. As the frequency increases, the circuit inductance changes from the low frequency limit, determined by the ratio of the resistance of the parallel current paths, to the high frequency value, determined by the inductance ratio of the current paths. At high signal frequencies, the inductive reactance dominates the interconnect impedance; therefore, the path of minimum inductance carries the largest share of the current, minimizing the overall impedance (see Fig. 2.10). Note that parallel current paths can be formed either by several physically distinct lines, as in multi-path current redistribution, or by different paths within the same line, as in the

proximity effect, as shown in Fig. 10.2. A thick line can be thought of
as being composed of multiple thin lines bundled together in parallel.
The proximity effect in such a thick line can be considered as a special
case of current redistribution among multiple thin lines forming a thick
line.

Fig. 10.2. A cross-sectional view of two parallel current paths (dark gray) sharing
the same current return path (light gray). The path closest to the return path,
path 1, has a lower inductance than the other path, path 2. The parallel paths can
be either two physically distinct lines, as shown by the dotted line, or two different
paths withing the same line, as shown by the dashed line.

Consider a simple case with two current paths with different inductive properties. The impedance characteristics are represented by the
circuit diagram shown in Fig. 10.3, where the inductive coupling between the two paths is neglected for simplicity. Assume that $L_1 < L_2$
and $R_1 > R_2$.

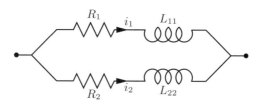

Fig. 10.3. A circuit model of two current paths with different inductive properties.

For the purpose of evaluating the variation of inductance with frequency, the electrical properties of the interconnect are characterized
by the inductive time constant $\tau = L/R$. The impedance magnitude
of these two paths is schematically shown in Fig. 10.4. The impedance of the first path is dominated by the inductive reactance above
the frequency $f_1 = \frac{1}{2\pi} \frac{R_1}{L_1} = \frac{1}{2\pi\tau_1}$. The impedance of the second path
is predominantly inductive above the frequency $f_2 = \frac{1}{2\pi} \frac{R_2}{L_2} = \frac{1}{2\pi\tau_2}$,
$f_2 < f_1$. At low frequencies, i.e., from DC to the frequency f_1, the ra-
tio of the two impedances is constant. The effective inductance at low
frequencies is therefore also constant, determining the low frequency

inductance limit. At high frequencies, *i.e.*, frequencies exceeding f_2, the ratio of the impedances is also constant, determining the high frequency inductance limit, $\frac{L_1 L_2}{L_1 + L_2}$. At intermediate frequencies from f_1 to f_2, the impedance ratio changes, resulting in a variation of the overall inductance from the low frequency limit to the high frequency limit. The frequency range of inductance variation is therefore determined by the two time constants, τ_1 and τ_2. The magnitude of the inductance variation depends upon both the difference between the time constants τ_1 and τ_2 and on the inductance ratio L_1 / L_2. Analogously, in the case of multiple parallel current paths, the frequency range and the magnitude of the variation in inductance is determined by the minimum and maximum time constants as well as the difference in inductance among the paths.

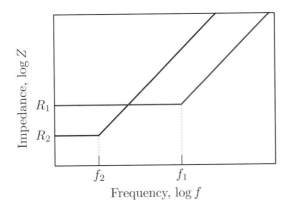

Fig. 10.4. Impedance magnitude versus frequency for two paths with dissimilar impedance characteristics.

The decrease in inductance begins when the inductive reactance $j\omega L$ of the path with the lowest R/L ratio becomes comparable to the path resistance R, $R \sim j\omega L$. The inductance, therefore, begins to decrease at a lower frequency if the minimum R/L ratio of the current paths is lower.

Due to this behavior, the proximity effect becomes significant at higher frequencies than multi-path current redistribution. Significant proximity effects occur in conductors containing current paths with significantly different inductive characteristics. That is, the inductive coupling of one edge of the line to the "return" current (*i.e.*, the current in the opposite direction) is substantially different from the inductive coupling of the other edge of the line to the same "return" current. In geometric terms, this characteristic means that the line width is larger than or comparable to the distance between the line and the return current. Consequently, the line with significant proximity effects is typically the immediate neighbor of the current return line. A narrower current loop is therefore formed with the current return path as compared to the other lines participating in the multi-path current redistribution. A smaller loop inductance L results in a higher R/L ratio. Referring to Fig. 2.10, current redistribution between paths one and two proceeds at frequencies lower than the onset frequency of the proximity effect in path one.

10.2.2 Analysis of inductance variation

The inductance of non-interdigitated grids versus signal frequency is shown in Fig. 10.5. At low frequencies, the forward and return currents are uniformly distributed among the lines. The two lines in the center of the grid form the smallest current loop while the lines at the periphery of the grid form wider current loops. The effective width of the current loop at low frequencies is relatively large, approximately half of the grid width. Non-interdigitated grids, therefore, have a relatively large inductance L and a low R/L ratio as compared to the other two grid types, interdigitated and paired. Consequently, the onset of a decrease in inductance occurs at a comparatively lower frequency, as illustrated in Figs. 10.5, 10.6, and 10.7. As the signal frequency increases, the current redistributes toward the center of the grid to decrease the grid inductance. Since the width of the grid is much larger than the width of the grid line, the decrease in inductance is primarily due to multi-path current redistribution among the different lines while current redistribution within the line cross sections (the proximity effect) is a secondary effect. The low frequency inductance of a non-interdigitated grid increases with grid width. As the grid width (*i.e.*, the number of lines) increases, the decrease in inductance with frequency becomes more significant and begins at a lower frequency [70].

In power grids with alternating power and ground lines (such as the interdigitated and paired grid structures illustrated in Figs. 10.1(b) and 10.1(c), respectively), each line has the same resistance and self inductance per length, and almost the same inductive coupling to the rest of the grid. As discussed in Chapter 11, long distance inductive coupling is

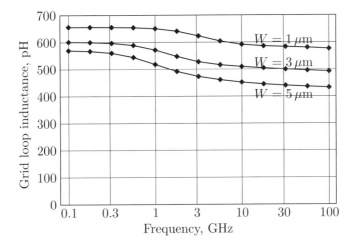

Fig. 10.5. Loop inductance of non-interdigitated grids versus signal frequency.

cancelled out in grids with a periodic structure, such that the lines are inductively coupled only to the immediate neighbors, making inductive coupling effectively a local phenomenon. As a result, the distribution of the current among the lines at low frequencies (where the current flows through the path of lowest resistance) practically coincides with the current distribution at high frequencies (where the current flows through the path of lowest inductance). That is, the line resistance has a negligible effect on the current distribution within the grid, *i.e.*, multi-path current redistribution is insignificant. Consequently, the decrease in inductance at high frequencies is caused primarily by the proximity effect which depends upon the line width, spacing, and material resistivity.

This situation is exemplified by paired grids, where multi-path current redistribution is insignificant and the proximity effect is more pronounced due to the small separation between adjacent power and ground lines. The loop inductance versus signal frequency for paired grids is shown in Fig. 10.6. The wider the line, the lower the frequency at which the onset of the proximity effect occurs and the larger the relative decrease in inductance [277], [278], as depicted in Fig. 10.6. Thus, the primary mechanism for a decrease in inductance in paired grids is the proximity effect.

The loop inductance versus signal frequency for interdigitated grids is shown in Fig. 10.7. As in paired grids, multi-path current

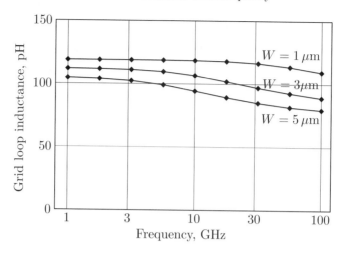

Fig. 10.6. Loop inductance of paired grids versus frequency.

redistribution is insignificant in interdigitated grids. However, the separation between grid lines is large as compared to the line width (unless the line width is comparable to the line pitch) and the proximity effect is, therefore, also insignificant [277], [278]. As shown in Fig. 10.7, the inductance of interdigitated grids is relatively constant with frequency, the decrease being limited to 10% to 12% of the low frequency inductance except for the case of very wide lines where the proximity effect becomes significant.

10.3 Summary

The variation of inductance with frequency in high performance power distribution grids is investigated in this chapter. The variation of inductance with frequency in three types of power grids is analyzed in terms of the mechanisms of inductance variation, as discussed in Section 2.2. These results support the design of area efficient and robust power distribution grids in high speed integrated circuits. The chapter results are summarized as follows.

- The inductance of power distribution grids decreases with increasing signal frequency

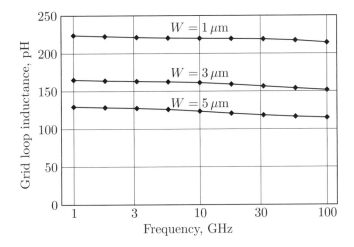

Fig. 10.7. Loop inductance of interdigitated grids versus frequency.

- The decrease in the inductance of non-interdigitated grids is primarily due to multi-path redistribution of the forward and return currents

- Multi-path current redistribution is greatly minimized in interdigitated and paired grids due to the periodic structure of these grids

- The smaller the separation between the power and ground lines and the wider the lines, the more significant the proximity effects become and the greater the relative decrease in inductance with frequency

- The wider the grid lines, the lower the frequency at which the onset of the decrease in inductance occurs

11

Inductance/Area/Resistance Tradeoffs

Tradeoffs among inductance, area, and resistance of power distribution grids are investigated in this chapter. As discussed in Section 1.3, design objectives, such as low impedance (low inductance and resistance), small area, and low current densities (for improved reliability), are typically in conflict. It is therefore important to make a balanced compromise among these design goals based upon application-specific constraints. A quantitative model of the inductance/area/resistance tradeoff in high performance power distribution networks is therefore necessary to achieve an efficient power distribution network. Another important goal is to provide quantitative guidelines to these tradeoffs and to bring intuition to the design of high performance power distribution networks.

Two tradeoff scenarios are considered in this chapter. The inductance versus resistance tradeoff under a constant grid area constraint in high performance power distribution grids is analyzed in Section 11.1. The inductance versus area tradeoff under a constant grid resistance constraint is analyzed in Section 11.2. The chapter concludes with a summary.

11.1 Inductance vs. resistance tradeoff under a constant grid area constraint

In the first tradeoff scenario, the fraction of the metal layer area dedicated to the power grid, called the grid area ratio and denoted as A, is assumed fixed, as shown in Fig. 11.1. The objective is to explore the tradeoff between grid inductance and resistance under the constraint of a constant area [276]. The area dedicated to the grid includes both

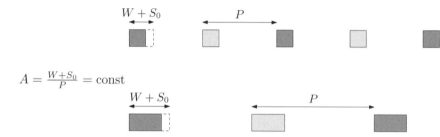

Fig. 11.1. Inductance versus resistance tradeoff scenario under a constant area constraint. As the line width varies, the grid area, including the minimum line spacing S_0, is maintained the same.

the line width W and the minimum spacing S_0 necessary to isolate the power line from any neighboring lines; therefore, the grid area ratio can be expressed as $A = \frac{W+S_0}{P}$, where P is the line pitch.

The inductance of paired grids is virtually independent of the separation between the power/ground line pairs. The effective current loop area in paired grids is primarily determined by the line spacing within each power/ground pair, which is much smaller than the separation between the power/ground pairs [69]. Therefore, only paired grids with an area ratio of 0.2 (*i.e.*, one fifth of the metal resources are allocated to the power and ground distribution) are considered here; the properties of paired grids with a different area ratio A (*i.e.*, different P/G separation) can be linearly extrapolated. In contrast, the dependence of the inductance of the interdigitated grids on the grid line pitch is substantial, since the effective current loop area is strongly dependent on the line pitch. Interdigitated grids with area ratios of 0.2 and 0.33 are analyzed here.

To investigate inductance tradeoffs in power distribution grids, the dependence of the grid inductance on line width is evaluated using FastHenry. Paired and interdigitated grids consisting of ten P/G lines are investigated. A line length of $1000\,\mu$m and a line thickness of $1\,\mu$m are assumed. The minimum spacing between the lines S_0 is $0.5\,\mu$m. The line width W is varied from $0.5\,\mu$m to $5\,\mu$m.

The grid inductance L_{grid} versus line width is shown in Fig. 11.2 for two signal frequencies: 1 GHz (the low frequency case) and 100 GHz (the high frequency case). The high frequency inductance is within 10% of the low frequency inductance for interdigitated grids, as mentioned

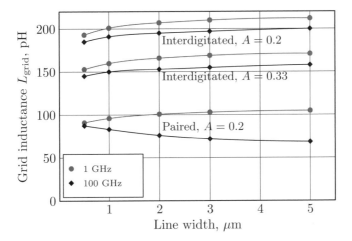

Fig. 11.2. The grid inductance versus line width under a constant grid area constraint for paired and interdigitated grids with ten P/G lines.

previously. The large change in inductance for paired grids is due to the proximity effect in closely spaced, relatively wide lines.

With increasing line width W, the grid line pitch P (and, consequently, the grid width) increases accordingly so as to maintain the desired grid area ratio $A = \frac{W+S_0}{P}$. Therefore, the inductance of a grid with a specific line width cannot be directly compared to the inductance of a grid with a different line width due to the difference in grid width. To perform a meaningful comparison of the grid inductance, the dimension specific data shown in Fig. 11.2 is converted to a dimension independent sheet inductance. The sheet inductance of a grid with a fixed area ratio A, $L_{\square}^{A}$, can be determined from L_{grid} through the following relationship,

$$L_{\square}^{A}(W) = L_{\text{grid}}\frac{NP}{l} = L_{\text{grid}}\frac{N}{l}\frac{W+S_0}{A}, \qquad (11.1)$$

where N is the number of lines (line pairs), P is the line (line pair) pitch in an interdigitated (paired) grid, and l is the grid length. For each of the six L_{grid} data sets shown in Fig. 11.2, a correspondent $L_{\square}^{A}$ versus line width data set is plotted in Fig. 11.3. As illustrated in Fig. 11.3, the sheet inductance $L_{\square}^{A}$ increases with line width; this increase with line width can be approximated as a linear dependence with high accuracy.

The low frequency sheet resistance of a grid is $R_{\square} = \rho_{\square}\frac{P}{W}$. The grid resistance under a constant area ratio constraint, $A = \frac{W+S_0}{P} = \text{const}$,

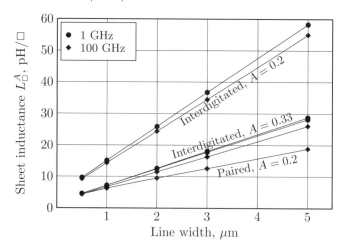

Fig. 11.3. The sheet inductance $L_\square^A$ versus line width under a constant grid area constraint.

can be expressed as a function of only the line width W,

$$R_\square^A = \frac{\rho_\square}{A} \frac{W + S_0}{W}. \tag{11.2}$$

This expression shows that as the line width W increases from the minimum line width $W_{\min} = S_0$ ($\frac{W+S_0}{W} = 2$) to a large width ($W \gg S_0$, $\frac{W+S_0}{W} \simeq 1$), the resistance decreases twofold. An intuitive explanation of this result is that at the minimum line width $W_{\min} = S_0$, only half of the grid area used for power routing is filled with metal lines (the other half is used for line spacing) while for large widths $W \gg S_0$, almost all of the grid area is metal.

In order to better observe the relative dependence of the grid sheet inductance and resistance on the line width, $L_\square^A$ and $R_\square^A$ are plotted in Fig. 11.4 normalized to the respective values at a minimum line width of $0.5\,\mu$m (such that $L_\square^A$ and $R_\square^A$ are equal to one normalized unit at $0.5\,\mu$m). As shown in Fig. 11.4, five out of six $L_\square^A$ lines have a similar slope. These lines depict the inductance of a paired grid at 1 GHz and the inductance of two interdigitated grids ($A = 0.2$ and $A = 0.33$) at 1 GHz and 100 GHz. The line with a lower slope represents a paired grid at 100 GHz. This different behavior is due to pronounced proximity effects in closely placed wide lines with very high frequency signals.

The dependence of the grid sheet inductance on line width is virtually linear and can be accurately approximated by

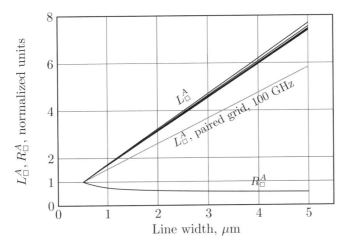

Fig. 11.4. Normalized sheet inductance and sheet resistance versus the width of the P/G line under a constant grid area constraint.

$$L_\square^A(W) = L_\square^A(W_{min}) \cdot \{1 + K \cdot (W - W_{min})\}, \qquad (11.3)$$

where $L_\square^A(W_{min})$ is the sheet inductance of a grid with a minimum line width and K is the slope of the lines shown in Fig. 11.4. Note that while $L_\square^A(W_{min})$ depends on the grid type and area ratio (as illustrated in Fig. 11.3), the coefficient K is virtually independent of these parameters (with the exception of the special case of paired grids at 100 GHz).

The grid inductance increases with line width, as shown in Fig. 11.4. The inductance increases eightfold (sixfold for the special case of a paired grid at 100 GHz) for a tenfold increase in line width [276]. The grid resistance decreases nonlinearly with line width. As mentioned previously, this decrease in resistance is limited to a factor of two.

The inductance versus resistance tradeoff has an important implication in the case where at the minimum line width the peak power noise is determined by the resistive voltage drop IR, but at the maximum line width the inductive voltage drop $L\frac{dI}{dt}$ is dominant. As the line width decreases, the inductive $L\frac{dI}{dt}$ noise becomes smaller due to the lower grid inductance L while the resistive IR noise increases due to the greater grid resistance R, as shown in Fig. 11.4. Therefore, a minimum total power supply noise, $IR + L\frac{dI}{dt}$, exists at some target line width. The line width that produces the minimum noise depends upon the ratio and relative timing of the peak current demand I and the peak transient current demand $\frac{dI}{dt}$. The optimal line width is, therefore,

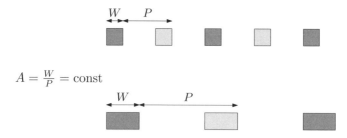

$$A = \frac{W}{P} = \text{const}$$

Fig. 11.5. Inductance versus area tradeoff scenario under a constant resistance constraint. As the line width varies, the metal area of the grid and, consequently, the grid resistance are maintained constant.

application dependent. This tradeoff provides guidelines for choosing the width of the power grid lines that produces the minimum noise.

11.2 Inductance vs. area tradeoff under a constant grid resistance constraint

In the second tradeoff scenario, the resistance of the power distribution grid is fixed (for example, by IR drop or electromigration constraints) [276], as shown in Fig. 11.5. The grid sheet resistance is

$$R_\square = \rho_\square \frac{P}{W} = \frac{\rho_\square}{M} = \text{const}, \tag{11.4}$$

where $\rho_\square$ is the sheet resistivity of the metal layer and $M = \frac{W}{P}$ is the fraction of the area filled with power grid metal, henceforth called the metal ratio of the grid. The constant resistance $R_\square$ infers a constant grid metal ratio M. The constraint of a constant grid resistance is similar to that of a constant grid area except that the line spacings are not considered as a part of the grid area. The objective is to explore tradeoffs between the grid inductance and area under the constraint of a constant grid resistance. This analysis is conducted similarly to the analysis described in the previous subsection. The grid inductance L_{grid}^R versus line width is shown in Fig. 11.6. The corresponding sheet inductance $L_\square^R$ versus line width data set is plotted in Fig. 11.7. The normalized sheet inductance and grid area data, analogous to the data shown in Fig. 11.4, is depicted in Fig. 11.8.

As shown in Fig. 11.8, under a constant resistance constraint, the grid inductance increases linearly with line width. Unlike in the first

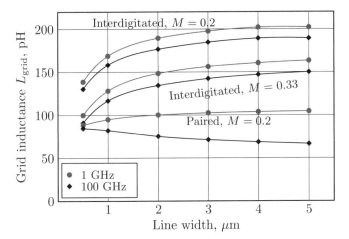

Fig. 11.6. The grid inductance versus line width under a constant grid resistance constraint for paired and interdigitated grids with ten P/G lines.

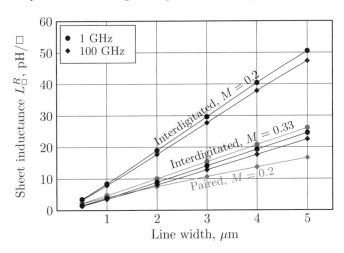

Fig. 11.7. The sheet inductance $L_\square^R$ versus line width under a constant grid resistance constraint.

scenario, the slope of the inductance increase with line width varies with the grid type and grid metal ratio. Paired grids have the lowest slope and interdigitated grids with a metal ratio of 0.33 have the highest slope. The lower slope of the inductance increase with line width is preferable, as, under a target resistance constraint, a smaller area and/or a less inductive power network implementation can be realized. The slope of the inductance increase with line width is independent

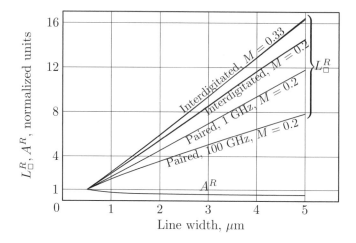

Fig. 11.8. Normalized sheet inductance $L_\square^R$ and grid area ratio A^R versus the width of the P/G line under a constant grid resistance (*i.e.*, constant grid metal ratio M) constraint.

of frequency in interdigitated grids (the lines for 1 GHz and 100 GHz coincide and are not discernible in the figure), while in paired grids the slope decreases significantly at high frequencies (100 GHz). The inductance increase varies from eight to sixteen fold, depending on grid type and grid resistance (*i.e.*, grid metal ratio), for a tenfold increase in line width. A reduction in the grid area is limited by a factor of two, similar to the decrease in resistance in the first tradeoff scenario.

11.3 Summary

Inductance/area/resistance tradeoffs in single layer power distribution grids are explored in this chapter. The primary conclusions can be summarized as follows.

- The grid inductance can be traded off against the grid resistance as the width of the grid lines is varied under a constant grid area constraint

- The grid inductance can be traded off against the grid area as the width of the grid lines is varied under a constant grid resistance constraint

- The grid inductance varies linearly with line width when either the grid resistance or the grid area is maintained constant

- The associated penalty in grid area (or resistance) is relatively small as long as the line width remains significantly greater than the minimum line spacing

12

Scaling Trends
of On-Chip Power Distribution Noise

A scaling analysis of the voltage drop across the on-chip power distribution networks is performed in this chapter. The design of power distribution networks in high performance integrated circuits has become significantly more challenging with recent advances in process technology. Insuring adequate signal integrity of the power supply has become a primary design issue in high performance, high complexity digital integrated circuits. A significant fraction of the on-chip resources is dedicated to achieve this objective.

State-of-the-art circuits consume higher current, operate at higher speeds, and have lower noise tolerance with the introduction of each new technology generation. CMOS technology scaling is forecasted to continue for at least another ten years [279]. The scaling trend of noise in high performance power distribution grids is, therefore, of practical interest. In addition to the constraints on the noise magnitude, electromigration reliability considerations limit the maximum current density in on-chip interconnect. The scaling of the peak current density in power distribution grids is also of practical interest. The results of this scaling analysis depend upon various assumptions. Existing scaling analyses of power distribution noise are reviewed and compared along with any relevant assumptions. The scaling of the inductance of an on-chip power distribution network as discussed here extends the existing material presented in the literature. Scaling trends of on-chip power supply noise in ICs packaged in high performance flip-chip packages are the focus of this investigation.

The chapter is organized as follows. Related existing work is reviewed in Section 12.1. The interconnect characteristics assumed in the analysis are discussed in Section 12.2. The model of the on-chip power

Table 12.1. Ideal scaling of CMOS circuits [280]

Parameter	Scaling factor
Device dimensions	$1/S$
Doping concentrations	S
Voltage levels	$1/S$
Current per device	$1/S$
Gate load	$1/S$
Gate delay	$1/S$
Device area	$1/S^2$
Device density	S^2
Power per device	$1/S^2$
Power density	1
Total capacitance	SS_C^2
Total power	S_C^2
Total current	SS_C^2

distribution noise used in the analysis is described in Section 12.3. The scaling of power noise is described in Section 12.4. Implications of the scaling analysis are discussed in Section 12.5. The chapter concludes with a summary.

12.1 Prior work

Ideal scaling of CMOS transistors was first described by Dennard *et al.* in 1974 [280]. Assuming a scaling factor S, where $S > 1$, all transistor dimensions uniformly scale as $1/S$, the supply voltage scales as $1/S$, and the doping concentrations scale as S. This "ideal" scaling maintains the electric fields within the device constant and ensures a proportional scaling of the I–V characteristics. Under the ideal scaling paradigm, the transistor current scales as $1/S$, the transistor power decreases as $1/S^2$, and the transistor density increases as S^2. The transistor switching time decreases as $1/S$, the power per circuit area remains constant, while the current per circuit area scales as S. The die dimensions increase by a chip dimension scaling factor S_C. The total capacitance of the on-chip devices and the circuit current both increase by SS_C^2 while the circuit power increases by S_C^2. The scaling of interconnect was first described by Saraswat and Mohammadi [281]. These ideal scaling relationships are summarized in Table 12.1.

Several research results have been published on the impact of technology scaling on the integrity of the IC power supply [132], [220], [282], [283]. The published analyses differ in the assumptions concerning the on-chip and package level interconnect characteristics. The analyses can be classified according to several categories: whether resistive IR or inductive $L\frac{dI}{dt}$ noise is considered, whether wire-bond or flip-chip packaging is assumed, and whether packaging or on-chip interconnect parasitic impedances are assumed dominant. Traditionally, the package-level parasitic inductance (the bond wires, lead frames, and pins) has dominated the total inductance of the power distribution system while the on-chip resistance of the power lines has dominated the total resistance of the power distribution system. The resistive noise has therefore been associated with the resistance of the on-chip interconnect and the inductive noise has been associated with the inductance of the off-chip packaging [283], [284], [285].

Scaling behavior of the resistive voltage drop in a wire bonded integrated circuit of constant size has been investigated by Song and Glasser in [282]. Assuming that the interconnect thickness scales as $1/S$, the ratio of the supply voltage to the resistive noise, *i.e.*, the signal-to-noise ratio (SNR) of the power supply voltage, scales as $1/S^3$ under ideal scaling (as compared to $1/S^4$ under constant voltage scaling). Song and Glasser proposed a multilayer interconnect stack to address this problem. Assuming that the top metal layer has a constant thickness, scaling of the power supply signal-to-noise ratio improves by a power of S as compared to standard interconnect scaling.

Bakoglu [132] investigated the scaling of both resistive and inductive noise in wire-bonded ICs considering the increase in die size by S_C with each technology generation. Under the assumption of ideal interconnect scaling (*i.e.*, the number of interconnect layers remains constant and the thickness of each layer is reduced as $1/S$), the SNR of the resistive noise decreases as $1/S^4S_C^2$. The SNR of the inductive noise due to the parasitic impedances of the packaging decreases as $1/S^4S_C^3$. These estimates of the SNR are made under the assumption that the number of interconnect levels increases as S. This assumption scales the on-chip capacitive load, average current, and, consequently, the SNR of both the inductive and resistive noise by a factor of S. Bakoglu also considered an improved scaling situation where the number of chip-to-package power connections increases as SS_C^2, effectively assuming flip-chip packaging. In this case, the resistive SNR_R scales as 1 assuming that the thickness of the upper metal levels is inversely scaled as S. The inductive SNR_L scales as $1/S$ under the assumption that the effective inductance per power connection scales as $1/S^2$.

A detailed overview of modeling and mitigation of package-level inductive noise is presented by Larsson [283]. The SNR of the inductive noise is shown to decrease as $1/S^2 S_C$ under the assumption that the number of interconnect levels remains constant and the number of chip-to-package power/ground connections increases as SS_C. The results and key assumptions of the power supply noise scaling analyses are summarized in Table 12.2.

The effect of the flip-chip pad density on the resistive drop in power supply grids has been investigated by Arledge and Lynch in [220]. All other conditions being equal, the maximum resistive drop is proportional to the square of the pad pitch. Based on this trend, a pad density of $4000\,\mathrm{pads/cm^2}$ is the minimum density required to assure an acceptable on-chip IR drop and I/O signal density at the 50 nm technology node [220].

Nassif and Fakhouri describe an analytic expression relating the maximum power distribution noise to the principal design and technology characteristics [286]. The expression is based on a lumped model similar to the model depicted in Fig. 12.3. The noise is shown to increase rapidly with technology scaling based on the ITRS predictions [287]. Assuming constant inductance, a reduction of the power grid resistance and an increase in the decoupling capacitance are predicted to be the most effective approaches to decreasing the power distribution noise.

12.2 Interconnect characteristics

The power noise scaling trends depend substantially on the interconnect characteristics assumed in the analysis. The interconnect characteristics are described in this section. The assumptions concerning the scaling of the global interconnect are discussed in Section 12.2.1. Variation of the grid inductance with interconnect scaling is described in Section 12.2.2. Flip-chip packaging characteristics are discussed in Section 12.2.3. The impact of the on-chip capacitance on the results of the analysis is discussed in Section 12.2.4.

Table 12.2. Scaling analyses of power distribution noise

Scaling analysis	Noise type	Noise scaling	SNR scaling	Analysis assumptions	
Glasser and Song [282]	On-chip IR noise	S^2	$1/S^3$	Ideal interconnect scaling	Wire-bond package, fixed die size
		S	$1/S^2$	Thickness of the top metal remains constant	
Bakoglu [132]	On-chip IR noise	$S^3 S_C^2$	$1/S^4 S_C^2$	Ideal interconnect scaling, wire-bond package (the number of power connections is constant)	Current and capacitance scale as $S^2 S_C^2$ (due to the scaling of the number of metal levels by S) as compared to $S S_C^2$
		$1/S$	1	Reverse interconnect scaling ($\propto S$), the number of power connections scale as $S S_C^2$ (flip-chip)	
	Package $L\frac{dI}{dt}$ noise	$S^3 S_C^3$	$1/S^4 S_C^3$	Number of power connections remains constant, inductance per connections increases as S_C	
		1	$1/S$	Number of power connections scale as $S S_C^2$ (flip-chip), inductance per connection scales as $1/S^2$	
Larsson [283]	Package $L\frac{dI}{dt}$ noise	$S S_C$	$1/S^2 S_C$	Wire-bond package, number of package connections increases as $S S_C$	
Mezhiba and Friedman	On-chip IR noise	1	$1/S$	Metal thickness remains constant	Area array flip-chip package, pad pitch scales as $1/\sqrt{S}$ (i.e., the number of power connections scales as $S S_C^2$)
		S	$1/S^2$	Ideal interconnect scaling	
	On-chip $L\frac{dI}{dt}$ noise	S	$1/S^2$	Metal thickness remains constant	
		1	$1/S$	Ideal interconnect scaling	

12.2.1 Global interconnect characteristics

The scaling of the cross-sectional dimensions of the on-chip global power lines directly affects the power distribution noise. Two scenarios of global interconnect scaling are considered here.

In the first scenario, the thickness of the top interconnect layers (where the conductors of the global power distribution networks are located) is assumed to remain constant. Through several recent technology generations, the thickness of the global interconnect layers has not been scaled in proportion to the minimum local line pitch due to power distribution noise and interconnect delay considerations. This behavior is in agreement with the 1997 edition of the International Technology Roadmap for Semiconductors (ITRS) [288], [289], where the minimum pitch and thickness of the global interconnect are assumed constant.

In the second scenario, the thickness and minimum pitch of the global interconnect layers are scaled in proportion to the minimum pitch of the local interconnect. This assumption is in agreement with the more recent editions of the ITRS [10], [287]. The scaling of the global interconnect in future technologies is therefore expected to evolve in the design envelope delimited by these two scenarios.

The number of metal layers and the fraction of the metal resources dedicated to the power distribution network are also assumed constant. The ratio of the diffusion barrier thickness to the copper interconnect core is assumed to remain constant with scaling. The increase in resistivity of the interconnect due to electron scattering at the interconnect surface interface (significant at line widths below 45 nm [10]) is neglected for relatively thick global power lines.

Under the aforementioned assumptions, in the constant metal thickness scenario, the effective sheet resistance of the global power distribution network remains constant with technology scaling. In the scenario of scaled metal thickness, the grid sheet resistance increases with technology scaling by a factor of S.

12.2.2 Scaling of the grid inductance

The inductive properties of power distribution grids are investigated in [69], [70]. It is shown that the inductance of the power grids with alternating power and ground lines behaves analogously to the grid resistance. That is, the grid inductance increases linearly with the grid length and decreases inversely linearly with the number of lines in the grid. This linear behavior is due to the periodic structure of the alternating power and ground grid lines. The long range inductive coupling of a specific (signal or power) line to a power line is cancelled out by the coupling to the ground lines adjacent to the power line, which carry current in the opposite direction [68], [70]. As described in Chapter 9, inductive coupling in periodic grid structures is effectively a short range interaction. Similar to the grid resistance, the grid inductance can be conveniently expressed as a dimension independent grid sheet inductance $L_\square$ [70], [276]. The inductance of a specific grid is obtained by multiplying the sheet inductance by the grid length and dividing by the grid width. As described in Section 9.6.4, the grid sheet inductance can be estimated as

$$L_\square = 0.8P \left(\ln \frac{P}{T+W} + \frac{3}{2} \right) \frac{\mu H}{\square}, \tag{12.1}$$

where W, T, and P are the width, thickness, and pitch of the grid lines, respectively. The sheet inductance is proportional to the line pitch P. The line density is reciprocal to the line pitch. A smaller line pitch means a higher line density and more parallel paths for the current to flow. The sheet inductance is however relatively insensitive to the cross-sectional dimensions of the lines, as the inductance of the individual lines is similarly insensitive to these parameters. Note that while the sheet resistance of the power grid is determined by the metal conductivity and the net cross-sectional area of the lines, the sheet inductance of the grid is determined by the line pitch and the ratio of the pitch to the line width and thickness.

In the constant metal thickness scenario, the sheet inductance of the power grid remains constant since the routing characteristics of the global power grid do not change. In the scaled thickness scenario, the line pitch, width, and thickness are reduced by S, increasing the line density and the number of parallel current paths. The sheet inductance therefore decreases by a factor of S, according to (12.1).

12.2.3 Flip-chip packaging characteristics

In a flip-chip package, the integrated circuit and package are interconnected via an area array of solder bumps mounted onto the on-chip I/O pads [103]. The power supply current enters the on-chip power distribution network from the power/ground pads. A view of the on-chip area array of power/ground pads is shown in Fig. 12.1.

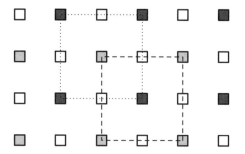

Fig. 12.1. An area array of on-chip power/ground I/O pads. The power pads are colored dark gray, the ground pads are colored light gray, and the signal pads are white. The current distribution area of the power pad (*i.e.*, the power distribution cell) in the center of the figure is delineated by the dashed line. The current distribution area of the ground pad in the center of the figure is delineated by the dotted line.

One of the main goals of this work is to estimate the significance of the *on-chip* inductive voltage drop in comparison to the on-chip resistive voltage drop. Therefore, all of the power/ground pads of a flip-chip packaged IC are assumed to be equipotential, *i.e.*, the variation in the voltage levels among the pads is considered negligible as compared to the noise within the on-chip power distribution network. For the purpose of this scaling analysis, a uniform power consumption per die area is assumed. Under these assumptions, each power (ground) pad supplies power (ground) current only to those circuits located in the area around the pad, as shown in Fig. 12.1. This area is referred to as a power distribution cell (or power cell). The edge dimensions of each power distribution cell are proportional to the pitch of the power/-ground pads. The size of the power cell area determines the effective distance of the on-chip distribution of the power current. The power distribution scaling analysis becomes independent of die size.

An important element of this analysis is the scaling of the flip-chip technology. The rate of decrease in the pad pitch and the rate of reduction in the local interconnect half-pitch are compared in Fig. 12.2, based on the ITRS [10]. At the 150 nm line half-pitch technology node, the pad pitch P is 160 μm. At the 32 nm node, the pad pitch is forecasted to be 80 μm. That is, the linear density of the pads doubles for a fourfold reduction in circuit feature size. The pad size and pitch P scale, therefore, as $1/\sqrt{S}$ and the area density ($\propto 1/P^2$) of the pads increases as S with each technology generation. Interestingly, one of the reasons given for this relatively infrequent change in the pad pitch (as compared with the introduction of new CMOS technology generations) is the cost of the test probe head [10]. The maximum density of the flip-chip pads is assumed to be limited by the pad pitch. Although the number of on-chip pads is forecasted to remain constant, some recent research has predicted that the number of on-chip power/-ground pads will increase due to electromigration and resistive noise considerations [220], [290].

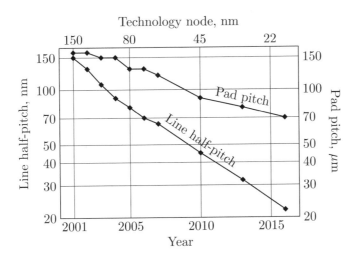

Fig. 12.2. Decrease in flip-chip pad pitch with technology generations as compared to the local interconnect half-pitch.

12.2.4 Impact of on-chip capacitance

On-chip capacitors are used to reduce the impedance of the power distribution grid lines as seen from the load terminals. A simple model of an on-chip power distribution grid with a power load and a decoupling capacitor is shown in Fig. 12.3. The on-chip loads are switched within tens of picoseconds in modern semiconductor technologies. The frequency spectrum of the load current therefore extends well beyond 10 GHz. The on-chip decoupling capacitors shunt the load current at the highest frequencies. The bulk of the power current bypasses the on-chip distribution network at these frequencies. At the lower frequencies, however, the capacitor impedance is relatively high and the bulk of the

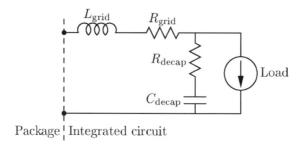

Fig. 12.3. A simplified circuit model of the on-chip power distribution network with a power load and a decoupling capacitance.

current flows through the on-chip power distribution network. The decoupling capacitors therefore serve as a low pass filter for the power current.

Describing the same effect in the time domain, the capacitors supply the (high frequency) current to the load during a switching transient. To prevent excessive power noise, the charge on the decoupling capacitor should be replenished by the (lower frequency) current flowing through the power distribution network before the next switching of the load, *i.e.*, typically within a clock period. The effect of the on-chip decoupling capacitors is therefore included in the model by assuming that the current transients within the on-chip power distribution network are characterized by the clock frequency of the circuit, rather than by the switching times of the on-chip load circuits. Estimates of the resistive voltage drop are based on the average power current, which is not affected by the on-chip decoupling capacitors.

12.3 Model of power supply noise

The following simple model is utilized in the scaling analysis of on-chip power distribution noise. A power distribution cell is modeled as a circle of radius r_c with a constant current consumption per area I_a, as described by Arledge and Lynch [220]. The model is depicted in Fig. 12.4. The total current of the cell is $I_{cell} = I_a \cdot \pi r_c^2$. The power network current is distributed from a circular pad of radius r_p at the center of the cell. The global power distribution network has an effective sheet resistance $\rho_\square$. The incremental voltage drop dV_R across the elemental circular resistance $\rho_\square \, dr / 2\pi r$ is due to the current $I_a(\pi r_c^2 - \pi r^2)$ flowing

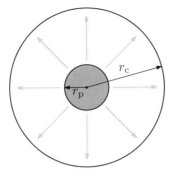

Fig. 12.4. A model of the power distribution cell. Power supply current spreads out from the power pad in the center of the cell to the cell periphery, as shown by the arrows.

through this resistance toward the periphery of the cell. The voltage drop at the periphery of the power distribution cell is

$$\Delta V_R = \int_{r_p}^{r_c} dV_R = \int_{r_p}^{r_c} I(r) \cdot dR(r)$$

$$= \int_{r_p}^{r_c} \pi(r_c^2 - r^2) I_a \cdot \rho_\square \frac{dr}{2\pi r}$$

$$= I_a \pi r_c^2 \rho_\square \cdot \frac{1}{2\pi} \left(\ln \frac{r_c}{r_p} + \frac{r_p^2}{2r_c^2} - \frac{1}{2} \right)$$

$$= I_{\text{cell}} \rho_\square \cdot C \left(\frac{r_c}{r_p} \right). \tag{12.2}$$

The resistive voltage drop is proportional to the product of the total cell current I_{cell} and the effective sheet resistance $\rho_\square$ with the coefficient C dependent only on the r_c/r_p ratio. The ratio of the pad pitch to the pad size is assumed to remain constant. The coefficient C, therefore, does not change with technology scaling.

The properties of the grid inductance are analogous to the properties of the grid resistance, as discussed in Section 12.2.1. Therefore, analogous to the resistive voltage drop ΔV_R discussed above, the inductive voltage drop ΔV_L is proportional to the product of the sheet inductance $L_\square$ of the global power grid and the magnitude of the cell transient current dI_{cell}/dt,

$$\Delta V_L = L_\square \frac{dI_{\text{cell}}}{dt} \cdot C \left(\frac{r_c}{r_p} \right). \qquad (12.3)$$

12.4 Power supply noise scaling

An analysis of the on-chip power supply noise is presented in this section. The interconnect characteristics assumed in the analysis are described in Section 12.2. The power supply noise model is described in Section 12.3. Ideal scaling of the power distribution noise in the constant thickness scenario is discussed in Section 12.4.1. Ideal scaling of the noise in the scaled thickness scenario is analyzed in Section 12.4.2. Scaling of the power distribution noise based on the ITRS projections is discussed in Section 12.4.3.

12.4.1 Analysis of constant metal thickness scenario

The scaling of a power distribution grid over four technology generations according to the constant metal thickness scenario is depicted in Fig. 12.5. The minimum feature size is reduced by $\sqrt{2}$ with each generation. The minimum feature size over four generations is therefore reduced by four, i.e., $\left(\sqrt{2} \right)^4 = 4$, while the size of the power distribution cell (represented by the size of the square grid) is halved ($\sqrt{4} = 2$). As the cross-sectional dimensions of the power lines are maintained constant in this scenario, both the sheet resistance $\rho_\square$ and sheet inductance $L_\square$ of the power distribution grid remain constant with scaling under these conditions.

The cell current I_{cell} is the product of the area current density I_a and the cell area πr_c^2. The current per area I_a scales as S; the area of the cell is proportional to P^2 which scales as $1/S$. The cell current I_{cell}, therefore, remains constant (i.e., scales as 1). The resistive drop ΔV_R, therefore, scales as $I_{\text{cell}} \cdot \rho_\square \propto 1 \cdot 1 \propto 1$. The resistive SNR_R^{I} of the power supply voltage, consequently, decreases with scaling as

$$\text{SNR}_R^{\text{I}} = \frac{V_{\text{dd}}}{\Delta V_R} \propto \frac{1/S}{1} \propto \frac{1}{S}. \qquad (12.4)$$

This scaling trend agrees with the trend described by Bakoglu in the improved scaling regime [132]. Faster scaling of the on-chip current as described by Bakoglu is offset by increasing the interconnect thickness by S which reduces the sheet resistance $\rho_\square$ by S. This trend is more

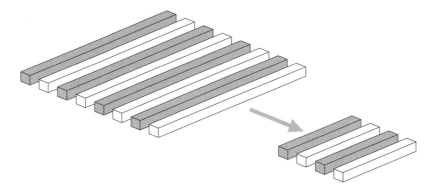

Fig. 12.5. The scaling of a power distribution grid over four technology generations according to the constant metal thickness scenario. The cross-sectional dimensions of the power lines remain constant. The size of the power distribution cell, represented by the size of the square grid, is halved.

favorable as compared to the $1/S^2$ dependence established by Song and Glasser [282]. The improvement is due to the decrease in the power cell area of a flip-chip IC by a factor of S whereas a wire-bonded die of constant area is assumed in [282].

The transient current dI_{cell}/dt scales as $I_{\mathrm{cell}}/\tau \propto 1/(1/S) \propto S$, where $\tau \propto 1/S$ is the transistor switching time. The inductive voltage drop ΔV_L, therefore, scales as $L_\square \cdot dI_{\mathrm{cell}}/dt \propto 1 \cdot S$. The inductive $\mathrm{SNR}_L^{\mathrm{I}}$ of the power supply voltage decreases with scaling as

$$\mathrm{SNR}_L^{\mathrm{I}} = \frac{V_{\mathrm{dd}}}{\Delta V_L} \propto \frac{1/S}{S} \propto \frac{1}{S^2}. \tag{12.5}$$

The relative magnitude of the inductive noise therefore increases by a factor of S faster as compared to the resistive noise. Estimates of the inductive and resistive noise described by Bakoglu also differ by a factor of S [132].

12.4.2 Analysis of the scaled metal thickness scenario

The scaling of a power distribution grid over four technology generations according to the scaled metal thickness scenario is depicted in Fig. 12.6. In this scenario, the cross-sectional dimensions of the power lines are reduced in proportion to the minimum feature size by a factor of four, while the size of the power distribution cell is halved. Under these conditions, the sheet resistance $\rho_\square$ of the power distribution grid increases by S, while the sheet inductance $L_\square$ of the power distribution grid decreases by S with technology scaling.

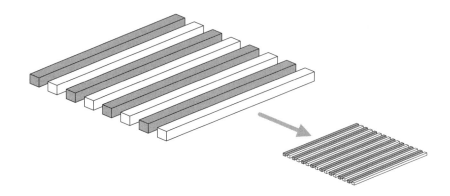

Fig. 12.6. The scaling of a power distribution grid over four technology generations according to the scaled metal thickness scenario. The cross-sectional dimensions of the power lines are reduced in proportion to the minimum feature size by a factor of four. The size of the power distribution cell, represented by the size of the square grid, is halved.

Analogous to the constant metal thickness scenario, the cell current I_{cell} remains constant. The resistive drop ΔV_R, therefore, scales as $I_{\text{cell}} \cdot \rho_\square \propto 1 \cdot S \propto S$. The resistive SNR_R^{II} of the power supply voltage, consequently, decreases with scaling as

$$\text{SNR}_R^{\text{II}} = \frac{V_{\text{dd}}}{\Delta V_R} \propto \frac{1/S}{S} \propto \frac{1}{S^2}. \tag{12.6}$$

As discussed in the previous section, the transient current dI_{cell}/dt scales as $I_{\text{cell}}/\tau \propto S$. The inductive voltage drop ΔV_L, therefore, scales as $L_\square \cdot dI_{\text{cell}}/dt \propto 1/S \cdot S \propto 1$. The inductive SNR_L^{II} of the power supply voltage decreases with scaling as

$$\text{SNR}_L^{\text{II}} = \frac{V_{\text{dd}}}{\Delta V_L} \propto \frac{1/S}{1} \propto \frac{1}{S}. \tag{12.7}$$

The rise of the inductive noise is mitigated if ideal interconnect scaling is assumed and the thickness, width, and pitch of the global power lines are scaled as $1/S$. In this scenario, the density of the global power lines increases as S and the sheet inductance $L_\square$ of the global power distribution grids decreases as $1/S$, mitigating the inductive noise and

SNR_L by S. The sheet resistance of the power distribution grid, however, increases as S, exacerbating the resistive noise and SNR_R by a factor of S. Currently, the parasitic resistive impedance dominates the total impedance of on-chip power distribution networks. Ideal scaling of the upper interconnect levels will therefore increase the overall power distribution noise. However, as CMOS technology approaches the nanometer range and the inductive and resistive noise becomes comparable, judicious tradeoffs between the resistance and inductance of the power networks will be necessary to achieve the minimum noise level (see Chapter 11) [206], [207].

12.4.3 ITRS scaling of power noise

Although the ideal scaling analysis allows the comparison of the rates of change of both resistive and inductive voltage drops, it is difficult to estimate the *ratio* of these quantities for direct assessment of their relative significance. Furthermore, practical scaling does not accurately follow the concept of ideal scaling due to material and technological limitations. An estimate of the ratio of the inductive to resistive voltage drop is therefore conducted in this section based on the projected 2001 ITRS data [10].

Forecasted demands in the supply current of high performance microprocessors are shown in Fig. 12.7. Both the average current and the transient current are rising exponentially with technology scaling. The rate of increase in the transient current is more than double the rate of increase in the average current as indicated by the slope of the trend lines depicted in Fig. 12.7. This behavior is in agreement with ideal scaling trends. The faster rate of increase in the transient current as compared to the average current is due to rising clock frequencies. The transient current of modern high performance processors is approximately one teraampere per second (10^{12} A/s) and is expected to rise, reaching hundreds of teraamperes per second. Such a high magnitude of the transient current is caused by switching hundreds of amperes within a fraction of a nanosecond.

In order to translate the projected current requirements into supply noise voltage trends, a case study interconnect structure is considered. The square grid structure shown in Fig. 12.8 is used here to serve as a model of the on-chip power distribution grid. The square grid consists of interdigitated power and ground lines with a $1\,\mu\text{m} \times 1\,\mu\text{m}$ cross section and a $1\,\mu\text{m}$ line spacing. The length and width of the grid are equal to

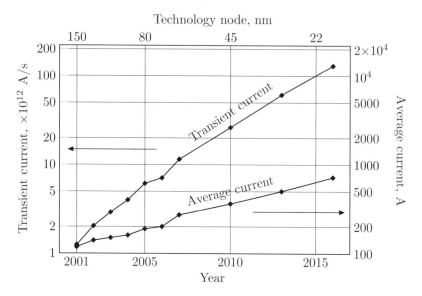

Fig. 12.7. Increase in power current demands of high performance microprocessors with technology scaling, according to the ITRS. The average current is the ratio of the circuit power to the supply voltage. The transient current is the product of the average current and the on-chip clock rate, $2\pi f_{\text{clk}}$.

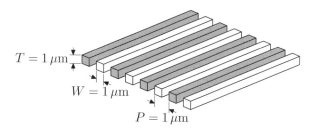

Fig. 12.8. Power distribution grid used to estimate trends in the power supply noise.

the size of a power distribution cell. The grid sheet inductance is 1.8 picohenrys per square, and the grid sheet resistance is 0.16 ohms per square. The size of the power cell is assumed to be twice the pitch of the flip-chip pads, reflecting that only half of the total number of pads are used for the power and ground distribution as forecasted by the ITRS for high performance ASICs.

The electrical properties of this structure are similar to the properties of the global power distribution grid covering a power distribution cell with the same routing characteristics. Note that the resistance and inductance of the *square* grid are independent of grid dimensions [276] (as long as the dimensions are severalfold greater than the line pitch). The average and transient currents flowing through the grid are, however, scaled from the IC current requirements shown in Fig. 12.7 in proportion to the area of the grid. The current flowing through the square grid is therefore the same as the current distributed through the power grid within the power cell. The power current enters and leaves from the same side of the grid, assuming the power load is connected at the opposite side. The voltage differential across this structure caused by the average and transient currents produces, respectively, on-chip resistive and inductive noise. The square grid has the same inductance to resistance ratio as the global distribution grid with the same line pitch, thickness, and width. Hence, the square grid has the same inductive to resistive noise ratio. The square grid model also produces the same rate of increase in the noise because the current is scaled proportionately to the area of the power cell.

The resulting noise trends under the constant metal thickness scenario are illustrated in Fig. 12.9. As discussed in Section 12.2, the area of the grid scales as $1/S$. The current area density increases as S. The total average current of the grid, therefore, remains constant. The resistive noise also remains approximately constant, as shown in Fig. 12.9. The inductive noise, alternatively, rises steadily and becomes comparable to the resistive noise at approximately the 45 nm technology node. These trends are in reasonable agreement with the ideal scaling predictions discussed in Section 12.4.1.

The inductive and resistive voltage drops in the scaled metal thickness scenario are shown in Fig. 12.10. The increase in inductive noise with technology scaling is limited, while the resistive noise increases by an order of magnitude. This behavior is similar to the ideal scaling trends for this scenario, as discussed in Section 12.4.2.

Note that the structure depicted in Fig. 12.8 has a lower inductance to resistance ratio as compared to typical power distribution grids because the power and ground lines are relatively narrow and placed adjacent to each other, reducing the area of the current loop and increasing the grid resistance [69], [276]. The width of a typical global power line varies from tens to a few hundreds of micrometers, resulting in a significantly higher inductance to resistance ratio. The results shown in Figs. 12.9 and 12.10 can be readily extrapolated to different

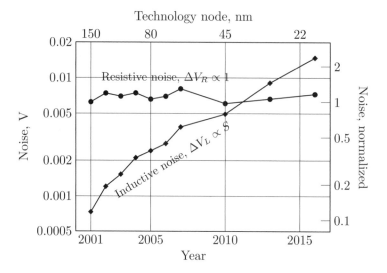

Fig. 12.9. Scaling trends of resistive and inductive power supply noise under the constant metal thickness scenario.

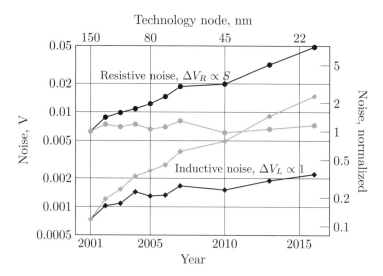

Fig. 12.10. Scaling trends of resistive and inductive power supply noise under the scaled metal thickness interconnect scaling scenario. The trends of the constant metal thickness scenario are also displayed in light gray for comparison.

grid configurations, using the expression for the grid sheet inductance (12.1).

Several factors offset the underestimation of the relative magnitude of the inductive noise due to the relatively low inductance to resistance ratio of the model shown in Fig. 12.8. If the global power distribution grid is composed of several layers of interconnect, the lines in the lower interconnect levels have a smaller pitch and thickness, significantly reducing the inductance to resistance ratio at high frequencies [291]. The transient current is conservatively approximated as the product of the average current I_{avg} and the angular clock frequency $2\pi f_{clk}$. This estimate, while serving as a useful scaling parameter, tends to overestimate the absolute magnitude of the current transients, increasing the ratio of the inductive and resistive voltage drops.

12.5 Implications of noise scaling

As described in the previous section, the amplitude of both the resistive and inductive noise relative to the power supply voltage increases with technology scaling. A number of techniques have been proposed to mitigate the unfavorable scaling of power distribution noise. These techniques are briefly summarized below.

To maintain a constant supply voltage to resistive noise ratio, the effective sheet resistance of the global power distribution grid should be reduced. There are two ways to allocate additional metal resources to the power distribution grid. One option is to increase the number of metalization layers. This approach adversely affects fabrication time and yield and, therefore, increases the cost of manufacturing. The ITRS forecasts only a moderate increase in the number of interconnect levels, from eight levels at the 130 nm line half-pitch node to eleven levels at the 32 nm node [10]. The second option is to increase the fraction of metal area per metal level allocated to the power grid. This strategy decreases the amount of wiring resources available for global signal routing and therefore can also necessitate an increase in the number of interconnect layers.

The sheet inductance of the power distribution grid, similar to the sheet resistance, can be lowered by increasing the number of interconnect levels. Furthermore, wide metal trunks typically used for power distribution at the top levels can be replaced with narrow interdigitated power/ground lines. Although this configuration substantially lowers the grid inductance, it increases the grid resistance and, consequently, the resistive noise [276].

Alternatively, circuit techniques can be employed to limit the peak transient power current demands of the digital logic. Current steering logic, for example, produces a minimal variation in the current demand between the transient response and the steady state response. In synchronous circuits, the maximum transient currents typically occur during the beginning of a clock period. Immediately after the arrival of a clock signal at the latches, a signal begins to propagate through the blocks of sequential logic. Clock skew scheduling can be exploited to spread in time the periods of peak current demand [36].

The constant metal thickness scaling scenario achieves a lower overall power noise until the technology generation is reached where the inductive and resistive voltage drops become comparable. Beyond this node, a careful tradeoff between the resistance and inductance of the power grid is necessary to minimize the on-chip power supply noise. The increase in the significance of the inductance of the power distribution interconnect is similar to that noted in signal interconnect [61], [292]. The trend, however, is delayed by several technology generations as compared to signal interconnect. As discussed in Section 12.2, the high frequency harmonics are filtered out by the on-chip decoupling capacitance and the power grid current has a comparatively lower frequency content as compared to the signal lines.

12.6 Summary

A scaling analysis of power distribution noise in flip-chip packaged integrated circuits is presented in this chapter. Published scaling analyses of power distribution noise are reviewed and various assumptions of these analyses are discussed. The primary conclusions can be summarized as follows.

- Under the constant metal thickness scenario, the relative magnitude (*i.e.*, the reciprocal of the signal-to-noise ratio) of the resistive noise increases by the scaling factor S, while the relative magnitude of the inductive noise increases by S^2

- Under the scaled metal thickness scenario, the scaling trend of the inductive noise improves by a factor of S, but the relative magnitude of the resistive noise increases by S^2

- The importance of on-chip inductive noise increases with technology scaling

- Careful tradeoffs between the resistance and inductance of power distribution networks in nanometer CMOS technologies will be necessary to achieve minimum power supply noise levels

13

Impedance Characteristics
of Multi-Layer Grids

The power distribution network spans many layers of interconnect with disparate electrical properties. The impedance characteristics of multi-layer power distribution grids and the relevant design implications are the subject of this chapter.

Decoupling capacitors are an effective technique to reduce the effect of the inductance on power distribution networks operating at high frequencies. The efficacy of decoupling capacitors depends on the impedance of the conductors connecting the capacitors to the power load and source. The optimal allocation of the on-chip decoupling capacitance depends on the impedance characteristics of the interconnect. Robust and area efficient design of multi-layer power distribution grids therefore requires a thorough understanding of the impedance properties of the power distributing interconnect structures.

Power distribution networks in high performance digital ICs are commonly structured as a multi-layer grid, as shown in Fig. 13.1. The inductive properties of single layer power grids have been described in Chapter 9. In grid layers with alternating power and ground lines, long distance inductive coupling is greatly diminished due to cancellation, turning inductive coupling in single layer power grids into, effectively, a local phenomenon. The grid inductance, therefore, behaves similarly to the grid resistance: increases linearly with grid length and decreases inversely linearly with grid width (*i.e.*, the number of lines in the grid). The electrical properties of power distribution grids can therefore be conveniently expressed by a dimension-independent sheet resistance $R_\square$ and sheet inductance $L_\square$ [276]. The inductance of the power grid layers can be efficiently estimated using simple models comprised of a few interconnect lines.

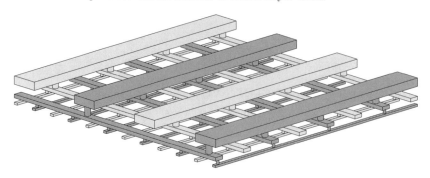

Fig. 13.1. A multi-layer power distribution grid. The ground lines are light grey, the power lines are dark grey.

Area/inductance/resistance tradeoffs in power distribution grids have also been investigated in Chapter 11. The sheet inductance of power distribution grids is shown to increase linearly with line width under two different tradeoff scenarios. Under the constraint of constant grid area, a tradeoff exists between the grid inductance and resistance. Under the constraint of a constant grid resistance, the grid inductance can be traded off against grid area.

The variation of inductance with frequency in single layer power grids has been characterized in Chapter 10. This variation is relatively moderate, typically less than 10% of the low frequency inductance. An exception from this behavior is power grids with closely spaced power and ground lines where the inductance variation with frequency is greater due to significant proximity effects.

Power distribution grids in modern integrated circuits typically consist of many grid layers, spanning an entire stack of interconnect layers. The objective of the present investigation is to characterize the electrical properties of these multi-layer grids, advancing the existing work beyond individual grid layers.

The chapter is organized as follows. The impedance characteristics of multi-layer power distribution grids are discussed in Section 13.1. A case study of a two layer power grid is presented in Section 13.2. The design implications of the impedance properties of a multi-layer grid are discussed in Section 13.3. The chapter concludes with a summary.

13.1 Electrical properties of multi-layer grids

A circuit model of multi-layer power distribution grids is developed in this section. The impedance characteristics of multi-layer grids are determined based on this model. The impedance characteristics of individual layers of multi-layer power distribution grids are discussed in Section 13.1.1. The variation with frequency of the impedance characteristics of several grid layers forming a multi-layer grid is analyzed in Section 13.1.2.

13.1.1 Impedance characteristics of individual grid layers

The power and ground lines within each layer of a multi-layer power distribution grid are orthogonal to the lines in the adjacent layers. Orthogonal lines have zero mutual partial inductance as there is no magnetic linkage [42]. Orthogonal grid layers can therefore be evaluated independently. A multi-layer grid can be considered to consist of two stacks of layers, with all of the lines in each stack parallel to each other, as shown in Fig. 13.2. Grid lines in one stack are orthogonal to the lines in the other stack. Grid layers in each stack only affect the grid inductance in the direction of the lines in the stack. This behavior is analogous to the properties of the grid resistance. The problem of characterizing a multi-layer grid is thereby reduced to determining the impedance characteristics of a stack of several individual grid layers with lines in the same direction.

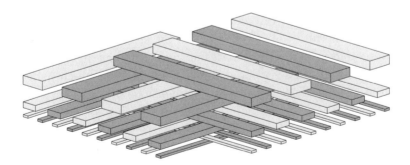

Fig. 13.2. A multi-layer grid consists of two stacks of layers. The lines in each stack are parallel to each other. The layers in one stack determine the resistive and inductive characteristics of the multi-layer grid in the direction of the lines in that stack, while the layers in the other stack determine the impedance characteristics in the orthogonal direction.

The power and ground lines in power distribution grids are connected to the lines in the adjacent layers through vias. The vias (or clusters of vias) are distributed along a power line at a pitch equal to the power line pitch in the adjacent layer. The line pitch is much larger than the via length. The inductance and resistance of the two close parallel power lines in different metal layers are therefore much larger than the resistance and inductance of the connecting vias. The effect of vias on the resistance and inductance of a power grid is therefore negligible. This property is a direct consequence of the characteristic that the distance of the lateral current distribution in power grids (hundreds or thousands of micrometers) is much larger than the distance of the vertical current distribution (several micrometers). The power current is distributed among the metal layers over a distance comparable to a line pitch. The power and ground lines are effectively connected in parallel.

Each layer of a typical multi-layer power distribution grid has significantly different electrical properties. Lines in the upper layers tend to be thick and wide, forming a low resistance global power distribution grid. Lines in the lower layers tend to be thinner, narrower, and have a smaller pitch. The lower the metal layer, the smaller the metal thickness, width, and pitch. The upper grid layers therefore have a relatively high inductance and low resistance, whereas the lower layers have a relatively low inductance and high resistance [70], [276]. The lower the layer, the higher the resistance and the lower the inductance. In those circuits employing flip-chip packaging with a high density area array of I/O contacts, the interconnect layers in the package are tightly coupled to the on-chip interconnect, effectively extending the on-chip interconnect hierarchy. The difference in the electrical properties across the interconnect hierarchy is particularly significant in nanoscale circuits. While the cross-sectional dimensions of local on-chip lines are measured in tens of nanometers, the dimensions of package lines are of the order of tens of micrometers. The three orders of magnitude difference in dimensions translates to six orders of magnitude difference in resistance (proportional to the cross-sectional area of the grid lines) and to three orders of magnitude difference in inductance (proportional to the grid line density).

The variation with frequency of the impedance of each layer in a grid stack comprised of N grid layers is schematically shown in Fig. 13.3. The layers are numbered from 1 (the uppermost layer) to N (the

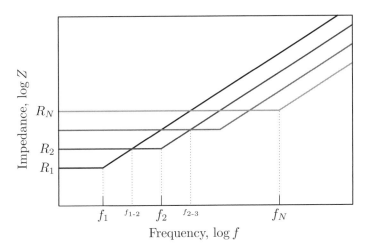

Fig. 13.3. Impedance of the individual grid layers comprising a multi-layer grid.

lowest layer). The grid layer resistance increases with layer number, $R_1 < R_2 < \ldots < R_N$, and the inductance decreases with layer number $L_1 > L_2 > \ldots > L_N$. At low frequencies, the uppermost layer has the lowest impedance as the layer with the lowest resistance. This layer, however, has the highest inductance and, consequently, the lowest transition frequency $f_1 = \frac{1}{2\pi}\frac{R_1}{L_1}$, as compared to the other layers (see Fig. 13.3). The transition frequency is the frequency at which the impedance of the grid layer changes in character from resistive to inductive. At this frequency, the inductive impedance of a grid layer is equal to the resistive impedance (neglecting skin and proximity effects), *i.e.*, $R_1 = \omega L_1$. The grid impedance increases linearly with frequency above f_1. The lowest grid layer has the highest resistance and the lowest inductance; therefore, this layer has the highest transition frequency f_N. As the inductance of the upper layers is higher than the lower layers, the impedance of an upper layer exceeds the impedance of any lower layer above a certain frequency. For example, the impedance of the first layer $R_1 + \omega L_1 \approx \omega L_1$ equals the magnitude of the second layer impedance $R_2 + \omega L_2 \approx R_2$ and exceeds the impedance of the second layer above frequency $f_{1\text{-}2} = \frac{1}{2\pi}\frac{R_2}{L_1}$, as shown in Fig. 13.3. Similarly, the impedance of layer k exceeds the impedance of layer l, $k < l$, at $f_{k-l} = \frac{1}{2\pi}\frac{R_l}{L_k}$.

13.1.2 Impedance characteristics of multi-layer grids

An entire stack of grid layers cannot be accurately described by a single RL circuit due to the aforementioned differences among the electrical properties of the individual grid layers. A stack of multiple grid layers can, however, be modeled by several parallel RL branches, each branch characterizing the electrical properties of one of the comprising grid layers, as shown in Fig. 13.4.

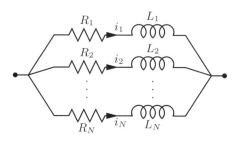

Fig. 13.4. Equivalent circuit of a stack of N grid layers.

Due to the difference in the electrical properties of the individual layers, the magnitude of the current in each grid layer varies significantly with frequency. At low frequencies, the low resistance uppermost layer is the path of lowest impedance, as shown in Fig. 13.3. The uppermost layer has the greatest effect on the low frequency resistance and inductance of the grid stack, as the largest share of the overall current flows through this layer. As the frequency increases to $f_{1\text{-}2} = \frac{1}{2\pi}\frac{R_2}{L_1}$ and higher, the impedance of the uppermost layer ωL_1 exceeds the impedance of the second uppermost layer R_2, as shown in Fig. 13.3. The second uppermost layer, therefore, carries the largest share of the overall current and most affects the inductance and resistance within this frequency range. As the frequency exceeds $f_{2\text{-}3} = \frac{1}{2\pi}\frac{R_3}{L_2}$, the next layer in the stack becomes the path of least impedance and so on. The process continues until at very high frequencies the lowest layer carries most of the overall current.

As the frequencies increase, the majority of the overall current is progressively transfered from the layers of low resistance and high inductance to the layers of high resistance and low inductance. The overall grid inductance, therefore, decreases with frequency and the overall grid resistance increases with frequency. A qualitative plot of the variation of the grid inductance and resistance with frequency is shown in

Fig. 13.5. At low frequency, all of the layers exhibit a purely resistive behavior and the current is partitioned among the layers according to the resistance of each layer. The share i_k of the overall current flowing through layer k is

$$i_k = \frac{I_k}{\sum_{n=1}^{N} I_n} = \frac{\prod_{n \neq k} R_n}{\sum_{m=1}^{N} \prod_{n \neq m} R_n}. \tag{13.1}$$

Note that $i_1 > i_2 > \ldots > i_N$ as $R_1 < R_2 < \ldots < R_N$. The resistance of a multi-layer grid R_0^{LF} at low frequency is therefore determined by the parallel connection of all of the individual layer resistances,

$$R_0^{LF} = R_1 \| R_2 \| \ldots \| R_N = \frac{\prod_{n=1}^{N} R_n}{\sum_{m=1}^{N} \prod_{n \neq m} R_n}. \tag{13.2}$$

The low frequency inductance of a multi-layer grid L_0^{LF} is, however,

$$L_0^{LF} = L_1 i_1^2 + L_2 i_2^2 + \ldots + L_N i_N^2 \approx L_1, \tag{13.3}$$

due to $L_1 > L_k$ and $i_1 > i_k$ for any $k \neq 1$.

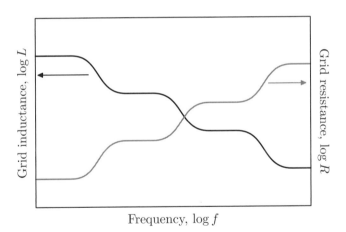

Fig. 13.5. Variation of the grid inductance and resistance of a multi-layer stack with frequency. As the signal frequency increases, the current flow shifts to the high resistance, low inductance layers, decreasing the inductance and increasing the resistance of the grid.

At very high frequencies, the resistance and inductance exchange roles. All of the grid layers exhibit a purely inductive behavior and the current is partitioned among the layers according to the inductance of each layer. The share of the overall current flowing through layer n is

$$i_k = \frac{I_k}{\sum_{n=1}^{N} I_n} = \frac{\prod_{n \neq k} L_n}{\sum_{m=1}^{N} \prod_{n \neq m} L_n}. \tag{13.4}$$

The relation among the currents of each layer is reversed as compared to the low frequency case: $i_1 < i_2 < \ldots < i_N$. The inductance of a multi-layer grid at high frequency L_0^{HF} is determined by the parallel connection of the individual layer inductances,

$$L_0^{HF} = L_1 \| L_2 \| \ldots \| L_N = \frac{\prod_{n=1}^{N} L_n}{\sum_{m=1}^{N} \prod_{n \neq m} L_n}. \tag{13.5}$$

The high frequency resistance of a multi-layer grid R_0^{HF} is

$$R_0^{HF} = R_1 i_1^2 + R_2 i_2^2 + \ldots + R_N i_N^2 \sim R_N, \tag{13.6}$$

due to $R_N > R_k$ and $i_N > i_k$ for any $k \neq N$.

The grid resistance and inductance vary with frequency between these limiting low and high frequency cases. If the difference in the electrical properties of the layers is sufficiently high, the variation of the grid inductance and resistance with frequency has a staircase-like shape, as shown in Fig. 13.5. As the frequency increases, the grid layers consecutively serve as the primary current path, dominating the overall grid impedance within a specific frequency range [291].

13.2 Case study of a two layer grid

The electrical properties of a two layer grid are evaluated in this section to quantitatively illustrate the concepts described in Section 13.1. The grid parameters are described in Fig. 13.6.

The analysis approach used to determine the electrical characteristics of a grid structure is described in Section 13.2.1. Magnetic coupling between grid layers is discussed in Section 13.2.2. The inductive characteristics of a two layer grid are discussed in Section 13.2.3. The resistive characteristics of a two layer grid are discussed in Section 13.2.3. The impedance characteristics of a two layer grid are summarized in Section 13.2.5.

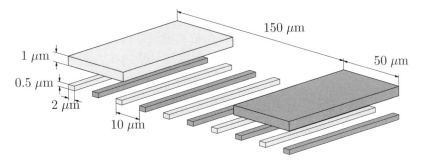

Fig. 13.6. General view of a two layer grid. The ground lines are white colored, the power lines are grey colored.

13.2.1 Simulation setup

The inductance extraction program FastHenry [67] is used to explore the inductive properties of grid structures. FastHenry efficiently calculates the frequency dependent impedance $R(\omega) + \omega L(\omega)$ of complex three-dimensional interconnect structures under a quasi-magnetostatic approximation. In the analysis, the lines are split into multiple filaments to account for skin and proximity effects, as discussed in Section 2.2. A conductivity of $58\,\mathrm{S}/\mu\mathrm{m} \simeq (1.72\,\mu\Omega \cdot \mathrm{cm})^{-1}$ is used in the analysis where an advanced process with copper interconnect is assumed [275].

When determining the loop inductance, all of the ground lines at one end of the grid are short circuited to form a ground terminal and all of the power lines at the same end of the grid are short circuited to form a power terminal. All of the lines at the other end of the grid are short circuited to complete the current loop. This configuration assumes that the power current loop is completed on-chip. This assumption is valid for high frequency signals which are effectively terminated through the on-chip decoupling capacitance which acts as a low impedance termination as compared to the inductive off-chip leads of the package. If the current loop is completed on-chip, the current in the power lines and the current in the ground lines always flow in opposite directions.

13.2.2 Inductive coupling between grid layers

An equivalent circuit diagram of a two layer power distribution grid is shown in Fig. 13.7. The partial mutual inductance between the lines in the two grid layers is significant as compared to the partial self inductance of the lines. Therefore, the two grid layers are, in general, magnetically coupled, as indicated in Fig. 13.7. It can be shown, however, that for practical geometries, magnetic coupling is significant only in interdigitated grids under specific conditions.

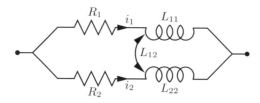

Fig. 13.7. An equivalent circuit diagram of a two layer grid.

The specific conditions are that the line pitch in both layers is the same and the separation between the two layers is smaller than the line pitch. The two layers with the same line pitch are spatially correlated, $i.e.$, the relative position of the lines in the two layers is repeated throughout the structure. The net inductance of such grids depends upon the mutual alignment of the two grid layers. For example, consider a two layer grid with each layer consisting of ten interdigitated power and ground lines with a $1\,\mu\text{m} \times 1\,\mu\text{m}$ cross section on an $8\,\mu\text{m}$ pitch, as in the cross section shown in Fig. 13.8. The separation between the layers is $4\,\mu\text{m}$. The variation of inductance of this two layer grid structure as a function of the physical offset between the two layers is shown in Fig. 13.9.

At $1\,\text{GHz}$, each of the layers has a loop inductance of $206\,\text{pH}$. The inductance of two identical parallel coupled inductors is

$$L_{1\|2} = \frac{L_{11}L_{22} - L_{12}^2}{L_{11} + L_{22} - 2L_{12}} = \frac{L_{11} + L_{12}}{2}. \tag{13.7}$$

In the case of zero coupling between the two grid layers, the net inductance of the two layer grid is approximately $206\,\text{pH}/2 = 103\,\text{pH}$ since the two grids are in parallel. If the ground lines in the top layer are placed immediately over the power lines of the bottom layer, as shown in Fig. 13.8(a), a close return path for the power lines is provided as compared to the neighboring ground lines of the bottom layer. The magnetic coupling between the two grid layers is negative in this case, resulting in a net inductance of $81\,\text{pH}$ for the two layer grid (which is lower than the uncoupled case of $103\,\text{pH}$), in agreement with (13.7). A two layer interdigitated grid effectively becomes a paired grid, as shown

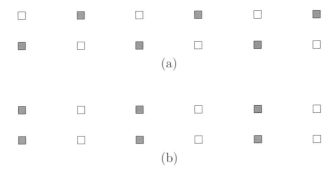

Fig. 13.8. Alignment of two layers with the same line pitch in a two layer grid resulting in the minimum and maximum grid inductance. The ground lines are white colored, the power lines are grey colored. a) The configuration with the minimum grid inductance: ground lines of one layer are aligned with the power lines of the other layer. b) The configuration with the maximum grid inductance: the ground lines of both layers are aligned with each other.

in Fig. 13.8(a). (Rather than equidistant line spacing, in paired grids the lines are placed in close power-ground line pairs [69].) If, alternatively, the ground lines of the top layer are aligned with the ground lines of the bottom layer, as shown in Fig. 13.8(b), the magnetic coupling between the two layers is positive and the net inductance of the two layer grid is 124 pH, higher than the uncoupled case, also in agreement with (13.7). As the offset between the grid layers changes between these two limits, the total inductance varies from a minimum of 81 pH to a maximum of 124 pH, passing a point where the effective coupling between the two layers is zero and the total inductance is 103 pH. The inductance at a 100 GHz signal frequency closely tracks this behavior at low frequencies.

If the layer separation is greater than the line pitch in either of the two layers, the net coupling from the lines in one layer to the lines in the other layer is insignificant. Coupling to the power lines is nearly cancelled by the coupling to the ground lines, carrying current in the opposite direction. This coupling cancellation is analogous to the cancellation of the long distance coupling within the same grid layer [69]. This cancellation also explains why two grid layers are effectively uncoupled if one of the layers is a paired grid. The power to ground line separation in a paired grid is smaller than the separation between two metalization layers with the grid lines in the same direction.

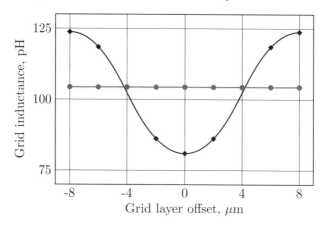

Fig. 13.9. Inductance of a two layer grid versus the physical offset between the two layers. The inductance of the grid with matched line pitch of the layers (black line) depends on the layer offset. (The low inductance alignment shown in Fig. 13.8(a) is chosen as the zero offset.) The inductance of the grid is constant where the line height, width, and pitch of the lower layer are twice as small as compared to the upper layer (the gray line).

It is possible to demonstrate that in the case where the line pitch is not matched, as shown in Fig. 13.10, the layer coupling is effectively cancelled, and the grid inductance is independent of the layer alignment, as shown in Fig. 13.9. Metalization layers in integrated circuits typically are of different thickness, line width, and line spacing. Therefore, unless intentionally designed otherwise, different grid layers typically have different line pitch and can be considered uncoupled, as has been implicitly assumed in Section 13.1.

Fig. 13.10. The cross section of a two layer grid with the line pitch of the upper layer a fractional multiple (5/4 in the case shown) of the line pitch in the bottom layer. Both effects illustrated in Fig. 13.8 occur at different locations (circled). The ground lines are white colored, the power lines are grey colored.

13.2.3 Inductive characteristics of a two layer grid

The variation of the sheet inductance with signal frequency in the two layer grid is shown in Fig. 13.11. Note that the inductance of the individual grid layers, also shown in Fig. 13.11, is virtually constant with frequency [69], [277]. The sheet inductance of the upper layer L_1 is $268\,\mathrm{pH}/\square$ at $1\,\mathrm{MHz}$ ($247\,\mathrm{pH}/\square$ at $100\,\mathrm{GHz}$). The sheet inductance of the bottom layer L_2 is $19.6\,\mathrm{pH}/\square$ at $1\,\mathrm{MHz}$ ($19\,\mathrm{pH}/\square$ at $100\,\mathrm{GHz}$). The inductance of the bottom grid layer is approximately fifteen times lower than the inductance of the upper grid layer. This difference in inductance is primarily due to the difference in the line density of the layers. The line density of the bottom layer is fifteen times higher, as determined by the line pitch of the layers ($150/10 = 15$). The inductance of a single line is relatively insensitive to the aspect ratio of the line cross section. The inductance of the two layer grid, however, varies significantly with signal frequency due to current redistribution, as discussed in Section 13.1.

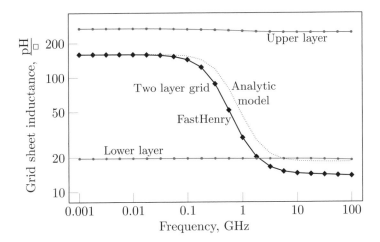

Fig. 13.11. Inductance of a two layer grid versus signal frequency. Both FastHenry data (solid line) and the analytic model data (dotted line) are shown. The individual inductance of the two comprising grid layers is shown for comparison (FastHenry data).

The inductive characteristics of a two layer grid can also be analytically determined based on the simple model shown in Fig. 13.7. Assuming $L_{12} = 0$ as discussed in Section 13.2.2, the loop inductance of a two layer grid is

$$L_0 = \frac{L_1(R_2^2 + \omega^2 L_1 L_2) + L_2(R_1^2 + \omega^2 L_1 L_2)}{(R_1 + R_2)^2 + \omega^2 (L_1 + L_2)^2}. \tag{13.8}$$

At high frequencies, where the resistance of the grid layers has no influence on the current distribution between the layers, the grid inductance described by (13.8) asymptotically approaches the inductance of two ideal parallel inductors,

$$L_0^{HF} = \frac{L_1 L_2}{L_1 + L_2}, \tag{13.9}$$

in agreement with (13.5). At low frequencies, the grid inductance described by (13.8) approaches the low frequency limit of the grid inductance,

$$L_0^{LF} = L_1 \left(\frac{R_2}{R_1 + R_2}\right)^2 + L_2 \left(\frac{R_1}{R_1 + R_2}\right)^2 = 160\,\text{pH}/\square, \tag{13.10}$$

in agreement with (13.3).

The variation of the grid inductance with frequency according to the analytic model described by (13.8) is also illustrated in Fig. 13.11 by the dotted line. The analytic model satisfactorily describes the variation of grid inductance with frequency. The discrepancy between the analytic and FastHenry data at high frequencies is due to proximity effects which are not captured by the model shown in Fig. 13.7.

13.2.4 Resistive characteristics of a two layer grid

The resistance of the two individual grid layers R_1 and R_2 and the resistance of the combined two layer grid R_0 are shown in Fig. 13.12. The resistance of the individual grid layers remains constant up to high frequencies. The resistance of the upper layer begins to moderately increase from approximately 0.5 GHz due to significant proximity effects in very wide lines. The resistance of both layers sharply increases above approximately 20 GHz due to significant skin effect. Note that the resistance of the grid comprised of the two layers exhibits significantly greater variation with frequency than either individual layer.

Similar to the grid inductance, the resistive characteristics of a two layer grid can be analytically determined from the properties of the comprising grid layers,

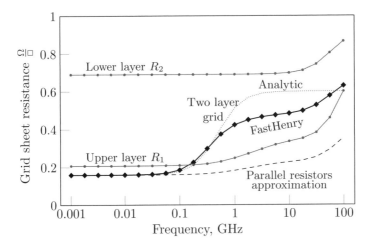

Fig. 13.12. Resistance of a two layer grid versus signal frequency. The individual resistance of the two comprising grid layers and the parallel resistance of the individual layer resistances are shown for comparison.

$$R_0 = \frac{R_1(R_1R_2 + \omega^2 L_2^2) + R_2(R_1R_2 + \omega^2 L_1^2)}{(R_1 + R_2)^2 + \omega^2(L_1 + L_2)^2}. \qquad (13.11)$$

The grid resistance versus frequency data based on the analytic model described by (13.11) is shown by the dotted line in Fig. 13.12. The analytic solution describes well the general character of the resistance variation with frequency. At low frequencies, the resistance of the two layer grid approaches the parallel resistance of two grid layers,

$$R_0^{LF} = R_1 \| R_2 = \frac{R_1 R_2}{R_1 + R_2} = 0.16\,\Omega/\square, \qquad (13.12)$$

in agreement with (13.2). The high frequency grid resistance asymptotically approaches

$$R_0^{HF} = R_1 \left(\frac{L_2}{L_1 + L_2}\right)^2 + R_2 \left(\frac{L_1}{L_1 + L_2}\right)^2 = 0.6\,\Omega/\square, \qquad (13.13)$$

in agreement with (13.6). This analytically calculated high frequency resistance overestimates the FastHenry extracted resistance of $0.48\,\Omega/\square$ (at 10 GHz). The discrepancy is due to pronounced proximity and skin effects at high frequencies.

13.2.5 Variation of impedance with frequency in a two layer grid

Having determined the variation with frequency of the resistance and inductance in the previous sections, it is possible to characterize the frequency dependent impedance characteristics of a two layer grid. The magnitude of the impedance calculated from the analytic models (13.8) and (13.11) is shown in Fig. 13.13 by the dotted line. Low frequency values of the individual layer inductance, L_1 and L_2, and resistance, R_1 and R_2, are used in the analytic model. The impedance magnitude based on FastHenry extracted data is shown by the solid line. The extracted impedance of the individual grid layers is also shown for comparison. Note that the impedance characteristics of the individual layers shown in Fig. 13.13 bear close resemblance to the schematic graph shown in Fig. 13.3.

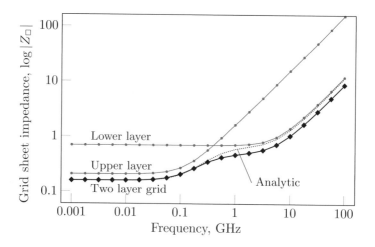

Fig. 13.13. Impedance magnitude of a two layer grid versus signal frequency. Both the extracted (solid line) and analytic (dotted line) data are shown. The impedance of the two comprising grid layers is also shown.

As discussed in Section 13.1, the low resistance upper grid dominates the impedance characteristics at low frequencies, while the low inductance lower grid determines the impedance characteristics at high frequencies [291]. The analytic model satisfactorily describes the frequency dependent impedance characteristics. The discrepancy between the analytic and extracted data at high frequencies is due to overestimation of the high frequency inductance by the analytic model, as shown in Fig. 13.11.

13.3 Design implications

The variation with frequency of the electrical properties of a multi-layer grid has several design implications. Modeling the resistance of a multi-layer grid as a parallel connection of individual layer resistances underestimates the high frequency resistance of the grid. The parallel resistance model, therefore, underestimates the resistive IR voltage drops during fast current transients. Representing a multi-layer grid inductance by the individual layer inductances connected in parallel is accurate only at very high frequencies. At lower frequencies, this model underestimates the grid inductance. Relatively low inductance and high resistance at high frequencies increase the damping factor of the power distribution grid (proportional to $R/\sqrt{L}$), thereby preventing resonant oscillations in power distribution networks at high frequencies. Conversely, resonant oscillations are more likely at lower frequencies, where the inductance is relatively high and the resistance is low.

Multi-layer grids with different grid layer impedance characteristics are well suited to distribute power in high speed integrated circuits. At low frequencies, where the grid impedance is dominated by the resistance, most of the current flows through the less resistive upper grid layers, decreasing the grid impedance. At high frequencies, where the grid impedance is dominated by the inductance, most of the current flows through the low inductance lower layers. Over the entire frequency range, the current flow changes so as to minimize the impedance of the grid. These properties of multi-layer power distribution grids support the design of power distribution networks with low impedance across a wide frequency range, necessary in high performance nanoscale integrated circuits.

The inductive properties of the interconnect changes the metal allocation strategy for global power distribution grids. In circuits based on resistance-only models, all of the metal area for the global power distribution is allocated in the upper layers with the lowest line resistance. The power interconnect in the lower metal layers connects the circuits to the global power grid and typically do not form continuous power grids. In multi-layer grids, however, significant metal resources are required to form continuous grids in the lower metal layers. This difference is a direct consequence of the inductive behavior of interconnect at high frequencies. A significant fraction of the high density lower metal layers should be used to lower the high frequency impedance of the power grid. In this manner, the frequency range of the grid impedance is extended to match the increased switching speeds of scaled transistors.

Redistribution of the grid current toward the lower layers at high frequencies increases the current density in the power and ground lines in the lower grid layers, degrading the electromigration reliability of the power distribution grid. The significance of these effects will increase as the frequency of the current delivered through the on-chip power distribution grid increases with higher operating speeds. An analysis of these effects is therefore necessary to ensure the integrity of high speed nanoscale integrated circuits.

13.4 Summary

The electrical characteristics of multi-layer power distribution grids are investigated in this chapter. The primary results are summarized as follows.

- The upper metal layers comprised of thicker and wider lines have low resistance and high inductance; the lower metal layers comprised of thinner and narrower lines have relatively high resistance and low inductance

- Inductive coupling between grid layers is shown to be insignificant in typical power distribution grids

- Due to this difference in electrical properties, the impedance characteristics of multi-layer grids vary significantly with frequency

- The current distribution among the grid layers changes with frequency, minimizing the overall impedance of the power grid

- As signal frequencies increase, the majority of the current flow shifts from the lower resistance upper layers to the lower inductance lower layers

- The inductance of a multi-layer grid decreases with frequency, while the resistance increases with frequency

- An analytic model describing the electrical properties of a multi-layer grid based on the inductive and resistive properties of the comprising grid layers is described

- A dense and continuous power distribution grid in the lower metal layers is essential to reduce the impedance of a power distribution grid operating at high frequencies

14

Multiple On-Chip Power Supply Systems

With recent developments in nanometer CMOS technologies, excessive power dissipation has become a limiting factor in integrating a greater number of transistors onto a single monolithic substrate. With the introduction of systems-on-chip and systems-in-package (SiP) technologies, the problem of heat removal has further worsened. Unless power consumption is dramatically reduced, packaging and performance of ultra large scale integration (ULSI) circuits will become fundamentally limited by heat dissipation.

Another driving factor behind the push for low power circuits is the growing market for portable electronic devices, such as PDAs, wireless communications, and imaging systems that demand high speed computation and complex functionality while dissipating as little power as possible [293]. Design techniques and methodologies for reducing the power consumed by an IC while providing high speed and high complexity systems are therefore required. These design technologies will support the continued scaling of the minimum feature size, permitting the integration of a greater number of transistors onto a single monolithic substrate.

The most effective way to reduce power consumption is to lower the supply voltage. Dynamic power currently dominates the total power dissipation, quadratically decreasing with supply voltage [294]. Reducing the supply voltage, however, increases the circuit delay. Chandrakasan *et al.* demonstrated in [295] that the increased delay can be compensated by shortening the critical paths using behavioral transformations such as parallelization and pipelining. The resulting circuit consumes less average power while satisfying global throughput constraints; albeit, at the cost of increased circuit area [296].

Power consumption can also be reduced by scaling the threshold voltage while simultaneously reducing the power supply [297]. This approach, however, results in significantly increased standby leakage current. To limit the leakage current during sleep mode, several techniques have been proposed, such as multi-threshold voltage CMOS [298], [299], variable threshold voltage schemes [300], [301], and circuits with an additional transistor behaving as a sleep switch [302]. These techniques, however, require additional process steps and/or additional circuitry to control the substrate bias or switch off portions of the circuit [301].

The total power dissipation can also be reduced by utilizing multiple power supply voltages [299], [303], [304]. In this scheme, a reduced voltage V_{dd}^L is applied to the non-critical paths, while a higher voltage V_{dd}^H is provided to the critical paths so as to achieve the specified delay constraints [299]. Multi-voltage schemes result in reduced total power without degrading the overall circuit performance. Multiple on-chip power supply systems are the subject of this chapter. Various circuit techniques exploiting multiple power supply voltages are presented in Section 14.1. Challenges to ICs with multiple supply voltages are discussed in Section 14.2. Choosing the optimum number and magnitude of the multi-voltage power supplies is discussed in Section 14.3. Some conclusions are offered in Section 14.4.

14.1 ICs with multiple power supply voltages

The strategy of exploiting multiple power supply voltages consists of two steps. Those logic gates with excessive slack (the difference between the required time and the arrival time of a signal) is first determined. A reduced supply voltage V_{dd}^L is provided to those gates to reduce power. Note that in most practical applications, the number of critical paths is only a small portion of the total number of paths in a circuit. Excess slack therefore exists in the majority of paths within a circuit. Determining those gates with excessive time slack is therefore an important and complex task [299]. A variety of computer-aided design (CAD) algorithms and tools have been developed to evaluate the delay characteristics of high complexity ICs such as microprocessors [305], [306]. Multi-voltage low power techniques are reviewed in this section. A low power technique with multiple power supply voltages is presented in Section 14.1.1. Clustered voltage scaling (CVS) is presented in

Section 14.1.2. Extended clustered voltage scaling (ECVS) is discussed in Section 14.1.3.

14.1.1 Multiple power supply voltage techniques

A critical delay path between flip flops FF_1 and FF_2 in a single supply voltage, synchronous circuit is shown in Fig. 14.1. Since the excessive slack remains in those paths located off the critical path, timing constraints are satisfied if the gates in the non-critical paths use a reduced supply voltage V_{dd}^L. A dual supply voltage circuit in which the original power supply voltage V_{dd}^H of each of the gates along the non-critical delay paths is replaced by a lower supply voltage V_{dd}^L is illustrated in Fig. 14.2. If a low voltage supply is available, the gates with V_{dd}^L can be selected to reduce the overall power using conventional algorithms such as gate resizing [307].

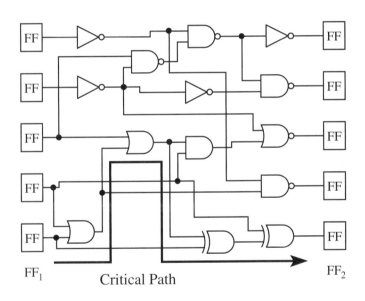

Fig. 14.1. An example single supply voltage circuit.

A circuit with multiple power supply voltages, however, can result in DC current flowing in a high voltage gate due to the direct connection between a low voltage gate and a high voltage gate. If a gate with a

reduced supply voltage is directly connected to a gate with the original
supply voltage, the "high" level voltage at node A is not sufficiently
high to turn off the PMOS device in a CMOS circuit, as shown in
Fig. 14.3. The PMOS device in the high voltage gate is therefore weakly
"ON," conducting static current from the power supply to ground.
These static currents significantly increase the overall power consumed
by an IC, wasting the savings in power achieved by utilizing a multi-
voltage power distribution system.

Level converters are typically inserted at node A to remove the
static current path [308]. A simple level converter circuit is illustrated in
Fig. 14.4. The level converter restores the full voltage swing from V_{dd}^L to
V_{dd}^H. Note that a great number of level converters is typically required,
increasing the area and power overhead. The problem of utilizing a dual
power supply voltage scheme is formulated as follows.

Problem formulation: For a given circuit, determine the gates and
registers to which a reduced power supply voltage V_{dd}^L should be applied such that the overall power and number of level converters are
minimized while satisfying system-level timing constraints [309].

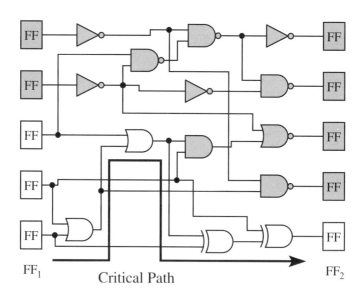

Fig. 14.2. An example dual supply voltage circuit. The gates operating at a lower
power supply voltage V_{dd}^L (located off the critical delay path) are shaded.

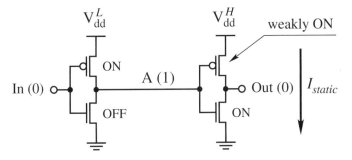

Fig. 14.3. Static current as a result of a direct connection between the V_{dd}^L gate and the V_{dd}^H gate.

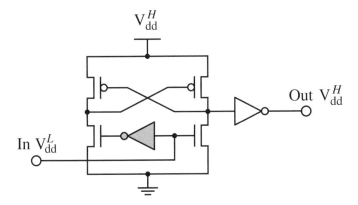

Fig. 14.4. Level converter circuit. The inverter operating at the reduced power supply voltage V_{dd}^L is shown in grey.

14.1.2 Clustered voltage scaling (CVS)

The number of level converters can be reduced by minimizing the connections between the V_{dd}^L gates and the V_{dd}^H gates. The CVS technique, proposed in [310], results in a circuit structure with a greatly reduced number of level converters, as shown in Fig. 14.5.

To avoid inserting level converters, the CVS technique exploits the specific connectivity patterns among the gates, such as a connection between V_{dd}^H gates, between V_{dd}^L gates, and between a V_{dd}^H gate and a V_{dd}^L gate. These connections do not require level converters to remove any static current paths. Level converters are only required at the interface between the output of a V_{dd}^L gate and the input of a V_{dd}^H gate. The number of required level converters in the CVS structure shown in

Fig. 14.5 is almost the same as the number of V_{dd}^L flip flops. The CVS technique therefore results in fewer level converters, reducing the overall power consumed by an integrated circuit.

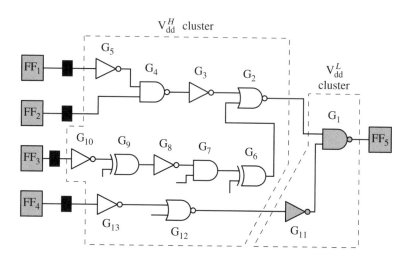

Fig. 14.5. A dual power supply voltage circuit with the clustered voltage scaling (CVS) technique [310]. The gates operating at a lower supply voltage are shaded. The level converters are shown as black rectangles.

14.1.3 Extended clustered voltage scaling (ECVS)

The number of gates with a lower power supply voltage can be increased by optimally choosing the insertion points of the level converters, further reducing overall power. As an example, in the CVS structure shown in Fig. 14.5, the path delay from flip flop FF_3 to gate G_2 is longer than the delay from FF_1 to G_2. Moreover, applying a lower power supply to gate G_2 can produce a timing violation. A high power supply should therefore be provided to G_2. From CVS connectivity patterns described in Section 14.1.2, note that G_3 also has to be supplied with V_{dd}^H. Alternatively, in a CVS structure, G_3 cannot be supplied with V_{dd}^L although excessive slack remains in the path from FF_1 to G_2. Similarly, G_4 and G_5 should be connected to V_{dd}^H to satisfy existing timing constraints. If the insertion point of the level converter adjacent to FF_1 is moved to the interface between G_3 and G_2, gates G_3, G_4, and G_5 can be connected to V_{dd}^L, as illustrated in Fig. 14.6. Note that the structure shown in Fig. 14.6 is obtained from the CVS network by relaxing existing limitations on the insertion positions of the level converters. Such a technique is often referred to as the extended clustered voltage scaling technique [309], [311].

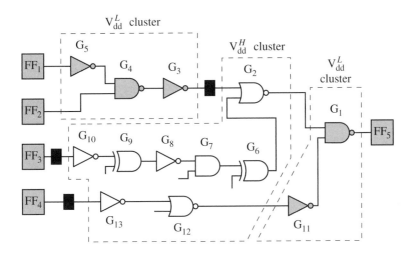

Fig. 14.6. A dual power supply voltage circuit with the extended clustered voltage scaling (ECVS) technique [309]. The gates operating at a lower supply voltage are shaded. The level converters are shown as black rectangles.

14.2 Challenges in ICs with multiple power supply voltages

The application of power reduction techniques with multiple supply voltages in modern high performance ICs is a challenging task. Circuit scheduling algorithms require complex computations, limiting the application of CVS and ECVS techniques to specific paths within an IC. Primary challenges of multi-voltage power reduction schemes are discussed in this section. The issues of area overhead and related trade-offs are introduced in Section 14.2.1. Power penalties are presented in Section 14.2.2. The additional design complexity associated with level converters and integrated DC–DC voltage converters is discussed in

Section 14.2.3. Several placement and routing strategies are described in Section 14.2.4.

14.2.1 Die area

As described in Section 14.1, level converter circuits are inserted at the interface between specific gates in power reduction schemes with multiple power supply voltages to reduce static current. Multi-voltage circuits require additional power connections, significantly increasing routing complexity and die area. Additional area results in greater parasitic capacitance of the signal lines, increasing the dynamic power consumed by an IC. As a result of the increased area, the time slack in the critical paths is often significantly smaller, reducing the power savings of a multi-voltage scheme. A tradeoff therefore exists between the power savings and area overhead in ICs with multiple power supply voltages. The critical paths should therefore be carefully determined in order to reduce the overall circuit power.

14.2.2 Power dissipation

Multi-voltage low power techniques require the insertion of level converters to reduce static current. The number of level converters depends upon the connectivity patterns at the interface between each critical and non-critical path. Improper scheduling of the critical paths can lead to an excessive number of level converters, increasing the power. The ECVS technique with relaxed constraints for level converters should therefore be used, resulting in a smaller number of level converters.

Note that the magnitude of the overall reduction in power is determined by the number and voltage of the available power supply voltages, as discussed in Section 14.3. It is therefore important to determine the optimum number and magnitude of the power supply voltages to maximize any savings in power. Also note that lower power supply voltages are often generated on-chip from a high voltage power supply using DC–DC voltage converters [312], [313]. The power and area penalties of the on-chip DC–DC voltage converters should therefore be considered to accurately estimate any savings in power.

Several primary factors, such as physical area, the number and magnitude of the power supply voltages, and the number of level converters contribute to the overall power overhead of any multi-voltage low power technique. Complex multi-variable optimization is thus required to determine the proper system parameters in order to achieve the greatest reduction in overall power [314].

14.2.3 Design complexity

Note that while significantly reducing power, a multiple power supply voltage scheme results in significantly increased design complexity. The complexity overhead of a multi-voltage low power technique is due to two aspects. The level converters not only dissipate power, but also dramatically increase the complexity of the overall design process. A level converter typically consists of both low voltage and high voltage gates, increasing the area and routing resources. Multiple level converters also increase the delay of the critical paths. High speed, low power level converters are therefore required to achieve a significant reduction in overall power while satisfying existing timing constraints [308], [315]. Standard logic gates with embedded level conversion as reported in [315] support the design of circuits without the addition of level converters, substantially reducing power, area, and complexity.

Monolithic DC – DC voltage converters are often integrated on-chip to enhance overall energy efficiency, improve the quality of the voltage regulation, decrease the number of I/O pads dedicated to power delivery, and reduce fabrication costs [316]. To lower the energy dissipated by the parasitic impedance of the circuit board interconnect, the passive components of a low frequency filter (e.g., the filter inductor and filter capacitor) are also placed on-chip, significantly increasing both the required area and design complexity. A great amount of on-chip decoupling capacitance is also often required to improve the quality of the on-chip power supply voltages [317]. The area and power penalty as well as the increased design complexity of the additional on-chip voltage converters should therefore be considered when determining the optimal number and magnitude of the multiple power supply voltages.

14.2.4 Placement and routing

To achieve the full benefit offered by multiple power supply voltage techniques, various design issues at both the high level and physical level should be simultaneously considered. Existing electronic design automation (EDA) placement and routing tools for conventional circuits with single power supply voltages, however, cannot be directly applied to low power techniques with multiple power supply voltages. Specific CAD tools, capable of placing and routing physical circuits with multiple power supplies based on high level gate assignment information, are therefore required. The placement and routing of ICs with multiple power supply voltages is a complex problem. Three widely utilized layout schemes are described in this section.

Area-by-area architecture

The simplest architecture for a circuit with dual power supply voltages is an area-by-area architecture [309], as shown in Fig. 14.7. In this architecture, the V_{dd}^L cells are placed in one area, while the V_{dd}^H cells are placed in a different area. The area-by-area technique iteratively generates a layout with existing placement and routing tools using one of the available power supply voltages. This architecture, however, results in a degradation in performance due to the substantially increased interconnect length between the V_{dd}^L and V_{dd}^H cells.

Row-by-row architecture

The layout architecture proposed in [318] is illustrated in Fig. 14.8. In this architecture, the V_{dd}^L cells and V_{dd}^H cells are placed in different rows. Each row only consists of V_{dd}^L cells *or* V_{dd}^H cells. This layout technique is therefore a row-by-row architecture. Note that in this architecture, a V_{dd}^L row is placed next to a V_{dd}^H row, reducing the interconnect length between the V_{dd}^L cells and the V_{dd}^H cells. The performance of a row-by-row layout architecture is therefore higher as compared to the performance of an area-by-area architecture. The row-by-row technique also results in smaller area, further improving system performance. Another advantage of this technique is that an original V_{dd}^H cell library can be used for the V_{dd}^L cells. Since the layout of the V_{dd}^L cells are the same as those of the V_{dd}^H cells, the original layout of the V_{dd}^H cells can be treated as V_{dd}^L cells. A lower power supply voltage can be provided to the V_{dd}^L cells.

In-row architecture

An improved row-by-row layout architecture is presented in [319]. This architecture is based on a modified cell library [319]. Unlike conventional standard cells, the new standard cell has two power rails and

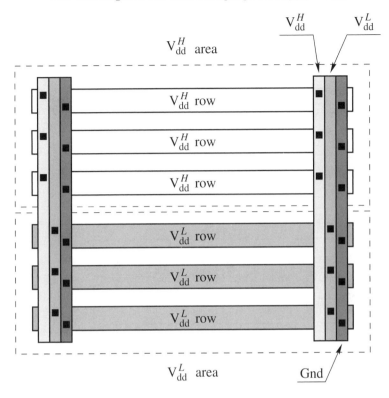

Fig. 14.7. Layout of an area-by-area architecture with a dual power supply voltage. In this architecture, the V_{dd}^L cells are placed in one area, while the V_{dd}^H cells are separately placed in a different area.

one ground rail. One of the power rails is connected to V_{dd}^L and the other power rail is connected to V_{dd}^H. The modified library supports the allocation of both V_{dd}^L cells and V_{dd}^H cells within the same row, as shown in Fig. 14.9. This layout scheme is therefore referred to as an in-row architecture. Note that the width of the power and ground lines in each cell is reduced, slightly increasing the overall area (a 2.7% area overhead as compared to the original cell) [319]. Since the number of V_{dd}^L cells is typically greater than the number of V_{dd}^H cells, the lower power supply provides higher current. The low voltage power rail is therefore wider than the high voltage power rail to maintain a similar voltage drop within each power rail. Note that the in-row architecture results in a significant reduction in the interconnect length between the V_{dd}^L and V_{dd}^H cells, as compared to a row-by-row scheme [319]. An

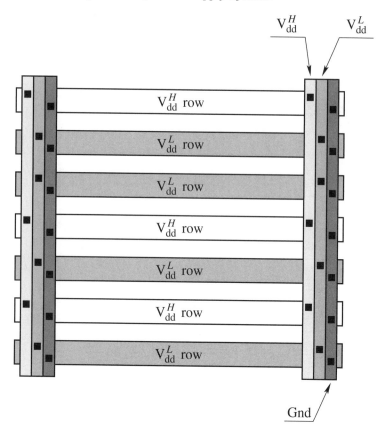

Fig. 14.8. Layout of a row-by-row architecture with a dual power supply voltage. In this architecture, the V_{dd}^L cells and V_{dd}^H cells are placed in different rows. Each row consists of only V_{dd}^L cells *or* V_{dd}^H cells.

in-row layout scheme should therefore be utilized in high performance, high complexity ICs to reduce overall power with minimal area and complexity penalties.

14.3 Optimum number and magnitude of available power supply voltages

In low power techniques with multiple power supply voltages, the power reduction is primarily determined by the number and magnitude of the available power supply voltages. The trend in power reduction with a

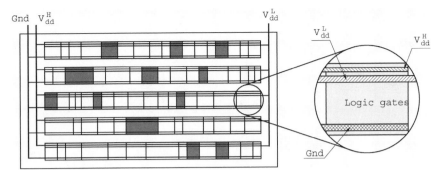

Fig. 14.9. In-row dual power supply voltage scheme. This architecture is based on a modified cell library with two power rails and one ground rail in each cell. The V_{dd}^H cells are shown in grey and the V_{dd}^L cells are white.

multi-voltage scheme as a function of the number of available supply voltages is illustrated in Fig. 14.10. Observe from Fig. 14.10 that if fewer power supplies than the optimum number are available ($n < n_{opt}$), the savings in power can be fairly small. The maximum power savings is achieved with the number of supply voltages close to the optimum number (represented by region $n = n_{opt}$ in Fig. 14.10). If more than the optimum number of power supplies are used, the savings in power becomes smaller, as depicted in Fig. 14.10 for $n > n_{opt}$. This decline in power reduction when the number of supply voltages is greater than the optimum number is due to the increased overhead of the additional power supplies (as a result of the increased area, number of level converters, and design complexity). Any savings in power is also constrained by the magnitude of the available power supplies. A tradeoff therefore exists between the number and magnitude of the available power supplies and the achievable power savings. A methodology is therefore required to estimate the optimum number and magnitude of the available power supply voltages in order to produce the greatest reduction in power. Design techniques for determining the optimum number and magnitude of the available power supplies are the subject of this section.

In systems with multiple power supply voltages (where $V_1 > V_2 > \cdots > V_n$), the power dissipation is [320]

$$P_n = f\left\{\left(C_1 - \sum_{i=2}^{n} C_i\right)V_1^2 + \sum_{i=2}^{n} C_i V_i^2\right\}, \qquad (14.1)$$

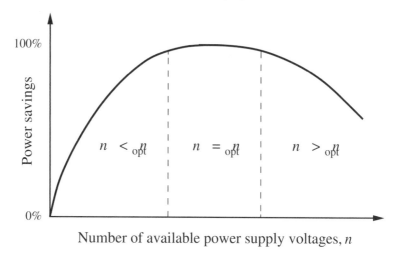

Fig. 14.10. Trend in power reduction with multi-voltage scheme as a function of the number of available supply voltages.

where C_i is the total capacitance of the logic gates and interconnects operating at a reduced supply voltage V_i and f is the operating frequency. The ratio of the power dissipated by a system with multiple power supply voltages as compared to the power dissipation in a single power supply system is

$$K_{V_{dd}} \equiv \frac{P_n}{P_1} = 1 - \sum_{i=2}^{n} \left[\left(\frac{C_i}{C_1} \right) \left\{ 1 - \left(\frac{V_i}{V_1} \right)^2 \right\} \right]. \qquad (14.2)$$

Since delay is proportional to the total capacitance, $\dfrac{C_i}{C_1}$ is

$$\frac{C_i}{C_1} = \frac{\int\limits_0^1 p(t)\, t_i\, dt}{\int\limits_0^1 p(t)\, t\, dt}, \qquad (14.3)$$

where $p(t)$ is the normalized path delay distribution function and t_i is the total delay of the circuits operating at V_i. For a path with a total delay $t_{i,0} < t < t_{i-1,0}$, where $t_{i,0}$ denotes the path delay at V_1 (equal to the cycle time when all of the circuits operate at V_i), the power dissipation is minimum when (V_i, V_{i-1}) are applied. In this case, t_i is

$$t_i = \begin{cases} \dfrac{t_{i,0}}{t_{i,0} - t_{i+1,0}}(t - t_{i+1,0}) & : \quad t_{i+1,0} \leq t \leq t_{i,0} \\[2ex] \dfrac{t_{i,0}}{t_{i-1,0} - t_{i,0}}(t_{i-1,0} - t) & : \quad t_{i,0} \leq t \leq t_{i-1,0}, \end{cases} \qquad (14.4)$$

where $t_{i,0}$ is

$$t_{i,0} = \left(\frac{V_1}{V_i}\right)\left(\frac{V_i - V_{\text{th}}}{V_1 - V_{\text{th}}}\right)^{\alpha}, \qquad (14.5)$$

V_{th} is the threshold voltage, and α is the velocity saturation index [321]. Note that $t_{n+1,0} = 0$. $K_{V_{\text{dd}}}$ can be determined from (14.1) – (14.5) for a specific $p(t)$, V_1, V_i, and V_{th}.

For a lambda-shaped normalized path delay distribution function $p(t)$ (see Fig. 14.11) as determined from post-layout static timing analysis, approximate rules of thumb for determining the optimum magnitude of the power supply voltages have been determined by Hamada *et al.* [320],

$$\text{for } \{V_1, V_2\} \qquad \frac{V_2}{V_1} = 0.5 + 0.5\frac{V_{\text{th}}}{V_1}, \qquad (14.6)$$

$$\text{for } \{V_1, V_2, V_3\} \qquad \frac{V_2}{V_1} = \frac{V_3}{V_2} = 0.6 + 0.4\frac{V_{\text{th}}}{V_1}, \qquad (14.7)$$

$$\text{for } \{V_1, V_2, V_3, V_4\} \qquad \frac{V_2}{V_1} = \frac{V_3}{V_2} = \frac{V_4}{V_3} = 0.7 + 0.3\frac{V_{\text{th}}}{V_1}. \qquad (14.8)$$

Criteria (14.6) – (14.8) can be used to determine the magnitude of each power supply voltage based on the total number of available power supply voltages. Note that these rules of thumb result in the optimum power supply voltages where the maximum difference in power reduction is less than 1% as compared to the absolute minimum (as determined from an analytic solution of the system of equations).

Note again that if a greater number of power supplies is used, the total power can be further reduced, reaching a constant power level at some number of power supplies (see Fig. 14.10). As determined in [320], up to three power supply voltages should be utilized to reduce the power consumed by an IC. The reduction in power diminishes as the power supply voltage is scaled and $\frac{V_{\text{th}}}{V_{\text{dd}}}$ increases.

A rule of thumb for two power supply voltages has been evaluated by simulations in [309]. For $V_{\text{dd}}^H = 3.3$ volts, a V_{dd}^L of 1.9 volts is estimated, exhibiting good agreement with (14.6). The dependence of the total power of a dual power supply media processor as a function of the lower

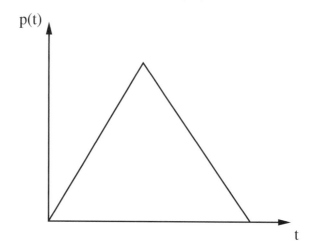

p(t)

t

Fig. 14.11. A lambda-shaped normalized path delay distribution function.

power supply V_{dd}^L is depicted in Fig. 14.12. Observe from Fig. 14.12 that
the minimum overall power is achieved at $V_{dd}^L = 1.9$ volts.

The minimum overall power of a dual power supply system can
be explained as follows. In a dual power supply system, the power
reduction is determined by two factors: the reduction in power of a
single logic gate due to scaling the power supply voltage from V_{dd}^H to
V_{dd}^L, and the number of original V_{dd}^H gates replaced with V_{dd}^L gates.
At lower V_{dd}^L, the power dissipated by a V_{dd}^L gate decreases, while the
number of original V_{dd}^H gates replaced with V_{dd}^L gates is reduced. This
behavior is due to the degradation in performance of the V_{dd}^L gates
at a lower V_{dd}^L. As a result, fewer gates can be replaced with lower
voltage gates without violating existing timing constraints. Conversely,
at a higher V_{dd}^L, the number of gates replaced with V_{dd}^L gates increases,
while the reduced power in a single V_{dd}^L gate decreases. The overall
power therefore has a minimum at a specific V_{dd}^L voltage, as shown in
Fig. 14.12.

Low power techniques with multiple power supply voltages and
a single fixed threshold voltage have been discussed in this chapter.
Enhanced results are achieved by simultaneously scaling the multiple
threshold voltages and the power supply voltages [299], [322], [323].
This approach results in reduced total power with low leakage currents.
The total power can also be lowered by simultaneously assigning threshold voltages during gate sizing. Nguyen *et al.* [324] demonstrated power

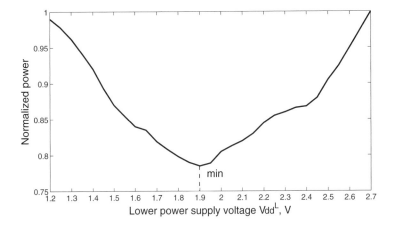

Fig. 14.12. Dependence of the total power of a dual power supply system on a lower power supply voltage V_{dd}^L [309]. The original high power supply voltage $V_{dd}^H = 3.3$ volts.

reductions approaching 32% on average (57% maximum) for the IS-CAS85 benchmark circuits. CVS with variable supply voltage schemes has been presented in [325]. In this scheme, the power supply voltage is gradually scaled based on an accurate model of the critical path delay. Up to a 70% power savings has been achieved as compared to the same circuit without these low power techniques. In [326], a column-based dynamic power supply has been integrated into a high frequency SRAM circuit. The power supply voltage is adaptively changed based on the read/write mode of the SRAM, reducing the total power.

As described in this chapter, power dissipation has become a major factor, limiting the performance of high complexity ICs. Multiple low power techniques should therefore be utilized to achieve significant power savings in modern nanoscale ICs.

14.4 Summary

The discussion of multiple on-chip power supply systems and different low power design techniques can be summarized as follows:

- The total power consumed by an IC can be reduced by utilizing multiple power supply voltages

- In multi-voltage low power techniques, a lower power supply voltage is applied to those logic gates with excessive slack to reduce power consumption

- In a multi-voltage scheme, the gates and flip flops with a lower power supply voltage should be determined such that the overall power and number of level converters are minimized while satisfying existing timing constraints

- CVS and ECVS techniques exploit specific connectivity patterns, reducing the number of level converters

- Various penalties, such as area, power, and design complexity, should be considered during the system design process so as to maximize the savings in power

- The in-row layout scheme reduces overall power with minimum area and design complexity

- A maximum of two or three supply voltages should be employed in low power applications

- Rules of thumb have been described for determining the optimum magnitude of the multiple power supply voltages

- A greater savings in power can be achieved by simultaneously scaling the multiple threshold voltages and power supply voltages

15

On-Chip Power Distribution Grids
with Multiple Supply Voltages

With the on-going miniaturization of integrated circuit feature size, the design of power and ground distribution networks has become a challenging task. With technology scaling, the requirements placed on the on-chip power distribution system have significantly increased. These challenges arise from shorter rise/fall times, lower noise margins, higher currents, and increased current densities. Furthermore, the power supply voltage has decreased to lower dynamic power dissipation. A greater number of transistors increases the total current drawn from the power supply. Simultaneously, the higher switching speed of a greater number of smaller transistors produces faster and larger current transients in the power distribution network [22]. The higher currents produce large IR voltage drops. Fast current transients lead to large $L\frac{dI}{dt}$ inductive voltage drops (ΔI noise) within the power distribution networks.

The lower voltage of the power supply level can be described as

$$V_{load} = V_{\text{dd}} - IR - L\frac{dI}{dt}, \tag{15.1}$$

where V_{load} is the voltage level seen by a current load, V_{dd} is the power supply voltage, I is the current drawn from the power supply, R and L are the resistance and inductance of the power distribution network, respectively, and dt is the rise time of the current drawn by the load. The power distribution networks must be designed to minimize voltage fluctuations, maintaining the power supply voltage as seen from the load within specified design margins (typically $\pm 5\%$ of the power supply level). If the power supply voltage drops too low, the performance (delay) and functionality of the circuit will be severely compromised. Excessive overshoots of the supply voltage can also affect circuit reliability and should therefore be reduced.

With a new era of nanometer scale CMOS circuits, power dissipation has become perhaps the critical design criterion. As described in Chapter 14, to manage the problem of high power dissipation, multiple on-chip power supply voltages have become commonplace [310]. This strategy has the advantage of permitting those modules along the critical paths to operate with the highest available voltage level (in order to satisfy target timing constraints) while permitting modules along the noncritical paths to use a lower voltage (thereby reducing energy consumption). In this manner, the energy consumption is decreased without affecting the circuit speed. This scheme is used to enhance speed in a smaller area as compared to the use of parallel architectures. Using multiple supply voltages for reducing power requirements has been investigated in the area of high level synthesis for low power [306], [327]. While it is possible to provide multiple supply voltages, in practical applications, such a scenario is expensive. Practically, a small number of voltage supplies (two or three) can be effective [299].

Power distribution networks in high performance ICs are commonly structured as a multi-layer grid [28]. In such a grid, straight power/ground lines in each metalization layer can span an entire die and are orthogonal to the lines in adjacent layers. Power and ground lines typically alternate in each layer. Vias connect a power (ground) line to another power (ground) line at the overlap sites. A typical on-chip power grid is illustrated in Fig. 15.1, where three layers of interconnect are depicted with the power lines shown in dark grey and the ground lines shown in light grey.

An on-chip power distribution grid in modern high performance ICs is a complex multi-level system. The design of on-chip power distribution grids with multiple supply voltages is the primary focus of this chapter. The chapter is organized as follows. Existing work on power distribution grids and related power distribution systems with multiple supply voltages is reviewed in Section 15.1. The structure of a power distribution grid and the simulation setup are reviewed in Section 15.2. The structure of a power distribution grid with dual supply voltages and dual grounds (DSDG) is discussed in Section 15.3. Interdigitated power distribution grids with DSDG are described in Section 15.4. Paired power distribution grids with DSDG are analyzed in Section 15.5. Simulation results are presented in Section 15.6. Circuit design implications are discussed in Section 15.7. Some specific conclusions are summarized in Section 15.8.

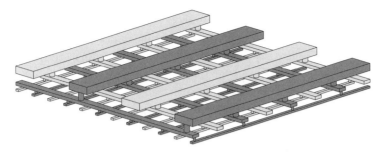

Fig. 15.1. A multi-layer on-chip power distribution grid [328]. The ground lines are light grey, the power lines are dark grey. The signal lines are not shown.

15.1 Background

On-chip power distribution grids have traditionally been analyzed as purely resistive networks [282]. In this early work, a simple model is presented to estimate the maximum on-chip IR drop as a function of the number of metal layers and the metal layer thickness. The optimal thickness of each layer produces the minimum IR drops. Design techniques are provided to maximize the available signal wiring area while maintaining a constant IR drop. These guidelines, however, have limited application to modern, high complexity power distribution networks. The inductive behavior of the on-chip power distribution networks has historically been neglected because the network inductance has been to date dominated by the off-chip parasitic inductance of the package. With the introduction of advanced packaging techniques and the increased switching speed of integrated circuits, this situation has changed. As noted in [208], by replacing wider power and ground lines with narrower interdigitated power and ground lines, the partial self-inductance of the power supply network can be reduced. The authors in [209] propose replacing the wide power and ground lines with an array of interdigitated narrow power and ground lines to decrease the characteristic impedance of the power grid. The dependence of the characteristic impedance on the separation between the metal lines and the metal ground plane is considered. The application of the proposed power delivery scheme, however, is limited to interdigitated structures.

Several design methodologies using multiple power supply voltages have been described in the literature. A row-by-row optimized power supply scheme, providing a different supply voltage to each cell row, is described in [318]. The original circuit is partitioned into two subcircuits by conventional layout methods. Another technique, presented in [319], decreases the total length of the on-chip power and ground lines by applying a multiple supply voltage scheme. A layout architecture exploiting multiple supply voltages in cell-based arrays is described in [309]. Three different layout architectures are analyzed. The authors show that the power consumed by an IC can be reduced, albeit with an increase in area. In previously reported publications, only power distribution systems with two power supply voltages and one common ground have been described. On-chip power distribution grids with multiple power supply voltages and multiple grounds are discussed in this chapter.

15.2 Simulation setup

The inductance extraction program FastHenry [67] is used to analyze the inductive properties of on-chip power grids. FastHenry efficiently calculates the frequency dependent self and mutual impedances, $R(\omega) + \omega L(\omega)$, in complex three-dimensional interconnect structures. A magneto-quasistatic approximation is utilized, meaning the distributed capacitance of the line and any related displacement currents associated with the capacitances are ignored. The accelerated solution algorithm employed in FastHenry provides approximately a 1% worst case accuracy as compared to directly solving the system of linear equations characterizing the system.

Copper is assumed as the interconnect material with a conductivity of $(1.72 \, \mu\Omega \cdot cm)^{-1}$. A line thickness of $1 \, \mu m$ is assumed for each of the lines in the grids. In the analysis, the lines are split into multiple filaments to account for the skin affect. The number of filaments are estimated to be sufficiently large so as to achieve a 1% accuracy. Simulations are performed assuming a 1 GHz signal frequency (modeling the low frequency case) and a 100 GHz signal frequency (modeling the high frequency case). The interconnect structures are composed of interdigitated and paired power and ground lines. Three different types of interdigitated power distribution grids are shown in Fig. 15.2. The total number of lines in each power grid is 24. Each of the lines is incorporated into a specific power distribution network and distributed equally between the power and ground networks. The maximum simulation time is under five minutes on a Sun Blade 100 workstation.

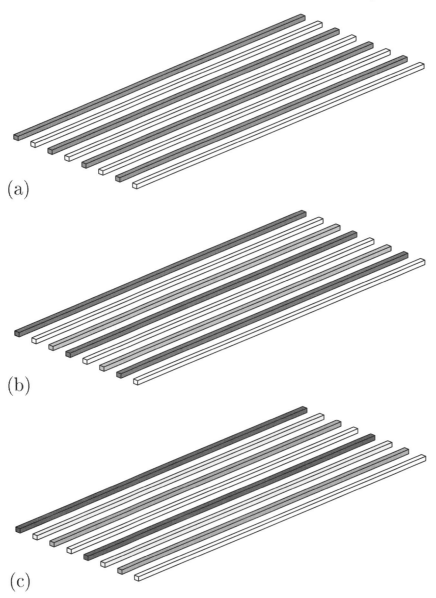

Fig. 15.2. Interdigitated power distribution grids under investigation. In all of the power distribution structures, the power lines are interdigitated with the ground lines. (a) A reference power distribution grid with a single supply voltage and a single ground (SSSG). The power lines are grey colored and the ground lines are white colored, (b) a power distribution grid with DSSG. The power lines are light and dark grey colored and the ground lines are white colored, (c) the power distribution grid with DSDG. The power lines are shown in black and dark grey colors and the ground lines are shown in white and light grey colors.

15.3 Power distribution grid with dual supply and dual ground

Multiple power supply voltages have been widely used in modern high performance ICs, such as microprocessors, to decrease power dissipation. Only power distribution schemes with dual supply voltages and a single ground (DSSG) have been reported in the literature [26], [28], [199], [309], [318], [319]. In such networks, both power supplies share the one common ground. The ground bounce produced by one of the power supplies therefore adds to the power noise in the other power supply. As a result, voltage fluctuations are significantly increased. To address this problem, an on-chip power distribution scheme with DSDG is presented. In this way, the power distribution system consists of two independent power delivery networks.

A power distribution grid with DSDG consists of two separate subnetworks with independent power and ground supply voltages and current loads. No electrical connection exists between the two power delivery subnetworks. In such a structure, the two power distribution systems are only coupled through the mutual inductance of the ground and power paths, as shown in Fig. 15.3.

The loop inductance of the current loop formed by the two parallel paths is

$$L_{loop} = L_{pp} + L_{gg} - 2M, \qquad (15.2)$$

where L_{pp} and L_{gg} are the partial self-inductance of the power and ground paths, respectively, and M is the mutual inductance between these paths. The current in the power and ground lines is assumed to always flow in opposite directions (a reasonable and necessary assumption in large power grids). The inductance of the current loop formed by the power and ground lines is therefore reduced by $2M$. The loop inductance of the power distribution grid can be further reduced by increasing the mutual inductive coupling between the power and ground lines. As described by Rosa in 1908 [43], the mutual inductance between two parallel straight lines of equal length is

$$M_{loop} = 0.2l \left(\ln\frac{2l}{d} - 1 + \frac{d}{l} - \ln\gamma + \ln k \right) \mu\text{H}, \qquad (15.3)$$

where l is the line length, and d is the distance between the line centers. This expression is valid for the case where $l \gg d$. The mutual inductance of two straight lines is a weak function of the distance between the lines [28].

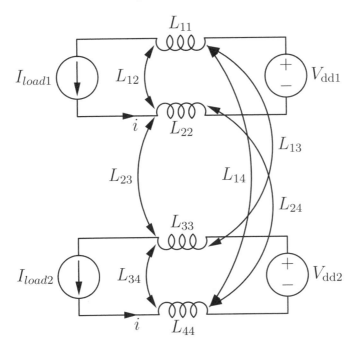

Fig. 15.3. Circuit diagram of the mutual inductive coupling of the DSDG power distribution grid. L_{11} and L_{33} denote the partial self-inductances of the power lines and L_{22} and L_{44} denote the partial self-inductances of the ground lines, respectively.

Analogous to inductive coupling between two parallel loop segments as described in [68], the mutual loop inductance of the two power distribution grids with DSDG is

$$M_{loop} = L_{13} - L_{14} + L_{24} - L_{23}. \tag{15.4}$$

Note that the two negative signs before the mutual inductance components in (15.4) correspond to the current in the power and ground paths flowing in opposite directions. Also note that since the mutual inductance M in (15.2) is negative, M_{loop} should be negative to lower the loop inductance. If M_{loop} is positive, the mutual inductive coupling between the power/ground paths is reduced and the effective loop inductance is therefore increased. If the distance between the lines making a loop is much smaller than the separation between the two loops, $L_{13} \approx L_{14}$ and $L_{23} \approx L_{24}$. This situation is the case for paired power distribution grids. In such grids, the power and ground lines are located in pairs in close proximity. For the interdigitated grid structure shown in Fig. 15.2(c), the distance between the lines d_{12} is the same as an off-set between the two loops d_{23}, as illustrated in Fig. 15.4. In this case, assuming $d_{12} = d_{23} = d$, from (15.3), M_{loop} between the two grids is approximately

$$M_{loop} = 0.2l \ln\frac{3}{4} \ \mu\text{H}. \tag{15.5}$$

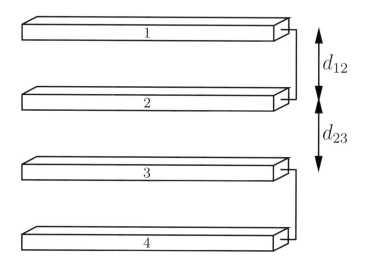

Fig. 15.4. Physical structure of an interdigitated power distribution grid with DSDG. The power delivery scheme consists of two independent power delivery networks.

Thus, M_{loop} between the two grids is negative (with an absolute value greater than zero) in DSDG grids. The loop inductance of the particular power distribution grid, therefore, can be further lowered by $2M$. Conversely, in grids with DSSG, currents in both power paths flow in the same direction. In this case, the resulting partial inductance of the current path formed by the two power paths is

$$L_{\|} = \frac{L_{pp}^1 L_{pp}^2 - M^2}{L_{pp}^1 + L_{pp}^2 - 2M}, \tag{15.6}$$

where L_{pp}^1 and L_{pp}^2 are the partial self-inductance of the two power paths, respectively, and M is the mutual inductance between these paths. The mutual inductance between the two loops is therefore increased. Thus, the loop inductance seen from a particular current load increases, producing larger power/ground $L\frac{dI}{dt}$ voltage fluctuations.

15.4 Interdigitated grids with DSDG

As shown in Section 15.3, by utilizing the power distribution scheme with DSDG, the loop inductance of the particular power delivery network is reduced. In power distribution grids with DSDG, the mutual inductance M between the power and ground paths in (15.2) includes two terms. One term accounts for the increase (or decrease) in the mutual coupling between the power and ground paths in a particular power delivery network due to the presence of the second power delivery network. The other term is the mutual inductance in the loop formed by the power and ground paths of the particular power delivery network. Thus, the mutual inductance in power distribution grids with DSDG is

$$M = M' + M_{loop}, \qquad (15.7)$$

where M' is the mutual inductance in the loop formed by the power and ground lines of the particular power delivery network and M_{loop} is the mutual inductance between the two power delivery networks. M' is always negative. M_{loop} can be either negative or positive.

The loop inductance of a conventional interdigitated power distribution grid with DSSG has recently been compared to the loop inductance of an example interdigitated power distribution grid with DSDG [329]. In general, multiple interdigitated power distribution grids with DSDG can be utilized, satisfying different design constraints in high performance ICs. Exploiting the symmetry between the power supply and ground networks, all of the possible interdigitated power distribution grids with DSDG can be characterized by two primary power delivery schemes. Two types of interdigitated power distribution grids with DSDG are described in this section. The loop inductance in the first type of power distribution grids is presented in Section 15.4.1. The loop inductance in the second type of power distribution grids is discussed in Section 15.4.2.

15.4.1 Type I interdigitated grids with DSDG

In the first type of interdigitated power distribution grid, the power and ground lines in each power delivery network and in different voltage domains (power and ground supply voltages) are alternated and equidistantly spaced, as shown in Fig. 15.5. In such power distribution grids, the distance between the lines inside the loop d_I^i is equal to the separation between the two loops s_I^i. Such power distribution grids are described here as *fully interdigitated* power distribution grids with DSDG.

Fig. 15.5. Physical structure of a fully interdigitated power distribution grid with DSDG. The distance between the lines making the loops d_I^i is equal to the separation between the two loops s_I^i.

Consistent with (15.4), the mutual inductive coupling of two current loops in fully interdigitated grids with DSDG is

$$M_{loop}^{\text{intI}} = L_{\text{Vdd1}-\text{Vdd2}} - L_{\text{Vdd1}-\text{Gnd2}} + L_{\text{Gnd1}-\text{Gnd2}} - L_{\text{Vdd2}-\text{Gnd1}}, \quad (15.8)$$

where L_{ij} is the mutual inductance between the power and ground paths in the two power distribution networks. In general, a power distribution grid with DSDG should be designed such that M_{loop} is negative with the absolute maximum possible value. Alternatively,

$$|L_{\text{Vdd1}-\text{Gnd2}}| + |L_{\text{Vdd2}-\text{Gnd1}}| > |L_{\text{Vdd1}-\text{Vdd2}}| + |L_{\text{Gnd1}-\text{Gnd2}}|. \quad (15.9)$$

For fully interdigitated power distribution grids with DSDG, the distance between the power and ground lines inside each loop d_I^i is the same as an offset between the two loops s_I^i. In this case, substituting the mutual inductances between the power and ground paths in the two voltage domains into (15.8), M_{loop}^{intI} between the two grids is determined by (15.5). Observe that M_{loop}^{intI} is negative. A derivation of the mutual coupling between the two current loops in fully interdigitated power distribution grids with DSDG is provided in Appendix A.

15.4.2 Type II interdigitated grids with DSDG

In the second type of interdigitated power distribution grid, a power/-ground line from one voltage domain is placed next to a power/ground line from the other voltage domain. Groups of power/ground lines are alternated and equidistantly spaced, as shown in Fig. 15.6. In such power distribution grids, the distance between the lines inside the loop d_{II}^i is two times greater than the separation between the lines. Since one loop is located inside the other loop, the separation between the two loops s_{II}^i is negative. Such power distribution grids are described here as *pseudo-interdigitated* power distribution grids with DSDG.

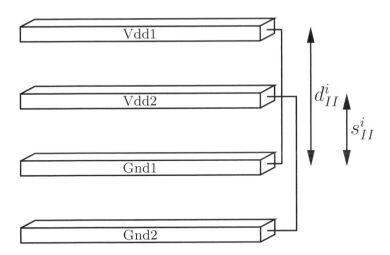

Fig. 15.6. Physical structure of a pseudo-interdigitated power distribution grid with DSDG. The distance between the lines making the loops d_{II}^i is two times greater than the separation between the lines.

The mutual inductive coupling of two current loops in pseudo-interdigitated grids with DSDG is determined by (15.8). For pseudo-interdigitated power distribution grids with DSDG, the distance between the power and ground lines inside each loop d^i_{II} is two time greater than the offset between the two loops s^i_{II}. In this case, substituting the mutual inductances between the power and ground paths in the different voltage domains into (15.8), the mutual inductive coupling between the two networks M^{intII}_{loop} is

$$M^{intII}_{loop} = 0.2l \left(\ln 3 - \frac{2d}{l} \right), \tag{15.10}$$

where d is the distance between the two adjacent lines. Observe that M^{intII}_{loop} is positive for $l \gg d$. The derivation of the mutual coupling between the two current loops in pseudo-interdigitated power distribution grids with DSDG is presented in Appendix B.

In modern high performance ICs, the inductive component of the power distribution noise has become comparable to the resistive noise [207]. In future nanoscale ICs, the inductive $L\frac{dI}{dt}$ voltage drop will dominate the resistive IR voltage drop, becoming the major component in the overall power noise. The partial self-inductance of the metal lines comprising the power distribution grid is constant for fixed parameters of a power delivery system (*i.e.*, the line width, line thickness, and line length). In order to reduce the power distribution noise, the total mutual inductance of a particular power distribution grid should therefore be negative with an absolute maximum value.

Comparing (15.5) to (15.10), note that for a line separation d much smaller than line length l, the mutual inductive coupling between different voltage domains in fully interdigitated grids M^{intI}_{loop} is negative with a nonzero absolute value, whereas the mutual inductive coupling between two current loops in pseudo-interdigitated grids M^{intII}_{loop} is positive. Moreover, since the distance between the lines comprising the loop in fully interdigitated power distribution grids is two times smaller than the line separation inside each current loop in pseudo-interdigitated power distribution grids, the mutual inductance inside the loop M'_{intI} is larger than M'_{intII}. Thus, the total mutual inductance as described by (15.7) in fully interdigitated grids is further increased by M^{intI}_{loop}. Conversely, the total mutual inductance in pseudo-interdigitated grids is reduced by M^{intII}_{loop}, as shown in Fig. 15.7. The total mutual inductance in fully interdigitated power distribution grids with DSDG is therefore greater than the total mutual inductance in pseudo-interdigitated grids with DSDG.

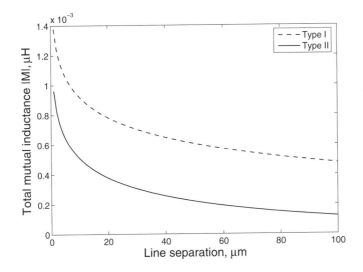

Fig. 15.7. Total mutual inductance of interdigitated power distribution grids with DSDG as a function of line separation. The length of the lines is $1000\,\mu$m.

15.5 Paired grids with DSDG

Another type of power distribution grid with alternating power and grounds lines is paired power distribution grids [28], [70]. Similar to interdigitated grids, the power and ground lines in paired grids are alternated, but rather than placed equidistantly, the lines are placed in equidistantly spaced pairs of adjacent power and ground lines. Analogous to the concepts presented in Section 15.3, the loop inductance of a particular power distribution network in paired power distribution grids with DSDG is affected by the presence of the other power distribution network.

In general, multiple paired power distribution grids with DSDG can be designed to satisfy different design constraints in high performance ICs. Exploiting the symmetry between the power and ground networks, each of the possible paired power distribution grids with DSDG can be characterized by the two main power delivery schemes. Two types

of paired power distribution grids with DSDG are presented in this
section. The loop inductance in the first type of power distribution
grid is described in Section 15.5.1. The loop inductance in the second
type of power distribution grid is discussed in Section 15.5.2.

15.5.1 Type I paired grids with DSDG

In the first type of paired power distribution grid with DSDG, the power
and ground lines of a particular power delivery network are placed in
equidistantly spaced pairs. The group of adjacent power and ground
lines from one voltage domain is alternated with the group of power
and ground lines from the other voltage domain, as shown in Fig. 15.8.
In such power distribution grids, the power and ground lines from a
specific power delivery network are placed in pairs. The separation
between the pairs is n times (where $n \geq 1$) larger than the separation
between the lines inside each pair. Such power distribution grids are
described here as *fully paired* power distribution grids with DSDG.
Note that in the case of $n = 1$, fully paired grids degenerate to fully
interdigitated grids.

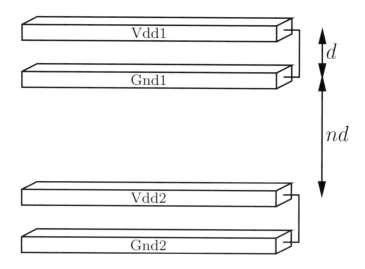

Fig. 15.8. Physical structure of a fully paired power distribution grid with DSDG.
In such a grid, each pair is composed of power and ground lines for a particular
voltage domain. The separation between the pairs is n times larger than the distance
between the lines making up the loop d.

Similar to the mutual inductance between the two loops in interdigitated power distribution grids as discussed in Section 15.4, the mutual inductive coupling of the two current loops in fully paired grids with DSDG is determined by (15.8). In fully paired power distribution grids with DSDG, the distance between the pairs is n times greater than the separation d between the power and ground lines making up the pair. Thus, substituting the mutual inductance between the power and ground lines for the different voltage domains into (15.8), the mutual inductive coupling between the two networks M_{loop}^{prdI} is

$$M_{loop}^{\mathrm{prdI}} = 0.2l \ln \left[\frac{(n+2)n}{(n+1)^2} \right]. \tag{15.11}$$

A derivation of the mutual coupling between the two current loops in fully paired power distribution grids with DSDG is presented in Appendix C. Note that M_{loop}^{prdI} is negative for $n \geq 1$ with an absolute value slightly greater than zero. Also note that the mutual inductance inside each current loop M'_{prdI} does not depend on n and is determined by (15.3).

15.5.2 Type II paired grids with DSDG

In the second type of paired power distribution grid with DSDG, a power/ground line from one voltage domain is placed in a pair with a power/ground line from the other voltage domain. The group of adjacent power lines alternates with the group of ground lines from different voltage domains, as shown in Fig. 15.9. In such power distribution grids, the power and ground lines from different power delivery networks are placed in pairs. The separation between the pairs is n times (where $n \geq 1$) larger than the separation between the lines within each pair. Such power distribution grids are described here as *pseudo-paired* power distribution grids with DSDG. Note that in the case of $n = 1$, pseudo-paired grids are identical to pseudo-interdigitated grids.

As discussed in Section 15.5.1, the mutual inductive coupling between the two power delivery networks in pseudo-paired grids with DSDG is determined by (15.8). In pseudo-paired power distribution grids with DSDG, the distance between the pairs is n times greater than the separation d between the power/ground lines making up the pair. The effective distance between the power and ground lines in a particular power delivery network is therefore $(n+1)d$. Substituting

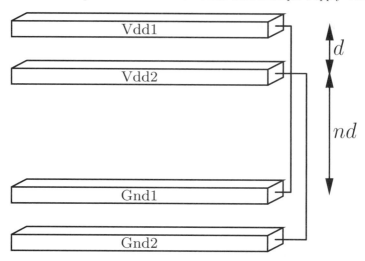

Fig. 15.9. Physical structure of a pseudo-paired power distribution grid with DSDG. In such a grid, each pair is composed of power or ground lines from the two voltage domains. The separation between the pairs is n times larger than the distance between the lines making up the loop d. The effective distance between the power and ground lines in a particular power delivery network is $(n + 1)d$.

the mutual inductance between the power and ground lines in the two different voltage domains into (15.8), the mutual inductive coupling between the two networks $M_{loop}^{\mathrm{prdII}}$ is

$$M_{loop}^{\mathrm{prdII}} = 0.2l \left[\ln \left(n^2 + 2n \right) - \frac{2nd}{l} \right]. \qquad (15.12)$$

A derivation of the mutual coupling between the two current loops in pseudo-paired power distribution grids with DSDG is provided in Appendix D. Note that $M_{loop}^{\mathrm{prdII}}$ is positive for $n \geq 1$. In contrast to fully paired grids, in pseudo-paired power distribution grids, the mutual inductance inside each current loop M'_{prdII} is a function of n,

$$M'_{\mathrm{prdII}} = 0.2l \left[\ln \frac{2l}{(n + 1)d} - 1 + \frac{(n + 1)d}{l} - \ln\gamma + \ln k \right]. \qquad (15.13)$$

Note that M'_{prdII} decreases with n, approaching zero for large n.

Comparing Fig. 15.8 to Fig. 15.9, note that the line separation inside each pair in the pseudo-paired power distribution grid is n times greater than the line separation between the power and ground lines making up a pair in fully paired power distribution grids. The mutual inductance within the power delivery network in fully paired power distribution grids M'_{prdI} is therefore greater than the mutual inductance within the power delivery network in pseudo-paired power distribution grids M'_{prdII}. Moreover, the distance between the lines in the particular voltage domain in fully paired power distribution grids does not depend on the separation between the pairs (no dependence on n). Thus, M'_{prdI} is a constant. The distance between the power/ground lines from the different voltage domains in pseudo-paired power distribution grids is smaller, however, than the distance between the power/ground lines from the different power delivery networks in fully paired power distribution grids. The magnitude of the mutual inductive coupling between the two current loops in pseudo-paired grids M_{loop}^{prdII} is therefore larger than the magnitude of the mutual inductive coupling between the two power delivery networks in fully paired grids M_{loop}^{prdI}. Note that the magnitude of M_{loop}^{prdII} increases with n and becomes much greater than zero for large n. Also note that M_{loop}^{prdI} is negative while M_{loop}^{prdII} is positive for all $n \geq 1$.

The total mutual inductance M as determined by (15.7) for two types of paired power distribution grids with DSDG is plotted in Fig. 15.10. Note that the total mutual inductance in fully paired grids is primarily determined by the mutual inductance inside each power delivery network M'_{prdI}. The absolute value of the total mutual inductance in fully paired grids is further increased by M_{loop}^{prdI}. As the separation between the pairs n increases, the mutual inductive coupling between the two current loops M_{loop}^{prdI} decreases, approaching zero at large n. Thus, the magnitude of the total mutual inductance in fully paired power distribution grids slightly drops with n. In pseudo-paired grids, however, the total mutual inductance is a non-monotonic function of n and can be divided into two regions. The total mutual inductance is determined by the mutual inductance inside each current loop M'_{prdII} for small n and by the mutual inductive coupling between the two voltage domains M_{loop}^{prdII} for large n. Since M'_{prdII} is negative and M_{loop}^{prdII} is positive for all n, the total mutual inductance in pseudo-paired grids is negative with a decreasing absolute value for small n. As n increases, M_{loop}^{prdII} begins to dominate and, at some n ($n = 8$ in Fig. 15.10), the total mutual inductance becomes positive with increasing absolute value. For large n, pseudo-paired grids with DSDG become identical to power distribution grids with DSSG. Similar to grids with DSSG, power and

ground paths in both voltage domains are strongly coupled, increasing
the loop inductance as seen from a specific power delivery network. The
resulting voltage fluctuations are therefore larger.

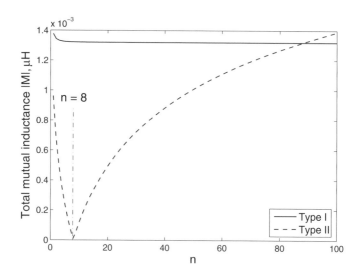

Fig. 15.10. Total mutual inductance of paired power distribution grids with DSDG
as a function of the ratio of the distance between the pairs to the line separation
inside each pair (n). The length of the lines is $1000\,\mu$m and the line separation inside
each pair d is $1\,\mu$m. Note that the total mutual inductance in pseudo-paired power
distribution grids becomes zero at $n = 8$.

15.6 Simulation results

To characterize the voltage fluctuations as seen at the load, both power
distribution grids are modeled as ten series RL segments. It is assumed
that both power delivery subnetworks are similar and source similar
current loads. Two equal current loads are applied to the power grid
with a single supply voltage and single ground. A triangular current
source with $50\,$mA amplitude, $100\,$ps rise time, and $150\,$ps fall time is
applied to each grid within the power distribution network. No skew
between the two current loads is assumed, modeling the worst case
scenario with the maximum power noise. For each grid structure, the
width of the lines varies from $1\,\mu$m to $10\,\mu$m, maintaining the line pair

pitch P at a constant value of 40 μm (80 μm in the case of paired grids). In paired power distribution grids, the line separation inside each pair is 1 μm. The decrease in the maximum voltage drop (or the voltage sag) from V_{dd} is estimated from SPICE for different line widths.

The resistance and inductance for the power distribution grids with SSSG operating at 1 GHZ and 100 GHz are listed in Table 15.1. The resistance and inductance for the power distribution grids with DSSG operating at 1 GHz and 100 GHz are listed in Table 15.2. Note that in the case of DSSG, only interdigitated grids can be implemented. The power grids with DSSG lack symmetry in both voltage domains which is necessary for paired grids. Also note that two types of interdigitated power distribution grids with DSSG can be implemented. Both types of interdigitated grids with DSSG are identical except for those power/ground lines located at the periphery of the power grid. Thus, the difference in loop inductance in both interdigitated grids with DSSG is negligible for a large number of power/ground lines comprising the grid. Only one interdigitated power distribution grid with DSSG is therefore analyzed. The impedance characteristics of the interdigitated and paired power distribution grids with DSDG are listed in Table 15.3, 15.4, and 15.5. The results listed in Tables 15.1 to 15.5 are discussed in Sections 15.6.1 to 15.6.4.

The performance of interdigitated power distribution grids is quantitatively compared to the power noise of a conventional power distribution scheme with DSSG in Section 15.6.1. The maximum voltage drop from V_{dd} for paired power distribution grids is evaluated in Section 15.6.2. Both types of power distribution grids are compared to the reference power distribution grid with SSSG. Power distribution schemes with decoupling capacitors are compared in Section 15.6.3. The dependence of the power noise on the switching frequency of the current loads is discussed in Section 15.6.4.

15.6.1 Interdigitated power distribution grids without decoupling capacitors

The maximum voltage drop for four interdigitated power distribution grids without decoupling capacitors is depicted in Fig. 15.11. For each of the power distribution grids, the maximum voltage drop decreases sublinearly as the width of the lines is increased. This noise voltage drop is caused by the decreased loop impedance. The resistance of the metal lines decreases linearly with an increase in the line width. The

Table 15.1. Impedance characteristics of power distribution grids with SSSG

Line cross section (μm $\times$ μm)	1 GHz				100 GHz			
	R_{pp}, R_{gg} (Ω)	L_{pp}, L_{gg} (nH)	L_{pg} (nH)	k	R_{pp}, R_{gg} (Ω)	L_{pp}, L_{gg} (nH)	L_{pg} (nH)	k
Interdigitated								
1×1	1.478	0.357	0.289	0.810	2.514	0.351	0.284	0.809
2×1	0.763	0.348	0.286	0.822	1.652	0.343	0.284	0.828
3×1	0.519	0.341	0.285	0.835	1.217	0.337	0.283	0.840
4×1	0.395	0.337	0.285	0.846	0.944	0.333	0.283	0.850
5×1	0.320	0.333	0.284	0.853	0.764	0.330	0.283	0.858
6×1	0.269	0.330	0.284	0.859	0.643	0.327	0.283	0.865
7×1	0.233	0.328	0.283	0.863	0.555	0.325	0.283	0.871
8×1	0.206	0.326	0.283	0.868	0.489	0.323	0.283	0.876
9×1	0.184	0.324	0.283	0.873	0.438	0.321	0.283	0.882
10×1	0.167	0.322	0.283	0.879	0.397	0.319	0.282	0.884
Paired								
1×1	1.467	0.357	0.332	0.930	2.652	0.352	0.329	0.935
2×1	0.747	0.349	0.324	0.928	1.728	0.344	0.323	0.939
3×1	0.504	0.343	0.319	0.930	1.274	0.338	0.319	0.944
4×1	0.382	0.339	0.315	0.929	0.987	0.333	0.315	0.846
5×1	0.309	0.335	0.312	0.931	0.798	0.330	0.312	0.845
6×1	0.260	0.332	0.309	0.931	0.671	0.327	0.310	0.948
7×1	0.225	0.330	0.307	0.930	0.580	0.325	0.308	0.948
8×1	0.199	0.328	0.305	0.930	0.510	0.322	0.306	0.950
9×1	0.179	0.326	0.303	0.929	0.456	0.321	0.304	0.949
10×1	0.163	0.324	0.301	0.929	0.413	0.319	0.303	0.950

Line pair pitch – 40 μm, grid length – 1000 μm,

and $k = \dfrac{L_{pg}}{\sqrt{L_{pp}L_{gg}}}$ – coupling coefficient

loop inductance increases slowly with increasing line width. As a result, the total impedance of each of the power distribution schemes decreases sublinearly, approaching a constant impedance as the lines become very wide.

As described in Section 15.3, the power distribution scheme with DSDG outperforms power distribution grids with DSSG. Fully inter-digitated grids with DSDG produce, on average, a 15.3% lower voltage drop as compared to the scheme with DSSG. Pseudo-interdigitated grids with DSDG produce, on average, a close to negligible 0.3% lower voltage drop as compared to the scheme with DSSG. The maximum improvement in noise reduction is 16.5%, which is achieved for an 8 μm

Table 15.2. Impedance characteristics of interdigitated power distribution grids with DSSG

Line cross section (μm × μm)	R_{pp}, R_{gg} (Ω)	L_{pp}^*, L_{gg}^* (nH)	L_{pg}^* (nH)	k^*	L_{pp}^{**}, L_{gg}^{**} (nH)	L_{pg}^{**} (nH)	k^{**}
			1 GHz				
1 × 1	2.180	0.397	0.289	0.728	0.396	0.285	0.720
2 × 1	1.109	0.385	0.287	0.745	0.383	0.283	0.738
3 × 1	0.748	0.377	0.286	0.759	0.375	0.282	0.752
4 × 1	0.566	0.370	0.286	0.773	0.368	0.281	0.764
5 × 1	0.456	0.365	0.285	0.781	0.363	0.281	0.774
6 × 1	0.383	0.361	0.285	0.789	0.359	0.280	0.780
7 × 1	0.330	0.358	0.285	0.796	0.355	0.280	0.789
8 × 1	0.290	0.355	0.285	0.804	0.352	0.280	0.795
9 × 1	0.260	0.352	0.285	0.810	0.349	0.280	0.802
10 × 1	0.235	0.349	0.285	0.817	0.346	0.279	0.806
			100 GHz				
1 × 1	3.603	0.391	0.285	0.729	0.389	0.281	0.722
2 × 1	2.357	0.379	0.285	0.752	0.377	0.280	0.743
3 × 1	1.730	0.372	0.285	0.766	0.369	0.280	0.759
4 × 1	1.338	0.366	0.285	0.779	0.363	0.280	0.771
5 × 1	1.081	0.361	0.285	0.789	0.358	0.280	0.782
6 × 1	0.908	0.357	0.284	0.796	0.354	0.279	0.788
7 × 1	0.784	0.354	0.284	0.802	0.350	0.279	0.796
8 × 1	0.691	0.351	0.284	0.809	0.347	0.279	0.803
9 × 1	0.618	0.348	0.284	0.816	0.345	0.279	0.809
10 × 1	0.560	0.346	0.284	0.821	0.342	0.279	0.816

Line pair pitch – 40 μm, grid length – 1000 μm,
* denotes coupling between V_{dd1} (V_{dd2}) and Gnd,
** denotes coupling between V_{dd1} and V_{dd2}

wide line, and 7.1%, which is achieved for a 1 μm wide line, for fully- and pseudo-interdigitated grids with DSDG, respectively. Note that pseudo-interdigitated power grids with DSDG outperform conventional power delivery schemes with DSSG for narrow lines. For wide lines, however, the power delivery scheme with DSSG results in a lower voltage drop. From the results depicted in Fig. 15.11, observe that the power delivery schemes with both DSDG and SSSG outperform the power grid with DSSG. The fully interdigitated power distribution grid with DSDG outperforms the reference power grid with SSSG by 2.7%. This behavior can be explained as follows. Since the number of lines

Table 15.3. Impedance characteristics of interdigitated power distribution grids with DSDG

Grid type	Cross section (μm×μm)	R_{pp}, R_{gg} (Ω)	L_{pp}^*, L_{gg}^* (nH)	L_{pg}^* (nH)	k^*	L_{pp}^{**}, L_{gg}^{**} (nH)	L_{pg}^{**} (nH)	k^{**}	$L_{pp}^\dagger, L_{gg}^\dagger$ (nH)	$L_{pg}^\dagger$ (nH)	$k^\dagger$
					1 GHz						
	1 × 1	2.887	0.439	0.293	0.667	0.439	0.279	0.636	0.438	0.284	0.648
	2 × 1	1.458	0.424	0.292	0.689	0.423	0.277	0.654	0.422	0.282	0.668
	3 × 1	0.979	0.414	0.291	0.703	0.413	0.276	0.668	0.410	0.281	0.685
	4 × 1	0.738	0.406	0.291	0.717	0.405	0.276	0.681	0.402	0.280	0.697
	5 × 1	0.594	0.400	0.290	0.725	0.398	0.275	0.691	0.395	0.280	0.709
	6 × 1	0.497	0.394	0.290	0.736	0.393	0.275	0.700	0.389	0.279	0.717
	7 × 1	0.428	0.390	0.290	0.744	0.388	0.275	0.709	0.384	0.279	0.727
	8 × 1	0.376	0.385	0.290	0.753	0.384	0.275	0.716	0.380	0.279	0.734
	9 × 1	0.336	0.382	0.290	0.759	0.380	0.275	0.724	0.376	0.279	0.742
	10 × 1	0.304	0.379	0.290	0.766	0.376	0.274	0.728	0.372	0.278	0.747
Type I					**100 GHz**						
	1 × 1	4.703	0.434	0.290	0.668	0.432	0.275	0.637	0.429	0.279	0.650
	2 × 1	3.070	0.419	0.290	0.692	0.417	0.275	0.659	0.413	0.279	0.676
	3 × 1	2.251	0.408	0.290	0.711	0.406	0.275	0.677	0.403	0.279	0.692
	4 × 1	1.739	0.401	0.290	0.723	0.399	0.275	0.689	0.395	0.279	0.706
	5 × 1	1.406	0.394	0.290	0.736	0.392	0.274	0.699	0.388	0.278	0.716
	6 × 1	1.179	0.389	0.290	0.746	0.387	0.274	0.708	0.383	0.278	0.726
	7 × 1	1.017	0.385	0.289	0.751	0.383	0.274	0.715	0.378	0.278	0.735
	8 × 1	0.896	0.381	0.289	0.759	0.379	0.274	0.723	0.374	0.278	0.743
	9 × 1	0.802	0.377	0.289	0.767	0.375	0.274	0.731	0.370	0.278	0.751
	10 × 1	0.727	0.374	0.289	0.773	0.372	0.274	0.737	0.367	0.278	0.757
					1 GHz						
	1 × 1	2.893	0.439	0.279	0.636	0.439	0.293	0.667	0.438	0.284	0.648
	2 × 1	1.466	0.423	0.277	0.655	0.424	0.292	0.689	0.422	0.282	0.668
	3 × 1	0.987	0.413	0.276	0.668	0.414	0.291	0.703	0.410	0.281	0.685
	4 × 1	0.747	0.405	0.276	0.681	0.406	0.291	0.717	0.402	0.280	0.697
	5 × 1	0.601	0.398	0.275	0.691	0.400	0.290	0.725	0.395	0.280	0.709
	6 × 1	0.504	0.393	0.275	0.700	0.394	0.290	0.736	0.389	0.279	0.717
	7 × 1	0.435	0.388	0.275	0.709	0.390	0.290	0.744	0.384	0.279	0.727
	8 × 1	0.383	0.384	0.275	0.716	0.386	0.290	0.751	0.380	0.279	0.734
	9 × 1	0.342	0.380	0.275	0.724	0.382	0.290	0.759	0.376	0.279	0.742
	10 × 1	0.310	0.377	0.274	0.727	0.379	0.290	0.765	0.372	0.278	0.747
Type II					**100 GHz**						
	1 × 1	4.756	0.432	0.275	0.637	0.434	0.290	0.668	0.429	0.279	0.650
	2 × 1	3.109	0.417	0.275	0.659	0.419	0.290	0.692	0.413	0.279	0.676
	3 × 1	2.281	0.406	0.275	0.677	0.408	0.290	0.711	0.403	0.279	0.692
	4 × 1	1.764	0.399	0.275	0.689	0.401	0.290	0.723	0.395	0.279	0.706
	5 × 1	1.425	0.392	0.274	0.699	0.394	0.290	0.736	0.388	0.278	0.716
	6 × 1	1.196	0.387	0.274	0.708	0.389	0.290	0.746	0.383	0.278	0.726
	7 × 1	1.031	0.383	0.274	0.715	0.385	0.290	0.753	0.378	0.278	0.735
	8 × 1	0.907	0.379	0.274	0.723	0.381	0.289	0.759	0.374	0.278	0.743
	9 × 1	0.812	0.375	0.274	0.731	0.377	0.289	0.767	0.370	0.278	0.751
	10 × 1	0.735	0.372	0.274	0.737	0.374	0.289	0.773	0.367	0.278	0.757

Line pair pitch – 40 μm, grid length – 1000 μm,

* denotes coupling between V_{dd1} (V_{dd2}) and Gnd_1 (Gnd_2),

** denotes coupling between V_{dd1} (Gnd_1) and V_{dd2} (Gnd_2),

$\dagger$ denotes coupling between Gnd_1 and V_{dd2}

Table 15.4. Impedance characteristics of Type I paired power distribution grids with DSDG

Line cross section ($\mu m \times \mu m$)	R_{pp}, R_{gg} (Ω)	L^{*}_{pp}, L^{*}_{gg} (nH)	L^{*}_{pg} (nH)	k^{*}	L^{**}_{pp}, L^{**}_{gg} (nH)	L^{**}_{pg} (nH)	k^{**}	$L^{\dagger}_{pp}, L^{\dagger}_{gg}$ (nH)	$L^{\dagger}_{pg}$ (nH)	$k^{\dagger}$	$L^{\ddagger}_{pp}, L^{\ddagger}_{gg}$ (nH)	$L^{\ddagger}_{pg}$ (nH)	$k^{\ddagger}$
					1 GHz								
1×1	2.883	0.439	0.389	0.886	0.439	0.279	0.636	0.439	0.279	0.636	0.439	0.278	0.633
2×1	1.450	0.425	0.376	0.885	0.423	0.277	0.655	0.424	0.278	0.656	0.423	0.277	0.655
3×1	0.972	0.415	0.366	0.882	0.413	0.276	0.668	0.413	0.277	0.671	0.413	0.276	0.668
4×1	0.733	0.407	0.359	0.882	0.405	0.276	0.681	0.405	0.276	0.681	0.404	0.275	0.681
5×1	0.590	0.400	0.353	0.883	0.398	0.275	0.691	0.398	0.276	0.693	0.398	0.275	0.691
6×1	0.495	0.395	0.348	0.881	0.392	0.275	0.702	0.393	0.276	0.702	0.392	0.274	0.699
7×1	0.428	0.390	0.344	0.882	0.388	0.275	0.709	0.388	0.276	0.711	0.387	0.274	0.708
8×1	0.378	0.386	0.340	0.881	0.383	0.275	0.718	0.384	0.276	0.719	0.383	0.274	0.715
9×1	0.339	0.382	0.336	0.880	0.379	0.274	0.723	0.380	0.276	0.726	0.379	0.274	0.723
10×1	0.308	0.379	0.333	0.879	0.376	0.274	0.729	0.377	0.276	0.732	0.375	0.273	0.728
					100 GHz								
1×1	5.121	0.434	0.388	0.894	0.431	0.275	0.638	0.431	0.275	0.638	0.431	0.275	0.638
2×1	3.324	0.417	0.376	0.902	0.414	0.275	0.664	0.414	0.275	0.664	0.413	0.275	0.666
3×1	2.441	0.405	0.367	0.906	0.402	0.275	0.684	0.402	0.275	0.684	0.402	0.274	0.682
4×1	1.887	0.397	0.361	0.909	0.394	0.274	0.695	0.394	0.275	0.698	0.393	0.274	0.697
5×1	1.525	0.390	0.355	0.910	0.387	0.274	0.708	0.387	0.275	0.711	0.387	0.274	0.708
6×1	1.279	0.385	0.350	0.909	0.381	0.274	0.719	0.382	0.275	0.720	0.381	0.274	0.719
7×1	1.102	0.380	0.246	0.911	0.377	0.274	0.727	0.377	0.275	0.729	0.376	0.274	0.729
8×1	0.970	0.376	0.343	0.912	0.372	0.274	0.737	0.373	0.275	0.737	0.372	0.273	0.734
9×1	0.867	0.372	0.339	0.911	0.369	0.274	0.743	0.369	0.275	0.745	0.368	0.273	0.742
10×1	0.785	0.369	0.336	0.911	0.365	0.274	0.751	0.366	0.275	0.751	0.365	0.273	0.748

Pairs pitch – 80 μm, grid length – 1000 μm, * denotes coupling between V_{dd1} and Gnd_1,
** denotes coupling between V_{dd1} and V_{dd2}, † denotes coupling between Gnd_1 and Gnd_2,
‡ denotes coupling between Gnd_1 and V_{dd2}

Table 15.5. Impedance characteristics of Type II paired power distribution grids with DSDG

Line cross section ($\mu m \times \mu m$)	R_{pp}, R_{gg} (Ω)	L_{pp}, L_{gg} (nH)	L_{pg}^* (nH)	k^*	L_{pp}^{**}, L_{gg}^{**} (nH)	L_{pg}^{**} (nH)	k^{**}	$L_{pp}^\dagger, L_{gg}^\dagger$ (nH)	$L_{pg}^\dagger$ (nH)	$k^\dagger$	$L_{pp}^\ddagger, L_{gg}^\ddagger$ (nH)	$L_{pg}^\ddagger$ (nH)	$k^\ddagger$
					1 GHz								
1×1	2.883	0.439	0.389	0.886	0.439	0.279	0.636	0.439	0.279	0.636	0.439	0.278	0.633
2×1	1.450	0.425	0.376	0.885	0.423	0.277	0.655	0.424	0.278	0.656	0.423	0.277	0.655
3×1	0.972	0.415	0.366	0.882	0.413	0.276	0.668	0.413	0.277	0.671	0.413	0.276	0.668
4×1	0.733	0.407	0.359	0.882	0.405	0.276	0.681	0.405	0.276	0.681	0.404	0.275	0.681
5×1	0.590	0.400	0.353	0.883	0.398	0.275	0.691	0.398	0.276	0.693	0.398	0.275	0.691
6×1	0.495	0.395	0.348	0.881	0.392	0.275	0.702	0.393	0.276	0.702	0.392	0.274	0.699
7×1	0.428	0.390	0.344	0.882	0.388	0.275	0.710	0.388	0.276	0.711	0.387	0.274	0.708
8×1	0.378	0.386	0.340	0.881	0.383	0.275	0.718	0.384	0.276	0.719	0.383	0.274	0.715
9×1	0.339	0.382	0.336	0.880	0.379	0.275	0.726	0.380	0.276	0.726	0.379	0.274	0.723
10×1	0.308	0.379	0.333	0.879	0.376	0.274	0.729	0.377	0.276	0.732	0.375	0.273	0.728
					100 GHz								
1×1	5.122	0.434	0.388	0.894	0.431	0.275	0.638	0.431	0.275	0.638	0.431	0.275	0.638
2×1	3.323	0.417	0.376	0.902	0.414	0.275	0.664	0.414	0.275	0.664	0.413	0.275	0.666
3×1	2.442	0.405	0.367	0.906	0.402	0.275	0.684	0.402	0.275	0.684	0.402	0.274	0.682
4×1	1.887	0.397	0.361	0.909	0.394	0.274	0.695	0.394	0.275	0.698	0.393	0.274	0.697
5×1	1.522	0.390	0.355	0.910	0.387	0.274	0.708	0.387	0.275	0.711	0.387	0.274	0.708
6×1	1.279	0.385	0.350	0.909	0.381	0.274	0.719	0.382	0.275	0.720	0.381	0.274	0.719
7×1	1.103	0.380	0.346	0.911	0.377	0.274	0.728	0.377	0.275	0.729	0.376	0.274	0.729
8×1	0.971	0.376	0.343	0.912	0.372	0.274	0.737	0.373	0.275	0.737	0.372	0.273	0.734
9×1	0.868	0.372	0.339	0.911	0.369	0.274	0.743	0.369	0.275	0.745	0.368	0.273	0.742
10×1	0.786	0.369	0.336	0.911	0.365	0.274	0.751	0.366	0.275	0.751	0.365	0.273	0.748

Pairs pitch – 80 μm, grid length – 1000 μm, * denotes coupling between $V_{dd1} - V_{dd2}$,
** denotes coupling between $V_{dd1} - Gnd_1$, † denotes coupling between V_{dd1} and Gnd_2,
‡ denotes coupling between Gnd_1 and V_{dd2}

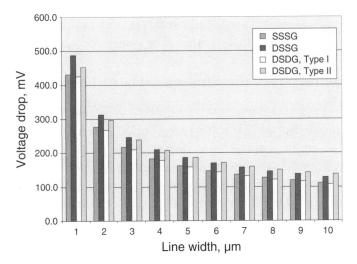

Fig. 15.11. Maximum voltage drop for the four interdigitated power distribution grids under investigation. No decoupling capacitors are added.

dedicated to each power delivery network in the grid with DSDG is two times smaller than the total number of lines in the reference grid, the resistance of each subnetwork is two times greater than the resistance of the reference power grid. The loop inductance of an interdigitated power distribution grid depends inversely linearly on the number of lines in the grid [70]. The loop inductance of each subnetwork is two times greater than the overall loop inductance of the grid with SSSG. Given two similar current loads applied to the reference power distribution scheme, the maximum voltage drop for both systems should be the same. However, from (15.4), the mutual inductive coupling in the power grid with DSDG increases due to the presence of the second subnetwork. As a result, the overall loop inductance of each network comprising the power grid with DSDG is lower, resulting in a lower power noise as seen from the current load of each subnetwork. Note from Fig. 15.7 that in pseudo-interdigitated power distribution grids with DSDG, the mutual inductance between two current loops M_{loop}^{intII} is positive, reducing the overall mutual inductance. The resulting loop inductance as seen from the load of the particular network is therefore increased, producing a larger inductive voltage drop. In many applications such as high performance microprocessors, mixed-signal circuits, and systems-on-chip, a power distribution network with DSDG is often utilized. In other applications, however, a fully interdigitated power distribution system with multiple voltages and multiple grounds can be a better alternative than distributing power with SSSG.

15.6.2 Paired power distribution grids without decoupling capacitors

The maximum voltage drop for three paired power distribution grids without decoupling capacitors is depicted in Fig. 15.12. Similar to interdigitated grids, the maximum voltage drop decreases sublinearly with increasing line width. Observe that fully paired power distribution grids with DSDG outperform conventional paired power distribution grids with SSSG by, on average, 2.3%. Note the information shown in Fig. 15.12, the ratio of the separation between the pairs to the distance between the lines in each pair (n) is eighty. Also note from Fig. 15.10 that the total mutual inductance in fully paired grids increases as n is decreased (the pairs are placed physically closer). Thus, better performance is achieved in fully paired grids with DSDG for densely placed pairs. In contrast to fully paired grids, in pseudo-paired grids with DSDG, the total mutual inductance is reduced by inductive coupling between the two current loops M_{loop}^{prdII}. For $n > 8$ (see Fig. 15.10), the mutual inductive coupling between the two current loops in pseudo-paired grid becomes comparable to the mutual inductive coupling between the two current loops in the conventional power grid with DSSG (the $-2M$ term in (15.2) becomes positive). As n further increases, the power and ground paths within the two voltage domains become strongly coupled, increasing the loop inductance.

To quantitatively compare interdigitated grids to paired grids, the maximum voltage drop for seven different types of power distribution grids without decoupling capacitors is plotted in Fig. 15.13. Note in Fig. 15.13 that the conventional power delivery scheme with DSSG results in larger voltage fluctuations as compared to fully interdigitated grids with DSDG. The performance of pseudo-interdigitated grids with DSDG is comparable to the performance of the conventional delivery scheme with DSSG. In pseudo-interdigitated grids, the positive mutual inductance between two current loops lowers the overall negative mutual inductance. The loop inductance in the specific power delivery network is therefore increased, resulting in greater power noise. Analogous to the conventional scheme, in pseudo-paired grids, the power and ground paths in different voltage domains are strongly coupled, producing the largest voltage drop. Both fully interdigitated and fully

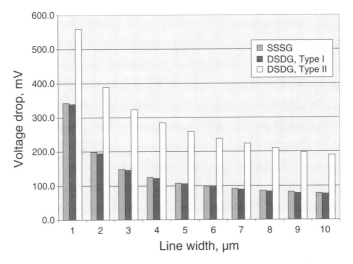

Fig. 15.12. Maximum voltage drop for the three paired power distribution grids under investigation. No decoupling capacitors are added.

paired power distribution grids with DSDG produce the lowest voltage fluctuations, slightly outperforming the reference power delivery network with SSSG. In these grids, the resulting loop inductance is reduced due to strong coupling between the power/ground pairs from different voltage domains (with currents flowing in opposite directions). Alternatively, the total mutual inductance is negative with large magnitude, reducing the loop inductance. Both fully interdigitated and fully paired power distribution grids with DSDG should be used in those systems with multiple power supply voltages. Fully interdigitated and fully paired power distribution grids with DSDG can also be a better alternative than a power distribution grid with SSSG.

15.6.3 Power distribution grids with decoupling capacitors

To lower the voltage fluctuations of on-chip power delivery systems, decoupling capacitors are placed on ICs to provide charge when the voltage drops [28]. The maximum voltage drop of seven power distribution schemes with decoupling capacitors operating at 1 GHz is shown in Fig. 15.14. All of the decoupling capacitors are assumed to be ideal, *i.e.*, no parasitic resistances and inductances are associated with the capacitor. Also, all of the decoupling capacitors are assumed to be effective (located inside the effective radius of an on-chip decoupling capacitor,

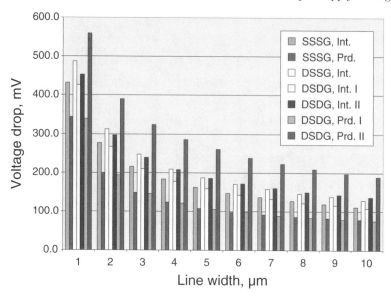

Fig. 15.13. Maximum voltage drop for interdigitated and paired power distribution grids under investigation. No decoupling capacitors are added.

as described in Chapter 18 [200]). The total budgeted capacitance is divided equally between the two supply voltages. The decoupling capacitor added to the power distribution grid with SSSG is two times larger than the decoupling capacitor in each subnetwork of the power delivery scheme with dual voltages. As shown in Fig. 15.14, the maximum voltage drop decreases as the lines become wider. The maximum voltage drop of the fully interdigitated power distribution scheme with DSDG is reduced by, on average, 9.2% (13.6% maximum) for a 30 pF decoupling capacitance as compared to a conventional power distribution scheme with DSSG. For a 20 pF decoupling capacitance, however, a fully interdigitated power distribution grid with DSDG produces about 55% larger power noise as compared to a conventional power distribution scheme with DSSG. This performance degradation is caused by on-chip resonances, as explained below.

Comparing the data shown in Fig. 15.13 to that shown in Fig. 15.14, note that the voltage drop of the power distribution grids with decoupling capacitors as compared to the case with no decoupling capacitances is greatly reduced for narrow lines and is higher for wider lines. This behavior can be explained as follows. For narrow lines, the grid resistance is high and the loop inductance is low. The grid

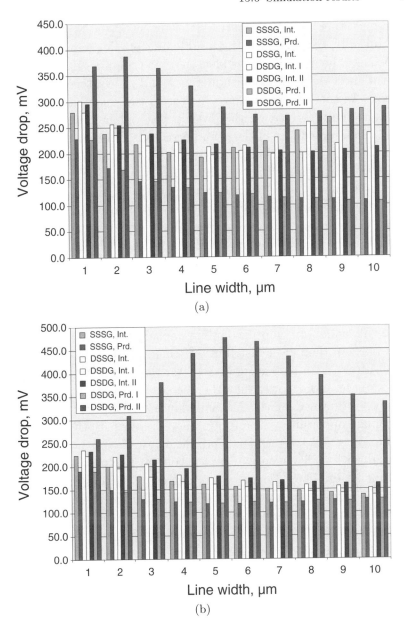

Fig. 15.14. Maximum voltage drop for seven types of power distribution grids with a decoupling capacitance of (a) 20 pF and (b) 30 pF added to each power supply. The switching frequency of the current loads is 1 GHz.

impedance, therefore, is primarily determined by the resistance of the lines. Initially, the system with an added decoupling capacitor is over-damped. As the lines become wider, the grid resistance decreases faster than the increase in the loop inductance and the system becomes less damped. As the loop inductance increases, the resonant frequency of an RLC circuit, formed by the on-chip decoupling capacitor and the parasitic RL impedance of the grid, decreases. This resonant frequency moves closer to the switching frequency of the current load. As a result, the voltage response of the overall system oscillates. Since the decoupling capacitance added to the power grid with SSSG is two times larger than the decoupling capacitance added to each power supply voltage in the dual voltage schemes, the system with a single supply voltage is more highly damped and the self-resonant frequency is significantly lower. Furthermore, the resonant frequency is located far from the switching frequency of the circuit.

For narrow lines propagating a signal with 1 GHz harmonics, the resulting power noise in fully interdigitated power grids with DSDG with 20 pF added on-chip decoupling capacitance is smaller than the power noise of the power distribution scheme with SSSG, as shown in Fig. 15.14(a). With increasing line width, the inductance of the power grids increases more slowly than the decrease in the grid resistance. An RLC system formed by the RL impedance of the power grid and the decoupling capacitance, therefore, is less damped. Both of the power distribution grids with DSDG and the conventional power distribution grid with SSSG result in larger voltage fluctuations as the line width increases. The self-resonant frequency of the fully interdigitated grid with DSDG is almost coincident with the switching frequency of the current load. The self-resonant frequency of the power grid with SSSG however is different from the switching frequency of the current source. Thus, for wide lines, a conventional power delivery scheme with SSSG outperforms the fully interdigitated power distribution grid with DSDG. Note that the loop inductance in pseudo-interdigitated power distribution grids with DSDG is greater than the loop inductance in fully interdigitated grids. As a result, the self-resonant frequency of a pseudo-interdigitated grid with DSDG is smaller than the switching frequency of the current load, resulting in smaller power noise as compared to power grids with SSSG and fully interdigitated grids with DSDG. Also note that the loop inductance in paired power distribution grids is further reduced as compared to interdigitated grids. In this case, the self-resonant frequency of all of the paired power distribution grids is greater than the circuit switching frequency. Thus, the power noise in paired power distribution grids gradually decreases as the line width increases (and is slightly higher in wide lines in the case of pseudo-paired grids).

Increasing the on-chip decoupling capacitance from 20 pF to 30 pF further reduces the voltage drop. For a 30 pF decoupling capacitance in a pseudo-paired power delivery scheme with DSSG, the self-resonant frequency is close to the switching frequency of the current load. Simultaneously, the grid resistance decreases much faster with increasing line width than the increase in the loop inductance. The system becomes underdamped with the self-resonant frequency equal to the circuit switching frequency. As a result, the system produces high amplitude voltage fluctuations. The maximum voltage drop in the case of a pseudo-paired power grid with DSDG therefore increases as the lines become wider. This phenomenon is illustrated in Fig. 15.14(b) for a line width of 5 μm.

With decoupling capacitors, the self-resonant frequency of an on-chip power distribution system is lowered. If the resonant frequency of an RLC system with intentionally added decoupling capacitors is sufficiently close to the circuit switching frequency, the system will produce high amplitude voltage fluctuations. Voltage sagging will degrade system performance and may cause significant failure. An excessively high power supply voltage can degrade the reliability of a system. The decoupling capacitors for power distribution systems with multiple supply voltages therefore have to be carefully designed. Improper choice (magnitude and location) of the on-chip decoupling capacitors can therefore worsen the power noise, further degrading system performance [199], [26].

15.6.4 Dependence of power noise on the switching frequency of the current loads

To model the dependence of the power noise on the switching frequency, the power grids are stimulated with triangular current sources with a 50 mA amplitude, 20 ps rise times, and 30 ps fall times. The switching frequency of each current source varies from 1 GHz to 10 GHz to capture the resonances in each power grid. For each grid structure, the width of the line is varied from 1 μm to 10 μm. The maximum voltage drop is determined from SPICE for different line widths at each frequency.

The maximum voltage drop for the power distribution grid with SSSG is illustrated in Fig. 15.15. The maximum voltage drop decreases slightly for wider lines. Note that with decoupling capacitors, the voltage drop is lower except for two regions. The significant increase in power noise at specific frequencies and line widths is due to the following two effects. As lines become wider, the resistance of the power grid is lower, whereas the inductance is slightly increased, decreasing the damping of the entire system. When the switching frequency of a current load approaches the self-resonant frequency of the power grid, the voltage drop due to the RLC system increases (due to resonances). As the width of the lines increases, the system becomes more underdamped, resulting in a sharper resonant peak. The amplitude of the resonant peak increases rapidly as the system becomes less damped. The maximum voltage drop occurs between 6 GHz and 7 GHz for a power grid with a 20 pF decoupling capacitance, as shown in Fig. 15.15(a).

The maximum voltage drop also increases at high frequencies in narrow lines. Decoupling capacitors are effective only if the capacitor is fully charged within one clock cycle. The effectiveness of the decoupling capacitor is related to the RC time constant, where R is the resistance of the interconnect connecting the capacitor to the power supply. For narrow resistive lines, the time constant is prohibitively large at high frequencies, $i.e$, the decoupling capacitor cannot be fully charged within one clock period. The effective magnitude of the decoupling capacitor is therefore reduced. The capacitor has the same effect on the power noise as a smaller capacitor [200].

By increasing the magnitude of the decoupling capacitor, the overall power noise can be further reduced, as shown in Fig. 15.15(b). Moreover, the system becomes more damped, producing a resonant peak with a smaller amplitude. The self-resonant frequency of the power delivery system is also lowered. Comparing Figs. 15.15(a) to 15.15(b), note that the resonant peak shifts in frequency from approximately 6 GHz to 7 GHz for a 20 pF decoupling capacitance to 5 GHz to 6 GHz for a 30 pF decoupling capacitance. Concurrently, increasing the decoupling capacitor increases the RC time constant, making the capacitor less effective at high frequencies in narrow resistive lines. Note the significant increase in the maximum voltage drop for a 1 μm wide line for a 30 pF decoupling capacitance as compared to the case of a 20 pF decoupling capacitance. Power distribution grids with DSSG and DSDG behave similarly. For the same decoupling capacitance and for

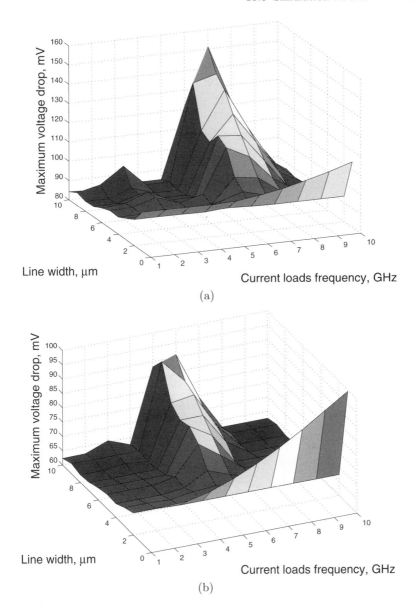

(a)

(b)

Fig. 15.15. Maximum voltage drop for the power distribution grid with SSSG as a function of frequency and line width for different values of decoupling capacitance: a) decoupling capacitance budget of 20 pF, b) decoupling capacitance budget of 30 pF.

the non-resonant case, both the fully- and pseudo-interdigitated power distribution schemes with DSDG result in a lower voltage drop than a power distribution scheme with DSSG. The magnitude of the decoupling capacitance needs to be carefully chosen to guarantee that the two prohibited regions are outside the operating frequency of the system for a particular line width. Also, for narrow lines, the magnitude of the decoupling capacitor is limited by the RC time constant. The amplitude of the resonant peak can be lowered by increasing the parasitic resistance of the decoupling capacitors.

15.7 Design implications

Historically, due to low switching frequencies and the high resistance of on-chip interconnects, resistive voltage drops have dominated the overall power noise. In modern high performance ICs, the inductive component of the power distribution noise has become comparable to the resistive noise [28]. It is expected that in future nanoscale ICs, the inductive $L\frac{dI}{dt}$ voltage drop will dominate the resistive IR voltage drop, becoming the primary component of the overall power noise [207]. As shown previously, the performance of the power delivery schemes with DSDG depends upon the switching frequency of the current load, improving with frequency (due to increased mutual coupling between the power and ground lines). It is expected that the performance of the power distribution grids with DSDG will increase with technology scaling.

As discussed in Section 15.6, fully interdigitated power distribution grids with DSDG outperform pseudo-interdigitated grids with DSDG. Moreover, in pseudo-interdigitated grids, the power/ground lines from different voltage domains are placed next to each over, increasing the coupling between the different power supply voltages. Pseudo-interdigitated power distribution grids with DSDG should therefore not be used in those ICs where high isolation is required between the power supply voltages (e.g., mixed-signal ICs, systems-on-chip). Rather, fully interdigitated power distribution grids with DSDG should be utilized.

Similar to interdigitated grids, fully paired power distribution grids with DSDG produce smaller power noise as compared to pseudo-paired power distribution grids with DSDG. In pseudo-paired grids, the separation between the power/ground lines from different voltage domains is much smaller than the distance between the power and ground lines inside each power delivery network (current loop). Different power supply voltages are therefore strongly coupled in pseudo-paired grids. Note that pseudo-paired grids have the greatest coupling between different power supplies among all of the power distribution schemes described in this chapter. Such grids, therefore, are not a good choice for distributing power in mixed-signal ICs. Later in the design flow, when it is prohibitively expensive to redesign the power distribution system, the spacing between the pairs in pseudo-paired grids with DSDG should be decreased. If the pairs are placed close to each over (n is small), as illustrated in Fig. 15.10, the loop inductance of a particular current loop is lowered, approaching the loop inductance in pseudo-interdigitated grids.

The self-resonant frequency of a system is determined by the power distribution network. For example, in power distribution grids with DSDG, the decoupling capacitance added to each power delivery network is two times smaller than the decoupling capacitance in the power delivery scheme with SSSG. The loop inductance of power distribution grids with DSDG is comparable however to the loop inductance of power distribution grids with SSSG. Assuming the same decoupling capacitance, the self-resonant frequency of power distribution grids with DSDG is higher than the self-resonant frequency of the reference power delivery scheme with SSSG, increasing the maximum operating frequency of the overall system. Note that for comparable resonant frequencies, the resistance of the power distribution grid with DSDG is two times greater than the resistance of a conventional power grid with SSSG. Thus, power distribution grids with DSDG are more highly damped, resulting in reduced voltage fluctuations at the resonant frequency. Also note that on-chip decoupling capacitors lower the resonant frequency of the system. On-chip power distribution grids with decoupling capacitors should therefore be carefully designed to avoid (and control) any on-chip resonances.

Power distribution grids operating at 1 GHz (the low frequency case) have been analyzed in this chapter. Comparing the results listed in Tables 15.1 – 15.5, the mutual inductive coupling at 100 GHz (the high frequency case) increases, reducing the loop inductance. Thus, for future generations of ICs operating at high frequencies [22], the performance of power distribution grids with DSDG is expected to improve by reducing the power distribution noise.

15.8 Summary

Power distribution grids with multiple power supply voltages are analyzed in this chapter. The primary results can be summarized as follows:

- Two types of interdigitated and paired on-chip power distribution grids with DSDG are presented

- Closed-form expressions to estimate the loop inductance in four types of power distribution grids with DSDG have been developed

- With no decoupling capacitors placed between the power supply and ground, fully- and pseudo-interdigitated power distribution grids outperform a conventional interdigitated power distribution grid with DSSG by 15.3% and 0.3%, respectively, in terms of lower power noise

- In the case of power grids with decoupling capacitors, the voltage drop is reduced by about 9.2% for fully interdigitated grids with a 30 pF additional decoupling capacitance and is higher by 55.4% in the case of an added 20 pF decoupling capacitance

- If no decoupling capacitors are added, the voltage drop of a fully interdigitated power distribution grid with DSDG is reduced by 2.7%, on average, as compared to the voltage drop of an interdigitated power distribution grid with SSSG

- In the case of a fully paired grid, the power noise is reduced by about 2.3% as compared to the reference paired power distribution grid with SSSG

- With on-chip decoupling capacitors added to the power delivery networks, both fully interdigitated and fully paired power distribution grids with DSDG slightly outperform the reference power distribution scheme with SSSG

- On-chip decoupling capacitors are shown to lower the self-resonant frequency of the on-chip power distribution grid, producing resonances. An improper choice of the on-chip decoupling capacitors can therefore degrade the overall performance of a system

- Fully interdigitated and fully paired power distribution grids with DSDG should be utilized in those ICs where high isolation is required between the power supply voltages so as to effectively decouple the power supplies

16
Decoupling Capacitors for Multi-Voltage Power Distribution Systems

Power dissipation has become a critical design issue in high performance microprocessors as well as battery powered and wireless electronics, multimedia and digital signal processors, and high speed networking. The most effective way to reduce power consumption is to lower the supply voltage. Reducing the supply voltage, however, increases the circuit delay [295], [297], [330]. The increased delay can be compensated by changing the critical paths with behavioral transformations such as parallelization or pipelining [331]. The resulting circuit consumes less power while satisfying global throughput constraints at the cost of increased circuit area.

Recently, the use of multiple on-chip supply voltages has become common practice [310]. This strategy has the advantage of permitting modules along the critical paths to operate with the highest available voltage level (in order to satisfy target timing constraints) while permitting modules along the non-critical paths to use a lower voltage (thereby reducing the energy consumption). A multi-voltage scheme lowers the speed of those circuits operating at a lower power supply voltage without affecting the overall frequency, thereby reducing power without decreasing the system frequency. In this manner, the energy consumption is decreased without affecting circuit speed. This scheme results in a smaller area as compared to parallel architectures. The problem of using multiple supply voltages for reducing the power requirements has been investigated in the area of high level synthesis for low power [306], [327]. While it is possible to provide many supply voltages, in practice such a scenario is expensive. Practically, the availability of a small number of voltage supplies (two or three) is reasonable, as discussed in Chapter 14.

The design of the power distribution system has become an increasingly difficult challenge in modern CMOS circuits [28]. As CMOS technologies are scaled, the power supply voltage is lowered. As clock rates rise and more functions are integrated on-chip, the power consumed has greatly increased. Assuming that only a small per cent of the power supply voltage (about 10%) is permitted as ripple voltage (noise), a target impedance for an example power distribution system is [107]

$$Z_{\text{target}} = \frac{V_{\text{dd}} \times \zeta}{I} = \frac{1.8 \text{ volts} \times 10\%}{100 \text{ ampers}} \approx 0.002 \text{ ohms}, \qquad (16.1)$$

where V_{dd} is the power supply voltage, ζ is the allowed ripple voltage, and I is the current. With general scaling theory [286], the current I is increasing and the power supply voltage is decreasing. The impedance of a power distribution system should therefore be decreased to satisfy power noise constraints. The target impedance of a power distribution system is falling at an alarming rate, a factor of five per computer generation [332]. The target impedance must be satisfied not only at DC, but also at all frequencies where current transients exist [134]. Several major components of a power delivery system are used to satisfy a target impedance over a broad frequency range. A voltage regulator module is effective up to about 1 kHz. Bulk capacitors supply current and maintain a low power distribution system impedance from 1 kHz to 1 MHz. High frequency ceramic capacitors maintain the power distribution system impedance from 1 MHz to several hundred MHz. On-chip decoupling capacitors can be effective above 100 MHz.

By introducing a second power supply, the power supplies are coupled through a decoupling capacitor effectively placed between the two power supply networks. Assuming a power delivery system with dual power supplies and only a small per cent of the power supply voltage is permitted as ripple voltage (noise), the following inequality for the magnitude of a voltage transfer function K_V should be satisfied,

$$|K_V| \leq \frac{\chi V_{\text{dd1}}}{V_{\text{dd2}}}, \qquad (16.2)$$

where V_{dd1} is a lower voltage power supply, χ is the allowed ripple voltage on a lower voltage power supply, and V_{dd2} is a higher voltage power supply. Since the higher voltage power supply is applied to the high speed paths, as for example a clock distribution network, V_{dd2} can be noisy. To guarantee that noise from the higher voltage supply does not affect the quiet power supply, (16.2) should be satisfied. For typical values of the power supply voltages and allowed ripple voltage for a CMOS $0.18\,\mu$m technology, $|K_V|$ is chosen to be less than or equal to 0.1 to effectively decouple a noisy power supply from a quiet power supply.

The design of a power distribution system with multiple supply voltages is the primary focus of this chapter. The influence of a second supply voltage on a system of decoupling capacitors is investigated. Noise coupling among multiple power distribution systems is also discussed in this chapter. A criterion for producing an overshoot-free voltage response is determined. It is shown that to satisfy a target specification in order to decouple multiple power supplies, it is necessary to maintain the magnitude of the voltage transfer function below 0.1. In certain cases, it is difficult to satisfy this criterion over the entire range of operating frequencies. In such a scenario, the frequency range of an overshoot-free voltage response can be traded off with the magnitude of the response [26]. Case studies are also presented in the chapter to quantitatively illustrate this methodology for designing a system of decoupling capacitors.

The chapter is organized as follows. The impedance of a power distribution system with multiple supply voltages is described in Section 16.1. A case study of the dependence of the impedance on the power distribution system parameters is presented in Section 16.2. The voltage transfer function of a power distribution system with multiple supply voltages is discussed in Section 16.3. Case studies examining the dependence of the magnitude of the voltage transfer function on the parameters of the power distribution system are illustrated in Section 16.4. Some specific conclusions are summarized in Section 16.5.

16.1 Impedance of a power distribution system

The impedance of a power distribution network is an important issue in modern high performance ICs such as microprocessors. The impedance should be maintained below a target level to guarantee the power and signal integrity of a system, as described in Chapter 5. The impedance of a power distribution system with multiple power supplies is described in Section 16.1.1. The antiresonance of capacitors connected in parallel is addressed in Section 16.1.2. The dependence of the impedance on the power distribution system is investigated in Section 16.1.3.

16.1.1 Impedance of a power distribution system

A model of the impedance of a power distribution system with two supply voltages is shown in Fig. 16.1. The impedance seen from the load of the power supply V_{dd1} is illustrated. The model of the impedance is applicable for the load of the power supply V_{dd2} if Z_1 is substituted for Z_2. The impedance of the power distribution system shown in Fig. 16.1 can be modeled as

$$Z = \frac{Z_1 Z_{12} + Z_1 Z_2}{Z_1 + Z_{12} + Z_2}.$$ (16.3)

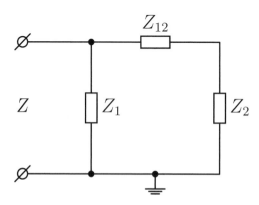

Fig. 16.1. Impedance of power distribution system with two supply voltages seen from the load of the power supply V_{dd1}.

Decoupling capacitors have traditionally been modeled as a series RLC network [132]. A schematic representation of a power distribution network with two supply voltages and the decoupling capacitors represented by RLC series networks is shown in Fig. 16.2.

In this case, the impedance of the power distribution network is

$$Z = \frac{a_4 s^4 + a_3 s^3 + a_2 s^2 + a_1 s + a_0}{b_3 s^3 + b_2 s^2 + b_1 s},$$ (16.4)

where

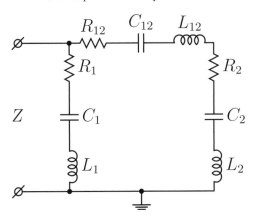

Fig. 16.2. Impedance of power distribution system with two supply voltages and the decoupling capacitors represented as series RLC networks.

$$a_4 = L_1(L_{12} + L_2), \tag{16.5}$$

$$a_3 = R_1 L_{12} + R_{12} L_1 + R_1 L_2 + R_2 L_1, \tag{16.6}$$

$$a_2 = R_1 R_{12} + R_1 R_2 + \frac{L_1}{C_{12}} + \frac{L_{12}}{C_1} + \frac{L_1}{C_2} + \frac{L_2}{C_1}, \tag{16.7}$$

$$a_1 = \frac{R_1}{C_2} + \frac{R_2}{C_1} + \frac{R_1}{C_{12}} + \frac{R_{12}}{C_1}, \tag{16.8}$$

$$a_0 = \frac{C_{12} + C_2}{C_1 C_{12} C_2}, \tag{16.9}$$

$$b_3 = L_1 + L_{12} + L_2, \tag{16.10}$$

$$b_2 = R_1 + R_{12} + R_2, \tag{16.11}$$

$$b_1 = \frac{1}{C_1} + \frac{1}{C_{12}} + \frac{1}{C_2}, \tag{16.12}$$

and $s = j\omega$ is a complex frequency.

The frequency dependence of the closed form expression for the impedance of a power distribution system with dual power supply voltages is illustrated in Fig. 16.3. The minimum power distribution system impedance is limited by the ESR of the decoupling capacitors. For on-chip applications, the ESR includes the parasitic resistance of the decoupling capacitor and the resistance of the power distribution network connecting a decoupling capacitor to a load. The resistance of the on-chip power distribution network is greater than the parasitic resistance of the on-chip decoupling capacitors. For on-chip applications, therefore, the ESR is represented by the resistance of the power delivery system. Conversely, for printed circuit board applications, the resistance of the decoupling capacitors dominates the resistance of the power delivery system. In this case, therefore, the ESR is primarily the resistance of the decoupling capacitors. In order to achieve a target impedance as described by (16.1), multiple decoupling capacitors are placed at different levels of the power grid hierarchy [107].

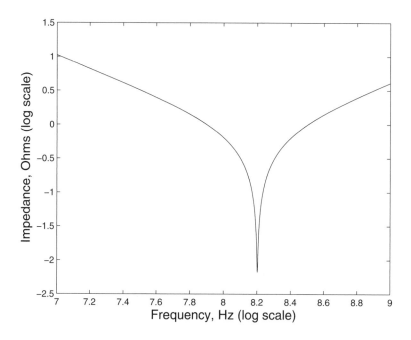

Fig. 16.3. Frequency dependence of the impedance of a power distribution system with dual supply voltages, $R_1 = R_{12} = R_2 = 10\,\mathrm{m\Omega}$, $C_1 = C_{12} = C_2 = 1\,\mathrm{nF}$, and $L_1 = L_{12} = L_2 = 1\,\mathrm{nH}$. Since all of the parameters of a power distribution system are identical, the system behaves as a single capacitor with one minimum at the resonant frequency. The minimum power distribution system impedance is limited by the ESR of the decoupling capacitors.

As described in [109], the ESR of the decoupling capacitors does not change the location of the poles and zeros of the power distribution system impedance, only the damping factor of the RLC system formed by the decoupling capacitor is affected. Representing a decoupling capacitor with a series LC network, the impedance of the power distribution system with dual power supply voltages is

$$Z = \frac{a_4 s^4 + a_2 s^2 + a_0}{b_3 s^3 + b_1 s}, \qquad (16.13)$$

where

$$a_4 = L_1(L_{12} + L_2), \qquad (16.14)$$

$$a_2 = \frac{L_1}{C_{12}} + \frac{L_{12}}{C_1} + \frac{L_1}{C_2} + \frac{L_2}{C_1}, \qquad (16.15)$$

$$a_0 = \frac{C_{12} + C_2}{C_1 C_{12} C_2}, \qquad (16.16)$$

$$b_3 = L_1 + L_{12} + L_2, \qquad (16.17)$$

$$b_1 = \frac{1}{C_1} + \frac{1}{C_{12}} + \frac{1}{C_2}. \qquad (16.18)$$

16.1.2 Antiresonance of parallel capacitors

To maintain the impedance of a power distribution system below a specified level, multiple decoupling capacitors are placed in parallel at different levels of the power grid hierarchy. The ESR affects the quality factor of the RLC system by acting as a damping element. The influence of the ESR on the impedance is therefore ignored. If all of the parameters of the circuit shown in Fig. 16.2 are equal, the impedance of the power distribution system can be described as a series RLC circuit. Expression (16.13) has four zeros and three poles. Two zeros are located at the same frequency as the pole when all of the parameters of the circuit are equal. The pole is therefore canceled for this special case and the circuit behaves as a series RLC circuit with one resonant frequency.

If the parameters of the power distribution system are not equal, the zeros of (16.13) are not paired. In this case, the pole is not canceled by a zero. For instance, in the case of two capacitors connected in parallel as shown in Fig. 16.4, in the frequency range from f_1 to f_2, the impedance of the capacitor C_1 has become inductive whereas the impedance of the capacitor C_2 remains capacitive. In this case, an LC tank will produce a peak at a resonant frequency located between f_1 and f_2. Such a phenomenon is called *antiresonance* [107] and is described in greater detail in Chapter 6.

The location of the antiresonant spike depends on the ratio of the ESL of the decoupling capacitors. Depending upon the parasitic inductance, the peak impedance caused by the decoupling capacitor is shifted to a different frequency, as shown in Fig. 16.4. For instance, if the parasitic inductance of C_1 is greater than the parasitic inductance of C_2, the antiresonance will appear at a frequency ranging from f_1 to f_2, *i.e.*, before the self-resonant frequency f_2 of the capacitor C_2. If the parasitic inductance of C_1 is lower than the parasitic inductance of C_2, the antiresonance will appear at a frequency ranging from f_2 to f_3, *i.e.*, after the self-resonant frequency of the capacitor C_2. The ESL of the decoupling capacitors, therefore, determines the frequency (location) of the antiresonant spike of the system [27].

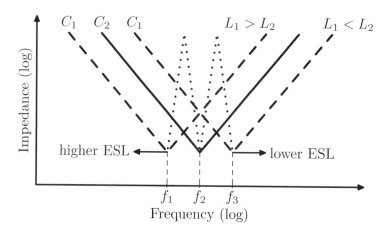

Fig. 16.4. Antiresonance of the two capacitors connected in parallel, $C_2 = C_1$. Two antiresonant spikes appear between frequencies f_1 and f_2 and f_2 and f_3 (dotted lines).

16.1.3 Dependence of impedance on power distribution system parameters

In practical applications, a capacitor C_{12} placed between V_{dd1} and V_{dd2} exists either as a parasitic capacitance or as a decoupling capacitor. Intuitively, from Fig. 16.2, by decreasing the impedance Z_{12} (increasing C_{12}), the greater part of Z_2 is connected in parallel with Z_1, reducing the impedance of the power distribution system as seen from the load of the power supply V_{dd1}. The value of a parasitic capacitance is typically much smaller than a decoupling capacitor such as C_1 and C_2. The decoupling capacitor C_{12} can be chosen to be equal to or greater than C_1 and C_2. Depending upon the placement of the decoupling capacitors, ESL can vary from 50 nH at the power supply to almost negligible values on-chip. The ESL includes both the parasitic inductance of the decoupling capacitors and the inductance of the power delivery system. For on-chip applications, the inductance of the decoupling capacitors is much smaller than the inductance of the power distribution network and can be ignored. At the board level, however, the parasitic inductance of the decoupling capacitors dominates the overall inductance of a power delivery system. For these reasons, the model depicted in Fig. 16.2 is applicable to any hierarchical level of a power distribution system from the circuit board to on-chip.

Assuming $C_1 = C_2$, if $C_{12} > C_1$, an antiresonance spike occurs at a lower frequency than the resonance frequency of an RLC series circuit. If $C_{12} < C_1$, the antiresonance spike occurs at a higher frequency than the resonance frequency of an RLC series circuit. This phenomenon is illustrated in Fig 16.5.

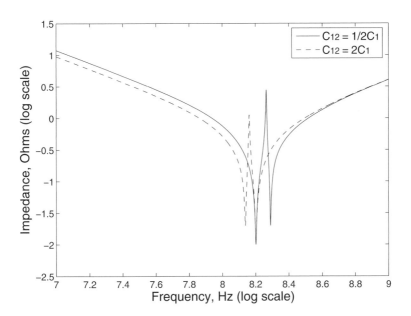

Fig. 16.5. Antiresonance of a power distribution system with dual power supply voltages, $R_1 = R_{12} = R_2 = 10\,\text{m}\Omega$, $C_1 = C_2 = 1\,\text{nF}$, and $L_1 = L_{12} = L_2 = 1\,\text{nH}$. Depending upon the ratio of C_{12} to C_1, the antiresonance appears before or after the resonant frequency of the system (the impedance minimum).

Antiresonance is highly undesirable because at a particular frequency, the impedance of a power distribution network can become unacceptably high. To cancel the antiresonance at a given frequency, a smaller decoupling capacitor is placed in parallel, shifting the antiresonance spike to a higher frequency. This procedure is repeated until the antiresonance spike appears at a frequency out of range of the operating frequencies of the system, as shown in Fig. 16.6.

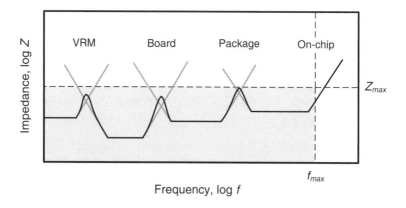

Fig. 16.6. Impedance of the power distribution system as a function of frequency. Decoupling capacitors are placed at different hierarchical levels to shift an antiresonant spike above the maximum operating frequency of the system.

Another technique for shifting the antiresonance spike to a higher frequency is to decrease the ESL of the decoupling capacitor. The dependence of the impedance of a power distribution system on the ESL is discussed below.

To determine the location of the antiresonant spikes, the roots of the denominator of (16.13) are evaluated. One pole is located at $\omega = 0$. Two other poles are located at frequencies,

$$\omega = \pm\sqrt{\frac{C_2 + C_1 C_2/C_{12} + C_1}{C_1 C_2 (L_1 + L_{12} + L_2)}}. \tag{16.19}$$

To shift the poles to a higher frequency, the ESL of the decoupling capacitors must be decreased. If the ESL of the decoupling capacitors is close to zero, the impedance of a power delivery network will not produce overshoots over a wide range of operating frequencies. Expression (16.19) shows that by minimizing the decoupling capacitor C_{12} between the two supply voltages, the operating frequency of the overshoot-free impedance of a power delivery network can be increased.

The dependence of the power distribution system impedance on the ESL of C_{12} is shown in Fig. 16.7(a). Note the strong dependence of the antiresonant frequency on the ESL of the decoupling capacitor located between V_{dd1} and V_{dd2}. As discussed above, the location of the antiresonant spike is determined by the ESL ratio of the decoupling capacitors. The magnitude of the antiresonance spike is determined by the total ESL of C_1, C_{12}, and C_2, as shown in Fig. 16.7(b).

By lowering the system inductance, the quality factor is decreased. The peaks become wider in frequency and lower in magnitude. The amplitude of the antiresonant spikes can be decreased by lowering the ESL of all of the decoupling capacitors within the power distribution system. As shown in Fig. 16.7(b), decreasing the parasitic inductance of all of the decoupling capacitors of the system reduces the peak magnitude. When the parasitic inductance of C_{12} is similar in magnitude to the other decoupling capacitors, from (16.4), the poles and zeros do not cancel, affecting the behavior of the circuit. The zero at the resonant frequency of a system (the minimum value of the impedance) decreases the antiresonant spike. The closer the location of an antiresonant spike is to the resonant frequency of a system, the greater the influence of a zero on the antiresonance behavior. From a circuits perspective, the more similar the ESL of each capacitor, the smaller the amplitude of the antiresonant spike. Decreasing the inductance of the decoupling capacitors has the same effect as increasing the resistance. Increasing the parasitic resistance of a decoupling capacitor is limited by the target impedance of the power distribution system. Decreasing the inductance of a power distribution system is highly desirable and, if properly designed, the inductance of a power distribution system can be significantly reduced [70].

16.2 Case study of the impedance of a power distribution system

The dependence of the impedance on the power distribution system parameters is described in this section to quantitatively illustrate the concepts presented in Section 16.1. An on-chip power distribution

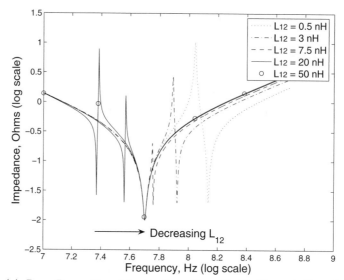

(a) $R_1 = R_{12} = R_2 = 10\,\mathrm{m\Omega}$, $C_1 = C_2 = 10\,\mathrm{nF}$, $C_{12} = 1\,\mathrm{nF}$, and $L_1 = L_2 = 1\,\mathrm{nH}$.

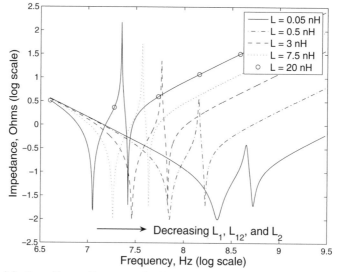

(b) $R_1 = R_{12} = R_2 = 10\,\mathrm{m\Omega}$, $C_1 = C_2 = 10\,\mathrm{nF}$, $C_{12} = 1\,\mathrm{nF}$, and $L_1 = L_{12} = L_2 = L$.

Fig. 16.7. Dependence of a dual V_{dd} power distribution system impedance on frequency for different ESL of the decoupling capacitors. The ESL of capacitors C_1, C_{12}, and C_2 is represented by L_1, L_{12}, and L_2, respectively.

system is assumed in this example. The total budgeted on-chip decoupling capacitance is distributed among the low voltage power supply ($C_1 = 10\,\mathrm{nF}$), high voltage power supply ($C_2 = 10\,\mathrm{nF}$), and the capacitance placed between the two power supplies ($C_{12} = 1\,\mathrm{nF}$). The ESR and ESL of the power distribution network are chosen to be equal to 0.1 ohms and 1 nH, respectively. The target impedance is 0.4 ohms.

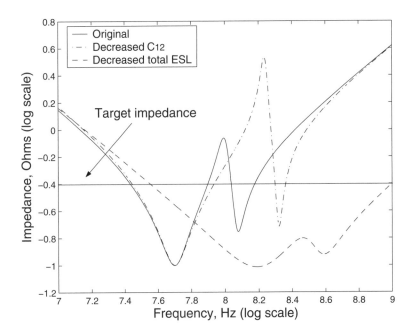

Fig. 16.8. The impedance of a power distribution system with dual power supply voltages as a function of frequency, $R_1 = R_{12} = R_2 = 100\,\mathrm{m\Omega}$, $C_1 = C_2 = 10\,\mathrm{nF}$, $C_{12} = 1\,\mathrm{nF}$, and $L_1 = L_{12} = L_2 = 1\,\mathrm{nH}$. The impedance of the example power distribution network produces an antiresonant spike with a magnitude greater than the target impedance (the solid line). The antiresonant spike is shifted to a higher frequency with a larger magnitude by decreasing C_{12} to 0.3 nF (the dashed-dotted line). By decreasing the total ESL of the system, the impedance can be maintained below the target impedance over a wide frequency range, from approximately 40 MHz to 1 GHZ (the dashed line).

For typical values of an example power distribution system, an antiresonant spike is produced at approximately 100 MHz with a magnitude greater than the target impedance, as shown in Fig. 16.8. According to (16.19), to shift the antiresonant spike to a higher frequency, the capacitor C_{12} should be decreased. As C_{12} is decreased to 0.3 nF,

Table 16.1. Case study of the impedance of a power distribution system

Tradeoff Scenario	Power Distribution System	Minimum frequency	Maximum frequency	Frequency range Δf
	Original	4 kHz	35.48 kHz	31.48 kHz
I	Decreased C_{12}	4 kHz	50.1 kHz	46.1 kHz
	Decreased L_1, L_{12}, L_2	4 kHz	1.26 MHz	1.256 MHz
	Original	100 kHz	1 MHz	900 kHz
II	Decreased C_{12}	100 kHz	2.82 MHz	2.72 MHz
	Decreased L_1, L_{12}, L_2	100 kHz	79 MHz	78.9 MHz
	Original	560 MHz	1 GHz	440 MHz
III	Decreased C_{12}	560 MHz	1.12 GHz	560 MHz
	Decreased L_1, L_{12}, L_2	890 MHz	7.9 GHz	7.01 GHz

Scenario I
Board

Original system: $R_1 = R_{12} = R_2 = 1\,\text{m}\Omega$, $L_1 = L_{12} = L_2 = 50\,\text{nH}$, $C_{12} = 100\,\mu\text{F}$, $C_1 = C_2 = 1\,\text{mF}$

Decreased C_{12}: $R_1 = R_{12} = R_2 = 1\,\text{m}\Omega$, $L_1 = L_{12} = L_2 = 50\,\text{nH}$, $C_{12} = 20\,\mu\text{F}$, $C_1 = C_2 = 1\,\text{mF}$

Decreased L_1, L_{12}, L_2: $R_1 = R_{12} = R_2 = 1\,\text{m}\Omega$, $L_1 = L_{12} = L_2 = 5\,\text{nH}$, $C_{12} = 100\,\mu\text{F}$, $C_1 = C_2 = 1\,\text{mF}$

Scenario II
Package

Original system: $R_1 = R_{12} = R_2 = 1\,\text{m}\Omega$, $L_1 = L_{12} = L_2 = 1\,\text{nH}$, $C_{12} = 3\,\mu\text{F}$, $C_1 = C_2 = 50\,\mu\text{F}$

Decreased C_{12}: $R_1 = R_{12} = R_2 = 1\,\text{m}\Omega$, $L_1 = L_{12} = L_2 = 1\,\text{nH}$, $C_{12} = 1\,\mu\text{F}$, $C_1 = C_2 = 50\,\mu\text{F}$

Decreased L_1, L_{12}, L_2: $R_1 = R_{12} = R_2 = 1\,\text{m}\Omega$, $L_1 = L_{12} = L_2 = 100\,\text{pH}$, $C_{12} = 3\,\mu\text{F}$, $C_1 = C_2 = 50\,\mu\text{F}$

Scenario III
On-chip

Original system: $R_1 = R_{12} = R_2 = 10\,\text{m}\Omega$, $L_1 = L_{12} = L_2 = 10\,\text{pH}$, $C_{12} = 1\,\text{nF}$, $C_1 = C_2 = 4\,\text{nF}$

Decreased C_{12}: $R_1 = R_{12} = R_2 = 10\,\text{m}\Omega$, $L_1 = L_{12} = L_2 = 10\,\text{pH}$, $C_{12} = 0.3\,\text{nF}$, $C_1 = C_2 = 4\,\text{nF}$

Decreased L_1, L_{12}, L_2: $R_1 = R_{12} = R_2 = 10\,\text{m}\Omega$, $L_1 = L_{12} = L_2 = 1\,\text{pH}$, $C_{12} = 1\,\text{nF}$, $C_1 = C_2 = 4\,\text{nF}$

the antiresonant spike appears at a higher frequency, approximately 158 MHz, and is of higher magnitude. To further decrease the impedance of a power distribution system with multiple power supply voltages, the total ESL of the decoupling capacitors should be decreased. As the total ESL of the system is decreased to 0.1 nH, the impedance of the power distribution system is below the target impedance over a wide frequency range, from approximately 40 MHz to 1 GHz. Three different tradeoff scenarios similar to the case study illustrated in Fig. 16.8 are summarized in Table 16.1. The design parameters for each scenario represent typical values of board, package, and on-chip power distribution systems with decoupling capacitors, as shown in Fig. 16.9. The minimum and maximum frequencies denote the frequency range in which the impedance of a power delivery network seen from the load of V_{dd1} does not exceed the target level of 400 mΩ. Note that by decreasing the decoupling capacitor placed between V_{dd1} and V_{dd2}, the range of operating frequencies, where the target impedance is met, is slightly increased. Alternatively, if the total ESL of the system is lowered by an order of magnitude, the frequency range Δf is increased by significantly more than an order of magnitude (for tradeoff scenario III, Δf increases from 560 MHz to 7.01 GHz).

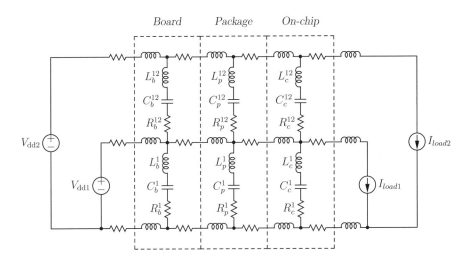

Fig. 16.9. Hierarchical model of a power distribution system with dual supply voltages and a single ground. The decoupling capacitors are represented by the series connected resistance, capacitance, and inductance. For simplicity, the decoupling capacitors placed between V_{dd2} and ground are not illustrated. Subscripts b, p, and c denote the board, package, and on-chip power delivery systems, respectively. Superscript 1 denotes the decoupling capacitors placed between V_{dd1} and ground and superscript 12 denotes the decoupling capacitors placed between V_{dd1} and V_{dd2}.

The design of a power distribution system with multiple power supply voltages is a complex task and requires many iterative steps. In general, to maintain the impedance of a power delivery system below a target level, the proper combination of design parameters needs to be determined. In on-chip applications, the ESL and C_{12} can be chosen to satisfy specific values. At the board level, the ESR and C_{12} can be adjusted to satisfy target impedance specifications. At the package level, the ESL, C_{12}, and ESR are the primary design parameters of the system. Usually, the total decoupling capacitance is constrained by the technology and application. In certain cases, it is possible to increase the decoupling capacitance. From (16.13), note that by increasing the decoupling capacitance, the overall impedance of a power distribution system with multiple power supply voltages can be significantly decreased.

16.3 Voltage transfer function of power distribution system

Classical methodologies for designing power distribution systems with a single power supply voltage typically only consider the target output impedance of the network. By introducing a second power supply voltage, a decoupling capacitor is effectively placed between the two power supply voltages [26], [199]. The problem of noise propagating from one power supply to the other power supply is aggravated if multiple power supply voltages are employed in a power distribution system. Since multiple power supplies are naturally coupled, the voltage transfer function of a multi-voltage power distribution network should be considered [333], [334]. The voltage transfer function of a power distribution system with dual power supplies is described in Section 16.3.1. The dependence of the magnitude of the voltage transfer function on certain parameters of the power distribution system is described in subsection 16.3.2.

16.3.1 Voltage transfer function of a power distribution system

A power distribution system with two power supply voltages and the decoupling capacitors represented by an RLC series network is shown in Fig. 16.10. All of the following formulae describing this system are symmetric in terms of the power supply voltages. The ESR and ESL of the three decoupling capacitors are represented by R_1, R_{12}, R_2 and L_1, L_{12}, L_2, respectively.

The voltage transfer function K_V of a power distribution system with two power supply voltages and decoupling capacitors, represented by an RLC network, is

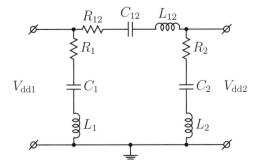

Fig. 16.10. Voltage transfer function of a power distribution network with two supply voltages and the decoupling capacitors represented as series RLC networks.

$$K_V = \frac{a_2 s^2 + a_1 s + a_0}{b_2 s^2 + b_1 s + b_0}, \tag{16.20}$$

where

$$a_2 = L_2 C_2, \tag{16.21}$$
$$a_1 = R_2 C_2, \tag{16.22}$$
$$a_0 = C_{12}, \tag{16.23}$$
$$b_2 = C_{12} C_2 (L_{12} + L_2), \tag{16.24}$$
$$b_1 = C_{12} C_2 (R_{12} + R_2), \tag{16.25}$$
$$b_0 = C_{12} + C_2. \tag{16.26}$$

Rearranging, (16.20) can be written as

$$K_V = \frac{1}{\dfrac{a_2 s^2 + a_1 s + a_0}{b_2 s^2 + b_1 s + b_0} + 1}, \tag{16.27}$$

where

$$a_2 = L_{12} C_{12} C_2, \tag{16.28}$$
$$a_1 = R_{12} C_{12} C_2, \tag{16.29}$$
$$a_0 = C_2, \tag{16.30}$$
$$b_2 = L_2 C_{12} C_2, \tag{16.31}$$
$$b_1 = R_2 C_{12} C_2, \tag{16.32}$$
$$b_0 = C_{12}. \tag{16.33}$$

Equations (16.20) and (16.27) are valid only for non-zero frequency, *i.e.*, for $s > 0$. Note from (16.20) that if all of the parameters of a power distribution system are identical, the transfer function equals 0.5 and is independent of frequency. The dependence of the voltage transfer function on the parameters of the power distribution system is discussed below.

16.3.2 Dependence of voltage transfer function on power distribution system parameters

In power distribution systems with two supply voltages, the higher power supply is usually provided for the high speed circuits while the lower power supply is used in the non-critical paths [316], [299]. The two power supplies are often strongly coupled, implying that voltage fluctuations on one power supply propagate to the other power supply. The magnitude of the voltage transfer function should be sufficiently small in order to decouple the noisy power supply from the quiet power supply. The objective is therefore to achieve a transfer function K_V such that the two power supplies are effectively decoupled.

The dependence of the magnitude of the voltage transfer function on frequency for different values of the ESR of the power distribution network with decoupling capacitors is shown in Fig. 16.11. Reducing the ESR of a decoupling capacitor decreases the magnitude and range of the operating frequency of the transfer function. Note that to maintain $|K_V|$ below or equal to 0.5, the following inequality has to be satisfied,

$$R_2 \leq R_{12}. \tag{16.34}$$

This behavior can be explained as follows. From (16.27), to maintain $|K_V|$ below or equal to 0.5,

$$\frac{L_{12}C_{12}C_2 s^2 + R_{12}C_{12}C_2 s + C_2}{L_2 C_{12}C_2 s^2 + R_2 C_{12}C_2 s + C_{12}} + 1 \geq 2. \tag{16.35}$$

For equal decoupling capacitors and parasitic inductances, (16.35) leads directly to (16.34). Generally, to maintain $|K_V|$ below or equal to 0.5,

$$L_2 C_2 C_3 s^2 + R_2 C_2 C_3 s + C_3 \geq L_3 C_2 C_3 s^2 + R_3 C_2 C_3 s + C_2. \tag{16.36}$$

From (16.36), in order to maintain the magnitude of the voltage transfer function below or equal to 0.5, the ESR and ESL of the decoupling capacitors should be chosen to satisfy (16.36).

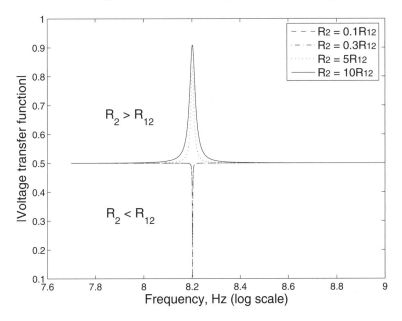

Fig. 16.11. Dependence of the magnitude of the voltage transfer function on frequency of a dual V_{dd} power distribution system for different values of ESR of the decoupling capacitors, $R_{12} = 10\,\text{m}\Omega$, $C_{12} = C_2 = 1\,\text{nF}$, and $L_{12} = L_2 = 1\,\text{nH}$.

To investigate the dependence of the magnitude of the voltage transfer function on the decoupling capacitors and associated parasitic inductances, the roots of the characteristic equation, the denominator of (16.20), should be analyzed. To produce an overshoot-free response, the roots of the characteristic equation must be real, yielding

$$R_{12} + R_2 \geq 2\sqrt{\frac{(L_{12} + L_2)(C_{12} + C_2)}{C_{12}C_2}}. \qquad (16.37)$$

In the case where $R_{12} = R_2 = R$, $L_{12} = L_2 = L$, and $C_{12} = C_2 = C$, (16.37) reduces to the well known formula [335],

$$R \geq 2\sqrt{\frac{L}{C}}. \qquad (16.38)$$

The dependence of the magnitude of the voltage transfer function on the ESL of a power distribution system is shown in Fig. 16.12. For the power distribution system parameters listed in Fig. 16.12, the critical value of L_2 to ensure an overshoot-free response is 0.49 nH. Therefore, in order to produce an overshoot-free response, the ESL of C_2 should be smaller than or equal to 0.49 nH.

Intuitively, if the ESL of a system is large, the system is under-damped and produces an undershoot and an overshoot. By decreasing L_2, the resulting inductance of the system in (16.37) is lowered and the system becomes more damped. As a result, the undershoots and overshoots of the voltage response are significantly smaller. If L_2 is decreased to the critical value, the system becomes overdamped, producing an overshoot-free voltage response.

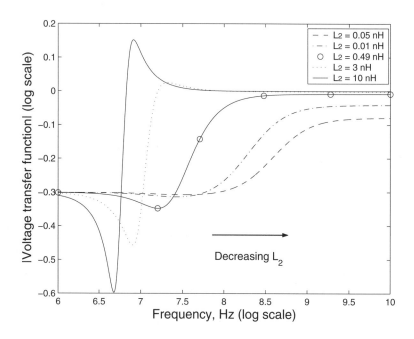

Fig. 16.12. Frequency dependence of the voltage transfer function of a dual V_{dd} power distribution system for different values of ESL of the decoupling capacitors, $R_{12} = R_2 = 100\,\mathrm{m\Omega}$, $C_{12} = C_2 = 100\,\mathrm{nF}$, and $L_{12} = 10\,\mathrm{pH}$.

As shown in Fig. 16.12, the magnitude of the voltage transfer function is strongly dependent on the ESL, decreasing with smaller ESL. It is highly desirable to maintain the ESL as low as possible to achieve a small overshoot-free response characterizing a dual V_{dd} power distribution system over a wide range of operating frequencies. Criterion (16.37) is strict and produces an overshoot-free voltage response. In most applications, if small overshoots (about 1%) are permitted, (16.37) is less strict, permitting the parameters of a power distribution network to vary over a wider range.

For the parameters listed in Fig. 16.12, the minimum overshoot-free voltage response equals 0.5. It is often necessary to maintain an extremely low magnitude voltage transfer function over a specific frequency range. This behavior can be achieved by varying one of the three design parameters (ESR, ESL or C) characterizing a decoupling capacitor while maintaining the other parameters at predefined values. In this case, for different decoupling capacitors, the magnitude of the voltage transfer function is maintained as low as 0.1 over the frequency range from DC to the self-resonant frequency of the decoupling capacitor induced by the RLC series circuit (hereafter called the *break frequency*).

The inductance of the decoupling capacitor has an opposite effect on the magnitude of the voltage transfer function. By increasing the ESL of a dual V_{dd} power distribution system, the magnitude of the voltage transfer function can be maintained below 0.1 from the self-resonant frequency (or break frequency) of the decoupling capacitor to the maximum operating frequency. From (16.27), for frequencies smaller than the break frequency, the magnitude of the voltage transfer function is approximately $\frac{C_{12}}{C_2}$. For frequencies greater than the break frequency, the magnitude of the voltage transfer function is approximately $\frac{L_2}{L_{12}}$. To maintain $|K_V|$ below 0.1, it is difficult to satisfy (16.37), and the range of operating frequency is divided by the break frequency into two ranges. This phenomenon is illustrated in Figs. 16.13(a) and 16.13(b).

16.4 Case study of the voltage response of a power distribution system

The dependence of the voltage transfer function on the parameters of a power distribution system is described in this section to quantitatively illustrate the concepts presented in Section 16.3. An on-chip power distribution system is assumed in this example. In modern high performance ICs, the total on-chip decoupling capacitance can exceed 300 nF, occupying about 20% of the total area of an IC [204]. In this example, the on-chip decoupling capacitance is assumed to be 160 nF. The total budgeted on-chip decoupling capacitance is arbitrarily distributed among the low voltage power supply ($C_1 = 100$ nF), high voltage power

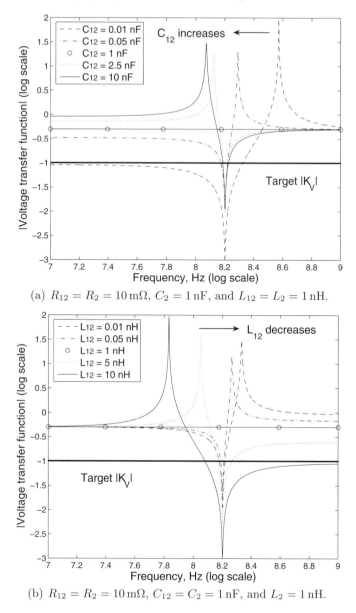

(a) $R_{12} = R_2 = 10\,\mathrm{m\Omega}$, $C_2 = 1\,\mathrm{nF}$, and $L_{12} = L_2 = 1\,\mathrm{nH}$.

(b) $R_{12} = R_2 = 10\,\mathrm{m\Omega}$, $C_{12} = C_2 = 1\,\mathrm{nF}$, and $L_2 = 1\,\mathrm{nH}$.

Fig. 16.13. Frequency dependence of the voltage transfer function of a dual V_{dd} power distribution system. The ESR and ESL of the decoupling capacitors for each power supply are represented by R_{12}, R_2 and L_{12}, L_2, respectively.

supply ($C_2 = 40\,\text{nF}$), and the capacitance placed between the two power
supplies ($C_{12} = 20\,\text{nF}$). The ESR and ESL of the decoupling capacitor
are chosen to be 0.1 ohms and 1 nH, respectively.

In designing a power distribution system with dual power supply
voltages, it is crucial to produce an overshoot-free voltage response
over the range of operating frequencies. Depending on the system parameters, it can be necessary to further decouple the power supplies, requiring the magnitude of the voltage transfer function to be decreased.
In this case, it is difficult to satisfy (16.37) and the range of operating
frequencies is therefore divided into two. There are two possible scenarios: 1) the two power supplies should be decoupled as much as possible
from DC to the break frequency, and 2) the two power supplies should
be decoupled as much as possible from the break frequency to infinity.

Note that infinite frequency is constrained by the maximum operating frequency of a specific system. Also note that the ESR, ESL,
and magnitude of the decoupling capacitors can be considered as design parameters. The ESR is limited by the target impedance of the
power distribution network. The ESL, however, can vary significantly.
The total budgeted decoupling capacitance is distributed among C_1,
C_{12}, and C_2. Note that C_{12} can range from zero (no decoupling capacitance between the two power supplies) to $C_{12} = C_{total} - C_1 - C_2$
(the maximum available decoupling capacitance between the two power
supplies), where C_{total} is the total budgeted decoupling capacitance.

16.4.1 Overshoot-free magnitude of a voltage transfer function

For typical values of an example power distribution system, (16.37)
is not satisfied and the response of the voltage transfer function produces an overshoot as shown in Fig. 16.14. To produce an overshoot-free
voltage response, the capacitor placed between the two power supplies
should be significantly increased, permitting the ESR and ESL to be
varied. Increasing the ESR of the decoupling capacitors to 0.5 ohms
produces an overshoot-free response. By decreasing the ESL of C_2, the
overshoot-free voltage response can be further decreased, also shown in
Fig. 16.14. As described in Section 16.3.2, at low frequency the magnitude of the voltage transfer function is approximately $\frac{C_{12}}{C_2}$. Note that
all curves start from the same point. By increasing the ESR, the system
becomes overdamped and produces an overshoot-free voltage response.
Since the ESR does not change the $\frac{L_2}{L_{12}}$ ratio, the voltage response of

the overdamped system is the same as the voltage response of the initial underdamped system. Note that the dashed line and solid line converge to the same point at high frequencies, where the magnitude of the voltage transfer function is approximately $\frac{L_2}{L_{12}}$. By decreasing L_2, the total ESL of the system is lowered and the system becomes overdamped, producing an overshoot-free voltage response. Also, since the $\frac{L_2}{L_{12}}$ ratio is lowered, the magnitude of the voltage response is significantly reduced at high frequencies.

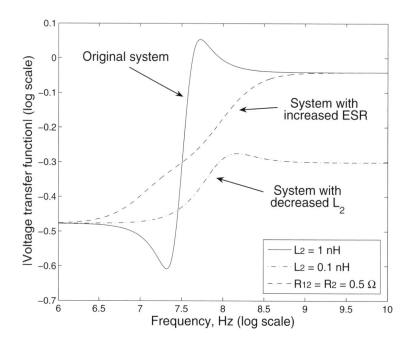

Fig. 16.14. Dependence of the magnitude of the voltage transfer function of a dual V_{dd} power distribution system on frequency for different values of the ESR and ESL of the decoupling capacitors, $R_{12} = R_2 = 0.1\,\Omega$, $C_{12} = 20\,\text{nF}$, $C_2 = 40\,\text{nF}$, and $L_{12} = L_2 = 1\,\text{nH}$. The initial system with $L_2 = 1\,\text{nH}$ produces an overshoot (solid line). To produce an overshoot-free voltage response, either the ESR of the system should be increased (dashed line) or the ESL should be decreased (dash-dotted line).

In general, a design methodology for producing an overshoot-free response of a power distribution system with dual power supply voltages is as follows. Based on the available decoupling capacitance for each power supply, the value of the decoupling capacitor placed between the two power supplies is determined by $C_{12} = C_{total} - C_1 - C_2$. The ESR is chosen to be less than or equal to the target impedance to satisfy the impedance constraint. The critical ESL of the capacitors C_{12} and C_2 is determined from (16.37). If the parasitic inductance of C_{12} and C_2 is less than or equal to the critical ESL, the system will produce an overshoot-free voltage response and no adjustment is required. Otherwise, the total decoupling capacitance budget should be redistributed among C_1, C_{12}, and C_2 until (16.37) is satisfied. In certain cases, the total budgeted decoupling capacitance should be increased to satisfy (16.37).

16.4.2 Tradeoff between the magnitude and frequency range

If it is necessary to further decouple the power supplies, the frequency range of the overshoot-free voltage response can be traded off with the magnitude of the voltage response, as described in Section 16.3.2. There are two ranges of interest. The magnitude of the voltage transfer function can be decreased over the frequency range from DC to the break frequency or from the break frequency to the highest operating frequency [26]. For the example power distribution system, as shown in Fig. 16.15(a), the magnitude of the voltage transfer function is overshoot-free from the break frequency to the highest operating frequency. To further decrease the magnitude of the voltage transfer function over a specified frequency range, the ESL of the decoupling capacitor placed between the two power supply voltages should be increased and C_{12} should be the maximum available decoupling capacitance, $C_{12} = C_{total} - C_1 - C_2$.

To decrease the magnitude of the voltage transfer function of a power distribution system with dual power supply voltages for frequencies less than the break frequency, the ESL of all of the decoupling capacitors and the value of C_{12} should be decreased, as shown in Fig. 16.15(b). If it is necessary to completely decouple the two power supply voltages, C_{12} should be minimized. This behavior can be explained as follows. The initial system produces an overshoot-free voltage response in the frequency range from DC to the highest operating frequency of the system. In order to satisfy the target $|K_V|$ at high frequencies, L_{12} should be increased in order to decrease the $\frac{L_2}{L_{12}}$ ratio. By increasing L_{12}, the magnitude of the voltage response falls below the target $|K_V|$ in the frequency range from the break frequency to the highest operating frequency of the system. At the same time, the

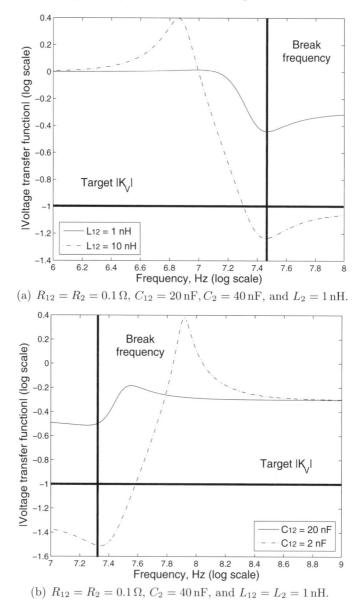

(a) $R_{12} = R_2 = 0.1\,\Omega$, $C_{12} = 20\,\mathrm{nF}$, $C_2 = 40\,\mathrm{nF}$, and $L_2 = 1\,\mathrm{nH}$.

(b) $R_{12} = R_2 = 0.1\,\Omega$, $C_2 = 40\,\mathrm{nF}$, and $L_{12} = L_2 = 1\,\mathrm{nH}$.

Fig. 16.15. Magnitude of the voltage transfer function of an example dual V_{dd} power distribution system as a function of frequency. The ESR and ESL of the decoupling capacitors are represented by R_{12} and R_2 and L_{12} and L_2, respectively.

system becomes underdamped and produces an overshoot as shown in Fig. 16.15(a). Similarly, by decreasing C_{12}, the $\frac{C_{12}}{C_2}$ ratio is lowered and the magnitude of the voltage response falls below the target $|K_V|$ in the frequency range from DC to the break frequency. The system becomes underdamped and produces an overshoot as shown in Fig. 16.15(b).

Table 16.2. Tradeoff between the magnitude and frequency range of the voltage response

Tradeoff Scenario	Power Distribution System	Minimum $\|K_V\|$	Maximum $\|K_V\|$	Minimum frequency	Maximum frequency
	Original	0.30	0.50	DC	∞
I	Increased L_{12}	0.09	0.56	63 kHz	∞
	Decreased C_{12}	0.05	0.60	DC	63 kHz
	Original	0.20	0.50	DC	∞
II	Increased L_{12}	0.09	0.50	3 MHz	∞
	Decreased C_{12}	0.03	0.60	DC	3 MHz
	Original	0.20	0.50	DC	∞
III	Increased L_{12}	0.09	0.50	3 GHz	∞
	Decreased C_{12}	0.05	0.45	DC	3 GHz

Scenario I Board   Original circuit: $R_{12} = R_2 = 2\,\text{m}\Omega$, $L_{12} = L_2 = 1\,\text{nH}$, $C_{12} = 2\,\text{mF}$, $C_2 = 4\,\text{mF}$

Increased L_{12}: $R_{12} = R_2 = 2\,\text{m}\Omega$, $L_{12} = 10\,\text{nH}$, $L_2 = 1\,\text{nH}$, $C_{12} = 2\,\text{mF}$, $C_2 = 4\,\text{mF}$

Decreased C_{12}: $R_{12} = R_2 = 2\,\text{m}\Omega$, $L_{12} = L_2 = 1\,\text{nH}$, $C_{12} = 200\,\mu\text{F}$, $C_2 = 4\,\text{mF}$

Scenario II Package   Original circuit: $R_{12} = R_2 = 10\,\text{m}\Omega$, $L_{12} = L_2 = 100\,\text{pH}$, $C_{12} = 10\,\mu\text{F}$, $C_2 = 40\,\mu\text{F}$

Increased L_{12}: $R_{12} = R_2 = 10\,\text{m}\Omega$, $L_{12} = 1\,\text{nH}$, $L_2 = 100\,\text{pH}$, $C_{12} = 10\,\mu\text{F}$, $C_2 = 40\,\mu\text{F}$

Decreased C_{12}: $R_{12} = R_2 = 10\,\text{m}\Omega$, $L_{12} = L_2 = 100\,\text{pH}$, $C_{12} = 1\,\mu\text{F}$, $C_2 = 40\,\mu\text{F}$

Scenario III On-chip   Original circuit: $R_{12} = R_2 = 10\,\text{m}\Omega$, $L_{12} = L_2 = 100\,\text{fH}$, $C_{12} = 20\,\text{nF}$, $C_2 = 40\,\text{nF}$

Increased L_{12}: $R_{12} = R_2 = 10\,\text{m}\Omega$, $L_{12} = 1\,\text{pH}$, $L_2 = 100\,\text{fH}$, $C_{12} = 20\,\text{nF}$, $C_2 = 40\,\text{nF}$

Decreased C_{12}: $R_{12} = R_2 = 10\,\text{m}\Omega$, $L_{12} = L_2 = 100\,\text{fH}$, $C_{12} = 2\,\text{nF}$, $C_2 = 40\,\text{nF}$

Three different tradeoff scenarios similar to the case study shown in Fig. 16.14 are summarized in Table 16.2. The design parameters for each scenario represent typical values of board, package, and on-chip decoupling capacitors, as shown in Fig. 16.9. The original system in each scenario produces an overshoot-free voltage response over a wide range of operating frequencies from DC to the highest operating frequency of the system. By increasing the ESL of the decoupling capacitor placed between the two power supplies, the system produces an overshoot and the range of operating frequencies is divided by two. The same phenomenon takes place if the value of the decoupling capacitor placed between the two power supplies is decreased. In the first case, when the ESL is increased by an order of magnitude, the magnitude of the voltage response is lowered by more than an order of magnitude from the break frequency to infinity. When C_{12} is decreased by an order of magnitude, the magnitude of the voltage response is lowered by more than an order of magnitude from DC to the break frequency. Note from the table that the location of the break point depends upon the particular system parameters. The break frequency of the board system occurs at a lower frequency as compared to the break frequency of the package power delivery network. Similarly, the break frequency of the package power distribution system is lower than the break frequency of the on-chip system. As previously mentioned, for typical power supplies values and allowed ripple voltage, $|K_V|$ should be less than 0.1 to decouple a noisy power supply from a quiet power supply. As listed in Table 16.2, this requirement is satisfied for the power distribution system if L_{12} is increased or C_{12} is decreased. The magnitude of the overshoot falls rapidly with decreasing ESL of the decoupling capacitors. Due to the extremely low value of the ESL in an on-chip power network, typically several hundred femtohenrys, the magnitude of the overshoot does not exceed the maximum magnitude of the overshoot-free voltage response.

Unlike the design methodology for producing an overshoot-free response as described in Section 16.4.1, a design methodology to trade off the magnitude of the voltage response of the power distribution system with the frequency range of an overshoot-free response is as follows. Based upon the available decoupling capacitance, the decoupling capacitances for each power supply are determined. Depending upon the target frequency range with respect to the break frequency, the ESL of the capacitor placed between the two power supplies and the decoupling capacitors should both be increased (above the break frequency). Otherwise, the capacitor placed between the two power supplies and the ESL of all of the decoupling capacitors should both be decreased (below the break frequency).

16.5 Summary

A system of decoupling capacitors used in power distribution systems with multiple power supply voltages is described in this chapter. The primary conclusions are summarized as follows:

- Multiple on-chip power supply voltages are often utilized to reduce power dissipation without degrading system speed

- To maintain the impedance of a power distribution system below a specified impedance, multiple decoupling capacitors are placed at different levels of the power grid hierarchy

- The decoupling capacitors should be placed both with progressively decreasing value to shift the antiresonance spike beyond the maximum operating frequency and with increasing ESR to control the damping characteristics

- The magnitude of the antiresonant spikes can also be limited by reducing the ESL of each of the decoupling capacitors

- To maintain the magnitude of the voltage transfer function below 0.5, the ESR and ESL of the decoupling capacitors should be carefully chosen to satisfy the overshoot-free voltage response criterion

- To further decouple the power supplies in frequencies ranging from DC to the break frequency, both the capacitor placed between the two power supply voltages and the ESL of each of the decoupling capacitors should be decreased

- To decouple the power supplies in frequencies ranging from the break frequency to infinity, both the ESL of the capacitor placed between the two power supply voltages and the decoupling capacitors should be increased

- The frequency range of an overshoot-free voltage response can be traded off with the magnitude of the response

17

On-chip Power Noise Reduction Techniques in High Performance ICs

Future generations of integrated circuit technologies are trending toward higher speeds and densities. The total capacitive load associated with the internal circuitry has been increasing for several generations of high complexity integrated circuits [165], [166]. As the operating frequencies increase, the average on-chip current required to charge and discharge these capacitances also has increased, while the switching time has decreased. As a result, a large change in the total on-chip current can occur within a brief period of time.

Due to the high slew rate of the currents flowing through the bonding wires, package pins, and on-chip interconnects, the ground and supply voltage can fluctuate (or bounce) due to the parasitic impedances associated with the package-to-chip and on-chip interconnects. These voltage fluctuations on the supply and ground rails, called ground bounce, ΔI noise, or simultaneous switching noise (SSN) [284], are larger since a significant number of the I/O drivers and internal logic circuitry switch close in time to the clock edges. SSN generates glitches on the ground and power supply wires, decreasing the effective current drive of the circuits, producing output signal distortion, thereby reducing the noise margins of a system. As a result, the performance and functionality of the system can be severely compromised.

In the past, research on SSN has concentrated on transient power noise caused by current flowing through the inductive bonding wires at the I/O buffers. SSN originating from the internal circuitry, however, has become an important issue in the design of VDSM high performance ICs, such as systems-on-chip, mixed-signal circuits, and microprocessors. This increased importance is due to fast clock rates, large on-chip switching activities and currents, and increased on-chip inductance, all of which are increasingly common characteristics of VDSM synchronous ICs.

Most of the work in this area falls into one of two categories: the first category includes analytic models that predict the behavior of the SSN, while the second category describes techniques to reduce ground bounce. A number of approaches have previously been proposed to analyze power and ground bounce and the effect of SSN on the performance of high complexity integrated circuits. Senthinathan *et al.* described an accurate technique for estimating the peak ground bounce noise by observing negative local feedback present in the current path of the driver [336]. This work suffers from the assumption that the switching currents of the output drivers are modeled as a triangular shape. In [337], Vaidyanath, Thoroddsen, and Prince relaxed this assumption by deriving an expression for the peak value of the ground bounce under the more realistic assumption that the ground bounce is a linear function of time during the output transition of the driver. Other research has considered short-channel effects in CMOS devices on the ground bounce waveform [338], [339], [340]. While most prior research has concentrated on the case where all of the drivers switch simultaneously, the authors in [339] consider the more realistic scenario when the drivers switch at different times. The idea of considering the effects of ground bounce on a tapered buffer has been presented in [341]. Tang and Friedman developed an analytic expression characterizing the on-chip SSN voltage based on a lumped RLC model characterizing the on-chip power supply rail rather than a single inductor to model a bonding wire [21]. In [342], Heydari and Pedram addressed ground bounce with no assumptions about the form of the switching current or noise voltage waveforms. The effect of ground bounce on the propagation delay and the optimum tapering factor of a multistage buffer is discussed. An analytic expression for the total propagation delay in the presence of ground bounce is also developed.

A number of techniques have been proposed to reduce SSN. In [343], a voltage controlled output buffer is described to control the slew rate. Ground bounce reduction is achieved by lowering the inductance in the power and ground paths by utilizing substrate conduction. An algorithm based on integer linear programming to skew the switching of the drivers to minimize ground bounce is presented in [344]. An architectural approach for reducing inductive noise caused by clock gating through gradual activation/deactivation units has been introduced in [345]. In [346], a routing method is described to distribute the ground bounce among the pads under a constraint of constant routing area. The total P/G noise of the system, however, is not reduced. Decoupling capacitors are often added to maintain the voltage on the P/G rails within specification, providing charge for the switching transients [342], [347]. Recently, various methods for reducing ground bounce have been introduced, such as bounce pre-generator circuits [348], supply current shaping, and clock frequency modulation [349].

Design techniques to reduce P/G noise in mixed-signal power distribution systems is the primary focus of this chapter. The efficiency of these techniques is based on the physical parameters of the system. The chapter is organized as follows. Ground noise reduction through the addition of a noise-free on-chip ground is described in Section 17.1. The efficiency of the technique as a function of the physical parameters of the system is investigated in Section 17.2. Some specific conclusions are summarized in Section 17.3.

17.1 Ground noise reduction through an additional low noise on-chip ground

An equivalent circuit of an SoC-based power delivery system is shown in Fig. 17.1. Traditionally, noisy digital circuits share the power and ground supply with noise sensitive analog circuits (see Chapter 15). If a number of digital blocks switch simultaneously, the current I_D drawn from the power distribution network can be significant. This large current passes through the parasitic resistance R_{Gnd}^p and inductance L_{Gnd}^p of the package, producing voltage fluctuations on the ground terminal (point A). As a result, ground bounce (or voltage fluctuations) appears at the ground terminal of the noise sensitive circuits.

To reduce voltage fluctuations at the ground terminal of the noise sensitive blocks, an on-chip low noise ground is added, as shown in Fig. 17.2. This approach utilizes a voltage divider formed by the impedance between the noisy ground terminal and the quiet ground terminal and the impedance of the path from the quiet ground terminal to the off-chip ground. The value of the capacitor is chosen to cancel the parasitic inductance of the additional low noise ground, $i.e.$, the ESL of the capacitor L_d and the on-chip and package parasitic inductances of the dedicated low noise ground L_c^3 and L_p^3, respectively. Alternatively, the capacitor is tuned in resonance with the parasitic inductances at a

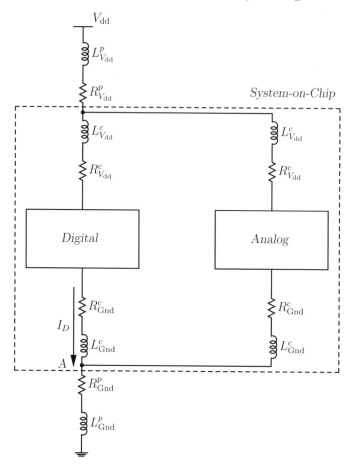

Fig. 17.1. An equivalent circuit for analyzing ground bounce in an SoC. The power distribution network is modeled as a series resistance and inductance. The superscripts p and c denote the parasitic resistance and inductance of the package and on-chip power delivery systems, respectively. The subscript V_{dd} denotes the power supply voltage and the superscript G_{nd} denotes the ground.

frequency that produces the greatest reduction in noise. The impedance of the additional ground path, therefore, behaves as a simple resistance.

The same technique can be used to reduce voltage fluctuations on the power supply. Based on the nature of the power supply noise, an additional ground path or power supply path can be provided. For instance, to ensure that the voltage does not drop below the power supply level, an on-chip path to the power supply is added. In the case of an overshoot, an additional ground path can be provided.

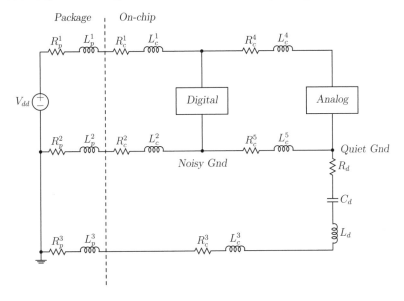

Fig. 17.2. Ground bounce reduction technique. The effective series resistance and effective series inductance of the decoupling capacitor are modeled by R_d and L_d, respectively. R_c^5 and L_c^5 represent the physical separation between the noisy and noise sensitive blocks. The impedance of the additional on-chip ground is modeled by R_c^3 and L_c^3, respectively.

17.2 Dependence of ground bounce reduction on system parameters

To determine the efficiency in reducing ground bounce, a simplified circuit model of the technique is used, as shown in Fig. 17.3. The ground bounce caused by simultaneously switching within the digital circuitry is modeled as a voltage source. A sinusoidal voltage source with an amplitude of 100 mV is used to determine the reduction in ground bounce at a single frequency. A triangular voltage source with an amplitude of 100 mV, rise time of 50 ps, and fall time of 200 ps is utilized to estimate the reduction in ground noise.

The dependence of the noise reduction technique on the physical separation between the noisy and noise sensitive circuits is presented in Section 17.2.1. The sensitivity of this technique to frequency and capacitance variations is discussed in Section 17.2.2. The dependence

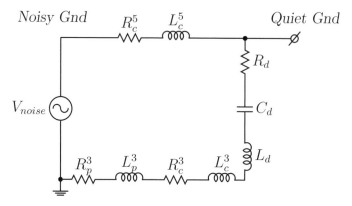

Fig. 17.3. Simplified circuit of the ground bounce reduction technique. The ground bounce due to simultaneously switching the digital circuits is modeled by a voltage source. The *Noisy Gnd* denotes an on-chip ground for the simultaneously switching digital circuits. The *Quiet Gnd* denotes a low noise ground for the noise sensitive circuits.

of ground noise on the impedance of an additional on-chip ground path is analyzed in Section 17.2.3.

17.2.1 Physical separation between noisy and noise sensitive circuits

To determine the dependence of the noise reduction technique on the physical separation between the noise source and noise receiver, the impedance of the ground path between the noisy and quiet terminals is modeled as a series RL, composed of the parasitic resistance and inductance per unit length. The peak voltage at the quiet ground is evaluated using SPICE where the distance between the digital and analog circuits is varied from one to ten unit lengths. The reduction in ground bounce as seen from the ground terminal of the noise sensitive circuit for sinusoidal and triangular noise sources is listed in Table 17.1.

Note that the reduction in ground noise increases linearly as the physical separation between the noisy and noise sensitive circuits becomes greater. A reduction in ground bounce of about 52% for a single frequency noise source and about 16% for a random noise source is achieved for a ground line (of ten unit lengths) between the digital and analog blocks. Enhanced results can be achieved if the impedance of the additional ground is much smaller than the impedance of the

Table 17.1. Ground bounce reduction as a function of the separation between the noisy and noise sensitive circuits

R_c^5	L_c^5	V_{quiet} (mV)		Noise Reduction (%)	
(mΩ)	(fH)	Sinusoidal	Triangular	Sinusoidal	Triangular
13	7	90.81	97.11	9.2	2.9
26	14	82.99	94.68	17.0	5.3
39	21	76.30	92.63	23.7	7.4
52	28	70.54	90.55	29.5	9.5
65	35	65.53	89.36	34.5	10.6
78	42	61.16	88.06	38.8	11.9
91	49	57.33	86.93	42.7	13.1
104	56	53.94	85.93	46.1	14.1
117	63	50.91	85.05	49.1	15.0
130	70	48.23	84.28	51.8	15.7

$$V_{noise} = 100\,\text{mV}, \ f = 1\,\text{GHz}, \ R_p^3 = 10\,\text{m}\Omega,$$
$$L_p^3 = 100\,\text{pH}, \ R_c^3 = 100\,\text{m}\Omega, \ L_c^3 = 100\,\text{fH}, \ R_d = 10\,\text{m}\Omega,$$
$$L_d = 10\,\text{fH}, \ C_d^{\text{Sin}} = 253\,\text{pF}, \ C_d^{\text{Triang}} = 63\,\text{pF}$$

interconnect between the noisy and noise sensitive modules. From a circuits perspective, the digital and analog circuits should be placed sufficiently distant and the additional low noise ground should be composed of multiple parallel lines. Moreover, the additional ground should be placed close to the multiple ground pins.

Note that since this noise reduction technique utilizes a capacitor tuned in resonance with the parasitic inductance of an additional ground path, this approach is frequency dependent and produces the best results for a single frequency noise source. In the case of a random noise source, the frequency harmonic with the highest magnitude should be significantly reduced, thereby achieving the greatest reduction in noise. For example, the second harmonic is selected in the case of a triangular noise source.

17.2.2 Frequency and capacitance variations

To determine the sensitivity of the ground bounce reduction technique on frequency and capacitance variations, the frequency is varied by ±50% from the resonant frequency and the capacitor is varied by ±10% from the target value. The range of capacitance variation is chosen based on typical process variations for a CMOS technology. The efficiency of the reduction in ground bounce for a sinusoidal noise source versus frequency and capacitance variations is illustrated in Figs. 17.4 and 17.5, respectively.

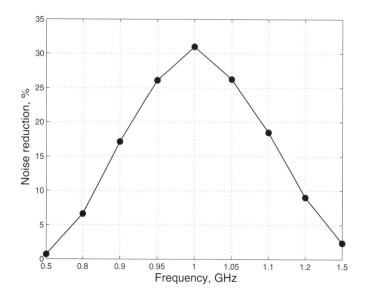

Fig. 17.4. Ground bounce reduction as a function of noise frequency. The reduction in noise drops linearly as the frequency varies from the target resonant frequency. The ground noise is modeled as a sinusoidal voltage source.

Note that the noise reduction drops linearly as the noise frequency varies from the target resonant frequency. The reduction in noise is slightly greater for higher frequencies. This phenomenon is due to the uncompensated parasitic inductance of the ground connecting the digital circuits to the analog circuits. As a result, at higher frequencies, the impedance of the ground path of a power delivery network increases, further reducing the noise. In general, the technique results in lower noise at higher frequencies. As illustrated in Fig. 17.5, the reduction in ground bounce is almost insensitive to capacitance variations. The efficiency of the technique drops by about 4% as the capacitance is varied by ±10%.

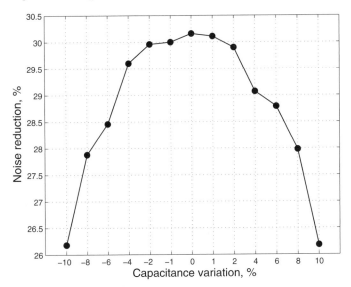

Fig. 17.5. Reduction in ground bounce as a function of capacitance variations. The reduction in ground bounce is almost insensitive to capacitance variations. The ground bounce is modeled as a sinusoidal voltage source.

17.2.3 Impedance of an additional ground path

As described in Section 17.1, the noise reduction technique utilizes a voltage divider formed by the ground of an on-chip power distribution system and an additional low noise ground. To increase the efficiency of the technique, the voltage transfer function of the voltage divider should be lowered, permitting a greater portion of the noise voltage to be diverted through the additional ground. As demonstrated in Section 17.2.1, placing the noisy and noise sensitive blocks farther from each other lowers the bounce at the ground terminal of the analog circuits. The ground noise can also be reduced by lowering the impedance of the low noise ground. The parasitic inductance of the additional ground is canceled by the capacitor tuned in resonance to the specific frequency. The impedance of the additional ground is therefore purely resistive at the resonant frequency. The reduction in noise for different values of the parasitic resistance of the low noise ground is listed in Table 17.2.

Note from Table 17.2 that by reducing the parasitic resistance of an on-chip low noise ground, the ground bounce can be significantly lowered. Noise reductions of about 68% and 22% are demonstrated for

Table 17.2. Ground bounce reduction for different values of parasitic resistance of the on-chip low noise ground

R_c^3	V_{quiet} (mV)		Noise Reduction (%)	
(mΩ)	Sinusoidal	Triangular	Sinusoidal	Triangular
100	60.54	87.88	39.5	12.1
80	56.52	86.57	43.5	13.4
60	51.67	84.98	48.3	15.0
40	45.79	83.03	54.2	17.0
20	38.59	80.60	61.4	19.4
10	34.37	79.15	65.6	20.9
5	32.08	78.37	67.9	21.6

$V_{noise} = 100\,\text{mV}$, $f = 1\,\text{GHz}$, $R_p^3 = 10\,\text{m}\Omega$, $L_p^3 = 100\,\text{pH}$,
$L_c^3 = 100\,\text{fH}$, $R_c^5 = 80\,\text{m}\Omega$, $L_c^5 = 40\,\text{fH}$, $R_d = 10\,\text{m}\Omega$,
$L_d = 10\,\text{fH}$, $C_d^{\text{Sin}} = 253\,\text{pF}$, $C_d^{\text{Triang}} = 63\,\text{pF}$

sinusoidal and triangular noise sources, respectively. The results listed in Table 17.2 are determined for an average resistance and inductance of the on-chip power distribution ground of five unit lengths (see Table 17.1). Thus, the ground bounce can be further reduced if the analog and digital circuits are placed farther from each other. Even better results can be achieved if the parasitic resistance of the package pins R_p^3 and decoupling capacitor R_d are lowered. From a circuits perspective, the low noise on-chip ground should be composed of many narrow lines connected in parallel to lower the parasitic resistance and inductance. A number of package pins should therefore be dedicated to the noise-free ground to lower the package resistance. A decoupling capacitor with a low ESR is also recommended.

17.3 Summary

Design techniques to reduce ground bounce in SoC and mixed-signal ICs are presented in this chapter and can be summarized as follows:

- A noise reduction technique with an additional on-chip ground is proposed to divert ground noise from the sensitive analog circuits

- The technique utilizes a decoupling capacitor tuned in resonance with the parasitic inductance of an additional low noise ground, making the technique frequency dependent

- The reduction in ground bounce, however, is almost independent of capacitance variations

- Noise reductions of 68% and 22% are demonstrated for a single frequency and random ground noise, respectively

- The noise reduction efficiency can be further enhanced by simultaneously lowering the impedance of the additional noise-free ground and increasing the impedance of the ground path between the digital (noisy) and analog (noise sensitive) circuits

Effective Radii of On-Chip Decoupling Capacitors

Decoupling capacitors are widely used to manage power supply noise. A decoupling capacitor acts as a reservoir of charge, which is released when the power supply voltage at a particular current load drops below some tolerable level. Alternatively, decoupling capacitors are an effective way to reduce the impedance of power delivery systems operating at high frequencies [27]. Since the inductance scales slowly [207], the location of the decoupling capacitors significantly affects the design of the P/G network in high performance ICs such as microprocessors. With increasing frequencies, a distributed hierarchical system of decoupling capacitors placed on-chip is needed to effectively manage power supply noise [334].

The efficacy of decoupling capacitors depends upon the impedance of the conductors connecting the capacitors to the current loads and power sources. During discharge, the current flowing from the decoupling capacitor to the current load results in resistive noise (IR drops) and inductive noise ($L\frac{dI}{dt}$ drops) due to the parasitic resistances and inductances of the power delivery network. The resulting voltage drop at the current load is therefore always greater than the voltage drop at the decoupling capacitor. Thus, a maximum parasitic impedance between the decoupling capacitor and the current load exists at which the decoupling capacitor is effective. Alternatively, to be effective, a decoupling capacitor should be placed close to a current load during discharge (within the maximum effective distance $d_Z^{\max}$), as shown in Fig. 18.1.

Once the switching event is completed, a decoupling capacitor has to be fully charged before the next clock cycle begins. During the charging phase, the voltage across the decoupling capacitor rises exponentially.

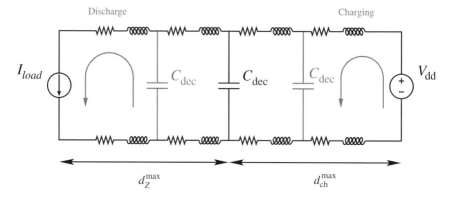

Fig. 18.1. Placement of an on-chip decoupling capacitor based on the maximum effective distance. To be effective, a decoupling capacitor should be placed close to the current load during discharge. During the charging phase, however, the decoupling capacitor should be placed close to the power supply to efficiently restore the charge on the capacitor. The specific location of a decoupling capacitor should therefore be determined to simultaneously satisfy the maximum effective distances $d_Z^{\max}$ during discharge and $d_{\mathrm{ch}}^{\max}$ during charging.

The charge time of a capacitor is determined by the parasitic resistance and inductance of the interconnect between the capacitor and the power supply. A design space for a tolerable interconnect resistance and inductance exists, permitting the charge on the decoupling capacitor to be restored within a target charge time. The maximum frequency at which the decoupling capacitor is effective is determined by the parasitic resistance and inductance of the metal lines and the size of the decoupling capacitor. A maximum effective distance based on the charge time, therefore, exists for each on-chip decoupling capacitor. Beyond this effective distance, the decoupling capacitor is ineffective. Alternatively, to be effective, an on-chip decoupling capacitor should be placed close to a power supply during the charging phase (within the maximum effective distance $d_{\mathrm{ch}}^{\max}$, see Fig. 18.1). The relative location of the on-chip decoupling capacitors is therefore of fundamental importance. A design methodology is therefore required to determine the location of an on-chip decoupling capacitor, simultaneously satisfying the maximum effective distances, $d_Z^{\max}$ and $d_{\mathrm{ch}}^{\max}$. This location is characterized by the effective radii of the on-chip decoupling capacitors and is the primary subject of this chapter. A design methodology to estimate the minimum required on-chip decoupling capacitance is also presented.

The chapter is organized as follows. Existing work on placing on-chip decoupling capacitors is reviewed in Section 18.1. The effective radius of an on-chip decoupling capacitor as determined by the target impedance is presented in Section 18.2. Design techniques to estimate the minimum magnitude of the required on-chip decoupling capacitance are discussed in Section 18.3. The effective radius of an on-chip decoupling capacitor based on the charge time is determined in Section 18.4. A design methodology for placing on-chip decoupling capacitors based on the maximum effective radii is presented in Section 18.5. A model of an on-chip power distribution network is developed in Section 18.6. Simulation results for typical values of on-chip parasitic resistances and inductances are presented in Section 18.7. Some circuit design implications are discussed in Section 18.8. Finally, some specific conclusions are summarized in Section 18.9.

18.1 Background

Decoupling capacitors have traditionally been allocated on a circuit board to control the impedance of a power distribution system and suppress EMI. Decoupling capacitors are also employed to provide the required charge to the switching circuits, enhancing signal integrity. Since the parasitic impedance of a circuit board-based power distribution system is negligible at low frequencies, board decoupling capacitors are typically modeled as ideal capacitors without parasitic impedances. In an important early work by Smith [262], the effect of a decoupling capacitor on the signal integrity in circuit board-based power distribution systems is presented. The efficacy of the decoupling capacitors is analyzed in both the time and frequency domains. Design criteria have been developed, however, which significantly overestimate the required decoupling capacitance. A hierarchical placement of decoupling capacitors has been presented by Smith et al. in [107]. The authors of [107] show that each decoupling capacitor is effective only within a narrow frequency range. Larger decoupling capacitors have a greater form factor (physical dimensions), resulting in higher parasitic impedances [150]. The concept of an effective series resistance and an effective series inductance of each decoupling capacitor is also introduced. The authors show that by hierarchically placing the decoupling capacitors from the voltage regulator module level to the package level, the impedance of the overall power distribution system can be maintained below a target impedance.

As the signal frequency increases to several megahertz, the parasitic impedance of the circuit board decoupling capacitors becomes greater than the target impedance. The circuit board decoupling capacitors therefore become less effective at frequencies above 10 to 20 MHz. Package decoupling capacitors should therefore be utilized in the frequency range from several megahertz to several hundred megahertz [107]. In modern high performance ICs operating at several gigahertz, only those decoupling capacitors placed on-chip are effective at these frequencies.

The optimal placement of on-chip decoupling capacitors has been discussed in [242]. The power noise is analyzed assuming an RLC network model, representing a multi-layer power bus structure. The current load is modeled by time-varying resistors. The on-chip decoupling capacitors are allocated to only those areas where the power noise is greater than the maximum tolerable level. Ideal on-chip decoupling capacitors are assumed in the algorithm proposed in [242]. The resulting budget of on-chip decoupling capacitance is therefore significantly overestimated. Another technique for placing on-chip decoupling capacitors has been described in [265]. The decoupling capacitors are placed based on activity signatures determined from microarchitectural simulations. The proposed technique produces a 30% decrease in the maximum noise level as compared to uniformly placing the on-chip decoupling capacitors. This methodology results in overestimating the capacitance budget due to the use of a simplified criterion for sizing the on-chip decoupling capacitors. Also, since the package level power distribution system is modeled as a single lumped resistance and inductance, the overall power supply noise is greatly underestimated.

An algorithm for automatically placing and sizing on-chip decoupling capacitors in application-specific integrated circuits is proposed in [272]. The problem is formulated as a nonlinear optimization and solved using a sensitivity-based quadratic programming solver. The proposed algorithm is limited to on-chip decoupling capacitors placed in rows of standard cells (in one dimension). The power distribution network is modeled as a resistive mesh, significantly underestimating the power distribution noise. In [268], the problem of on-chip decoupling capacitor allocation is investigated. The proposed technique is integrated into a power supply noise-aware floorplanning methodology. Only the closest power supply pins are considered to provide the

switching current drawn by the load. Additionally, only the shortest
and second shortest paths are considered between a decoupling capacitor and the current load. It is assumed that the current load is
located at the center of a specific circuit block. The technique does
not consider the degradation in effectiveness of an on-chip decoupling
capacitor located at some distance from the current load. Moreover,
only the discharge phase is considered. To be effective, a decoupling
capacitor should be fully charged before the following switching cycle.
Otherwise, the charge on the decoupling capacitor will be gradually
depleted, making the capacitor ineffective. The methodology described
in [268] therefore results in underestimating the power supply noise and
overestimating the required on-chip decoupling capacitance.

The problem of on-chip decoupling capacitor allocation has historically been considered as two independent tasks. The location of an
on-chip decoupling capacitor is initially determined. The decoupling
capacitor is next appropriately sized to provide the required charge
to the current load. As discussed in [200], the size of the on-chip decoupling capacitors is determined by the impedance (essentially, the
physical separation) between a decoupling capacitor and the current
load (or power supply).

Proper sizing and placement of the on-chip decoupling capacitors
however should be determined simultaneously. As shown in this chapter, on-chip decoupling capacitors are only effective in close vicinity
to the switching circuit. The maximum effective distance for both the
discharge and charging phase is determined. It is also shown that the
on-chip decoupling capacitors should be placed both close to the current load to provide the required charge and to the power supply to be
fully recharged before the next switching event. A design methodology
for placing and sizing on-chip decoupling capacitors based on a maximum effective distance as determined by the target impedance and
charge time is presented in this chapter.

18.2 Effective radius of on-chip decoupling capacitor based on a target impedance

Neglecting the parasitic capacitance [350], the impedance of a unit
length wire is $Z'(\omega) = r + j\omega l$, where r and l are the resistance and
inductance per length, respectively, and ω is an effective frequency,
as determined by the rise time of the current load. The inductance l

is the effective inductance per unit length of the power distribution grid, incorporating both the partial self-inductance and mutual coupling among the lines [70]. The target impedance of the metal line of a particular length is therefore

$$Z(\omega) = Z'(\omega) \times d, \tag{18.1}$$

where $Z'(\omega)$ is the impedance of a unit length metal line, and d is the distance between the decoupling capacitor and the current load. Substituting the expression for the target impedance Z_{target} presented in [26] into (18.1), the maximum effective radius $d_Z^{\max}$ between the decoupling capacitor and the current load is

$$d_Z^{\max} = \frac{Z_{target}}{Z'(\omega)} = \frac{V_{dd} \times Ripple}{I \times \sqrt{r^2 + \omega^2 l^2}}, \tag{18.2}$$

where $\sqrt{r^2 + \omega^2 l^2}$ denotes the magnitude of the impedance of a unit length wire, Z_{target} is the maximum impedance of a power distribution system, resulting in a power noise lower than the maximum tolerable level, and $Ripple$ is the maximum tolerable power noise (the ratio of the magnitude of the maximum tolerable voltage drop to the power supply level). Note that the maximum effective radius as determined by the target impedance is inversely proportional to the magnitude of the current load and the impedance of a unit length line. Also note that the per length resistance r and inductance l account for the ESR and ESL of an on-chip decoupling capacitor. The maximum effective radius as determined by the target impedance decreases rapidly with each technology generation (a factor of 1.4, on average, per computer generation), as shown in Fig. 18.2 [22]. Also note that in a meshed structure, multiple paths between any two points are added in parallel. The maximum effective distance corresponding to Z_{target} is, therefore, larger than the maximum effective distance of a single line, as discussed in Section 18.7. The maximum effective radius is defined in this chapter as follows:

Definition 1: The effective radius of an on-chip decoupling capacitor is the maximum distance between the current load (power supply) and the decoupling capacitor for which the capacitor is capable of providing sufficient charge to the current load, while maintaining the overall power distribution noise below a tolerable level.

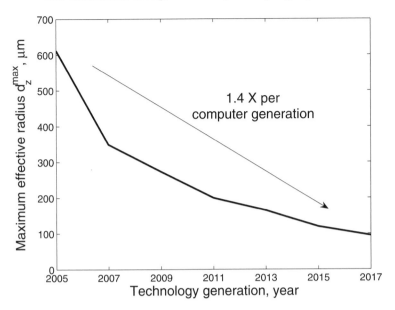

Fig. 18.2. Projection of the maximum effective radius as determined by the target impedance d_Z^{max} for future technology generations: $I_{\mathrm{max}} = 10\,\mathrm{mA}$, $V_{\mathrm{dd}} = 1\,\mathrm{V}$, and $Ripple = 0.1$. Global on-chip interconnects are assumed, modeling the highly optimistic scenario. The maximum effective radius as determined by the target impedance is expected to decrease at an alarming rate (a factor of 1.4 on average per computer generation).

18.3 Estimation of required on-chip decoupling capacitance

Once the specific location of an on-chip decoupling capacitor is determined as described in Section 18.2, the minimum required magnitude of the on-chip decoupling capacitance should be determined, satisfying the expected current demands. Design expressions for determining the required magnitude of the on-chip decoupling capacitors based on the dominant power noise are presented in this section. A conventional approach with dominant resistive noise is described in Section 18.3.1. Techniques for determining the magnitude of on-chip decoupling capacitors in the case of dominant inductive noise are developed in Section 18.3.2. The critical length of the P/G paths connecting the decoupling capacitor to the current load is presented in Section 18.3.3.

18.3.1 Dominant resistive noise

To estimate the on-chip decoupling capacitance required to support a specific local current demand, the current load is modeled as a triangular current source. The magnitude of the current source increases linearly, reaching the maximum current I_{max} at peak time t_p. The magnitude of the current source decays linearly, becoming zero at t_f, as shown in Fig. 18.3. The on-chip power distribution network is modeled as a series RL circuit. To qualitatively illustrate the methodology for placing on-chip decoupling capacitors based on the maximum effective radii, a single decoupling capacitor with a single current load is assumed to mitigate the voltage fluctuations across the P/G terminals.

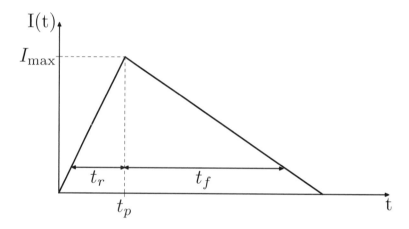

Fig. 18.3. Linear approximation of the current demand of a power distribution network by a current source. The magnitude of the current source reaches the maximum current I_{max} at peak time t_p. t_r and t_f denote the rise and fall time of the current load, respectively.

The total charge Q_{dis} required to satisfy the current demand during a switching event is modeled as the sum of the area of two triangles (see Fig. 18.3). Since the required charge is provided by an on-chip decoupling capacitor, the voltage across the capacitor during discharge drops below the initial power supply voltage. The required charge during the entire switching event is thus[1]

---

[1] In the general case with an *a priori* determined current profile, the required charge can be estimated as the integral of $I_{load}(t)$ from 0 to t_f.

$$Q_{\text{dis}}^f = \frac{I_{\max} \times (t_r + t_f)}{2} = C_{\text{dec}} \times (V_{\text{dd}} - V_C^f), \qquad (18.3)$$

where $I_{\max}$ is the maximum magnitude of the current load of a specific circuit block for which the decoupling capacitor is allocated, t_r and t_f are the rise and fall time, respectively, C_{dec} is the decoupling capacitance, V_{dd} is the power supply voltage, and V_C^f is the voltage across the decoupling capacitor after the switching event. Note that since there is no current after switching, the voltage at the current load is equal to the voltage across the decoupling capacitor.

The voltage fluctuations across the P/G terminals of a power delivery system should not exceed the maximum level (usually 10% of the power supply voltage [211]) to guarantee fault-free operation. Thus,

$$V_C^f \equiv V_{load}^f \geq 0.9\, V_{\text{dd}}. \qquad (18.4)$$

Substituting (18.4) into (18.3) and solving for C_{dec}, the minimum on-chip decoupling capacitance required to support the current demand during a switching event is

$$C_{\text{dec}}^f \geq \frac{I_{\max} \times (t_r + t_f)}{0.2\, V_{\text{dd}}}, \qquad (18.5)$$

where C_{dec}^f is the decoupling capacitance required to support the current demand during the entire switching event.

18.3.2 Dominant inductive noise

Note that (18.5) is applicable only to the case where the voltage drop at the end of the switching event is larger than the voltage drop at the peak time t_p ($IR \gg L\frac{dI}{dt}$). Alternatively, the minimum voltage at the load is determined by the resistive drop and the parasitic inductance can be neglected. This phenomenon can be explained as follows. The voltage drop as seen at the current load is caused by current flowing through the parasitic resistance and inductance of the on-chip power distribution system. The resulting voltage fluctuations are the sum of the ohmic IR voltage drop, inductive $L\frac{dI}{dt}$ voltage drop, and the voltage drop across the decoupling capacitor at t_p. A critical parasitic RL impedance, therefore, exists for any given set of rise and fall times. Beyond this critical impedance, the voltage drop at the load is primarily caused by the inductive noise ($L\frac{dI}{dt} \gg IR$), as shown in Fig. 18.4.

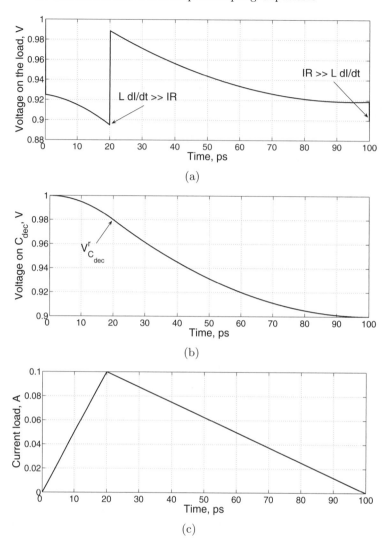

Fig. 18.4. Power distribution noise during discharge of an on-chip decoupling capacitor: $I_{max} = 100\,\text{mA}$, $V_{dd} = 1\,\text{V}$, $t_r = 20\,\text{ps}$, $t_f = 80\,\text{ps}$, $R = 100\,\text{m}\Omega$, $L = 15\,\text{pH}$, and $C_{dec} = 50\,\text{pF}$. a) Voltage across the terminals of the current load. b) Voltage across the decoupling capacitor. c) Current load modeled as a triangular current source. For these parameters, the parasitic impedance of the metal lines connecting the decoupling capacitor to the current load is larger than the critical impedance. The inductive noise therefore dominates the resistive noise and (18.5) underestimates the required decoupling capacitance. The resulting voltage drop on the power terminal of a current load is therefore larger than the maximum tolerable noise.

The decoupling capacitor should therefore be increased in the case of dominant inductive noise to reduce the voltage drop across the capacitor during the rise time V_C^r, lowering the magnitude of the power noise.

The charge Q_{dis}^r required to support the current demand during the rise time of the current load is equal to the area of the triangle formed by I_{max} and t_r. The required charge is provided by the on-chip decoupling capacitor. The voltage across the decoupling capacitor drops below the power supply level by ΔV_C^r. The required charge during t_r is[2]

$$Q_{dis}^r = \frac{I_{max} \times t_r}{2} = C_{dec} \times \Delta V_C^r, \qquad (18.6)$$

where Q_{dis}^r is the charge drawn by the current load during t_r and ΔV_C^r is the voltage drop across the decoupling capacitor at t_p. From (18.6),

$$\Delta V_C^r = \frac{I_{max} \times t_r}{2\,C_{dec}}. \qquad (18.7)$$

By time t_p, the voltage drop as seen from the current load is the sum of the ohmic IR drop, the inductive $L\frac{dI}{dt}$ drop, and the voltage drop across the decoupling capacitor. Alternatively, the power noise is further increased by the voltage drop ΔV_C^r. In this case, the voltage at the current load is

$$V_{load}^r = V_{dd} - I \times R - L\frac{dI}{dt} - \Delta V_C^r, \qquad (18.8)$$

where R and L are the parasitic resistance and inductance of the P/G lines, respectively. Linearly approximating the current load, dI is assumed equal to I_{max} and dt to t_r. Note that the last term in (18.8) accounts for the voltage drop ΔV_C^r across the decoupling capacitor during the rise time of the current at the load.

Assuming that $V_{load}^r \geq 0.9\,V_{dd}$, substituting (18.7) into (18.8), and solving for C_{dec}, the minimum on-chip decoupling capacitance to support the current demand during t_r is

$$C_{dec}^r \geq \frac{I_{max} \times t_r}{2\left(0.1\,V_{dd} - I \times R - L\frac{dI}{dt}\right)}. \qquad (18.9)$$

Note that if $L\frac{dI}{dt} \gg IR$, C_{dec} is excessively large. The voltage drop at the end of the switching event is hence always smaller than the maximum tolerable noise.

---

[2] In the general case with a given current profile, the required charge can be estimated as the integral of $I_{load}(t)$ from 0 to t_r.

Also note that, as opposed to (18.5), (18.9) depends upon the parasitic impedance of the on-chip power distribution system. Alternatively, in the case of the dominant inductive noise, the required charge released by the decoupling capacitor is determined by the parasitic resistance and inductance of the P/G lines connecting the decoupling capacitor to the current load.

18.3.3 Critical line length

Assuming the impedance of a single line, the critical line length d_{crit} can be determined by setting C_{dec}^{r} equal to C_{dec}^{f},

$$\frac{I_{\max} \times t_r}{\left(0.1\,V_{\text{dd}} - I\,r\,d_{\text{crit}} - l\,d_{\text{crit}}\,\frac{dI}{dt}\right)} = \frac{I_{\max} \times (t_r + t_f)}{0.1\,V_{\text{dd}}}. \qquad (18.10)$$

Solving (18.10) for d_{crit},

$$d_{\text{crit}} = \frac{0.1\,V_{\text{dd}}\left(1 - \dfrac{t_r}{t_r + t_f}\right)}{I\,r + l\,\dfrac{dI}{dt}}. \qquad (18.11)$$

For a single line connecting a current load to a decoupling capacitor, the minimum required on-chip decoupling capacitor is determined by (18.5) for lines shorter than d_{crit} and by (18.9) for lines longer than d_{crit}, as illustrated in Fig. 18.5. Note that for a line length equal to d_{crit}, (18.5) and (18.9) result in the same required capacitance. Also note that the maximum length of a single line is determined by (18.2). A closed-form solution for the critical line length has not been developed for the case of multiple current paths existing between the current load and a decoupling capacitor. In this case, the impedance of the power grid connecting a decoupling capacitor to a current load is extracted and compared to the critical impedance. Either (18.5) or (18.9) is utilized to estimate the required on-chip decoupling capacitance.

The dependence of the critical line length d_{crit} on the rise time t_r of the current load as determined by (18.11) is depicted in Fig. 18.6. From Fig. 18.6, the critical line length decreases sublinearly with shorter rise times. Hence, the critical line length will decrease in future nanometer technologies as transition times become shorter, significantly increasing the required on-chip decoupling capacitance. Also note that d_{crit} is determined by $\frac{t_r}{t_f}$, increasing with larger fall times.

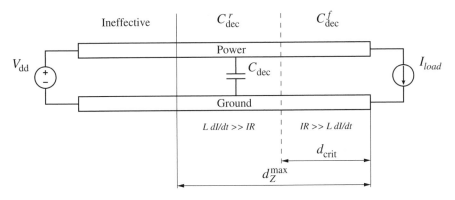

Fig. 18.5. Critical line length of an interconnect between a decoupling capacitor and a current load. The minimum required on-chip decoupling capacitance is determined by (18.5) for lines shorter than d_{crit} and by (18.9) for lines longer than d_{crit}. The decoupling capacitor is ineffective beyond the maximum effective radius as determined by the target impedance $d_Z^{\max}$.

Observe in Fig. 18.5 that the design space for determining the required on-chip decoupling capacitance is broken into two regions by the critical line length. The design space for determining the required on-chip decoupling capacitance (C_{dec}^r and C_{dec}^f) is depicted in Fig. 18.7. For the example parameters shown in Fig. 18.7, the critical line length is 125 μm. Note that the required on-chip decoupling capacitance C_{dec}^r depends upon the parasitic impedance of the metal lines connecting the decoupling capacitor to the current load. Thus, for lines longer than d_{crit}, C_{dec}^r increases exponentially as the separation between the decoupling capacitor and the current load increases, as shown in Fig. 18.7(a). Also note that for lines shorter than d_{crit}, the required on-chip decoupling capacitance does not depend upon the parasitic impedance of the power distribution grid. Alternatively, in the case of the dominant resistive drop, the required on-chip decoupling capacitance C_{dec}^f is constant and greater than C_{dec}^r (see region 1 in Fig. 18.7(b)). If $L\frac{dI}{dt}$ noise dominates the IR noise (the line length is greater than d_{crit}), the required on-chip decoupling capacitance C_{dec}^r increases substantially with line length and is greater than C_{dec}^f (see region 2 in Fig. 18.7(b)). Conventional techniques therefore significantly underestimate the required decoupling capacitance in the case of the dominant inductive noise. Note that in region 1, the parasitic impedance of the metal lines connecting a decoupling capacitor to the current load is not important.

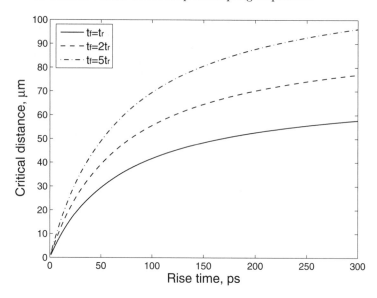

Fig. 18.6. Dependence of the critical line length d_{crit} on the rise time of the current load: $I_{\max} = 0.1\,\text{A}$, $V_{\text{dd}} = 1\,\text{V}$, $r = 0.007\,\Omega/\mu\text{m}$, and $l = 0.5\,\text{pH}/\mu\text{m}$. Note that d_{crit} is determined by $\frac{t_r}{t_f}$, increasing with larger t_f. The critical line length will shrink in future nanometer technologies as transition times become shorter.

In region 2, however, the parasitic impedance of the P/G lines should be considered. A tradeoff therefore exists between the size of C^r_{dec} and the distance between the decoupling capacitor and the current load. As C^r_{dec} is placed closer to the current load, the required capacitance can be significantly reduced.

18.4 Effective radius as determined by charge time

Once discharged, a decoupling capacitor must be fully charged to support the current demands during the following switching event. If the charge on the capacitor is not fully restored during the relaxation time between two consecutive switching events (the charge time), the decoupling capacitor will be gradually depleted, becoming ineffective after several clock cycles. A maximum effective radius, therefore, exists for an on-chip decoupling capacitor as determined during the charging phase for a target charge time. Similar to the effective radius based on the target impedance presented in Section 18.2, an on-chip decoupling

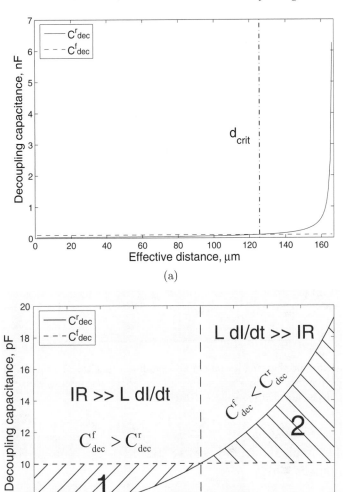

Fig. 18.7. Design space for determining minimum required on-chip decoupling capacitance: $I_{max} = 50\,\text{mA}$, $V_{dd} = 1\,\text{V}$, $r = 0.007\,\Omega/\mu\text{m}$, $l = 0.5\,\text{pH}/\mu\text{m}$, $t_r = 100\,\text{ps}$, and $t_f = 300\,\text{ps}$. a) The design space for determining the minimum required on-chip decoupling capacitance is broken into two regions by d_{crit}. b) The design space around d_{crit}. For the example parameters, the critical line length is $125\,\mu\text{m}$. In region 1, C_{dec}^f is greater than C_{dec}^r and does not depend upon the parasitic impedance. In region 2, however, C_{dec}^r dominates, increasing rapidly with distance between the decoupling capacitor and the current load.

capacitor should be placed in close proximity to the power supply (the power pins) to be effective.

To determine the current flowing through a decoupling capacitor during the charging phase, the parasitic impedance of a power distribution system is modeled as a series RL circuit between the decoupling capacitor and the power supply, as shown in Fig. 18.8. When the discharge is completed, the switch is closed and the charge is restored on the decoupling capacitor. The initial voltage V_C^0 across the decoupling capacitor is determined by the maximum voltage drop during discharge.

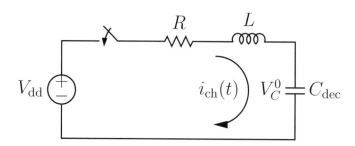

Fig. 18.8. Circuit charging an on-chip decoupling capacitor. The parasitic impedance of the power distribution system connecting the decoupling capacitor to the power supply is modeled by a series RL circuit.

For the circuit shown in Fig. 18.8, the KVL equation for the current in the circuit is [351]

$$L \frac{di_{\mathrm{ch}}}{dt} + R\,i_{\mathrm{ch}} + \frac{1}{C_{\mathrm{dec}}} \int i_{\mathrm{ch}}\, dt = V_{\mathrm{dd}}. \qquad (18.12)$$

Differentiating (18.12),

$$L \frac{d^2 i_{\mathrm{ch}}}{dt^2} + R \frac{di_{\mathrm{ch}}}{dt} + \frac{1}{C_{\mathrm{dec}}} i_{\mathrm{ch}} = 0. \qquad (18.13)$$

Equation (18.13) is a second order linear differential equation with the characteristic equation,

$$s^2 + \frac{R}{L} s + \frac{1}{L C_{\mathrm{dec}}} = 0. \qquad (18.14)$$

The general solution of (18.13) is

$$i_{ch}(t) = K_1 e^{s_1 t} + K_2 e^{s_2 t}, \qquad (18.15)$$

where s_1 and s_2 are the roots of (18.14),

$$s_{1,2} = -\frac{R}{2L} \pm \sqrt{\left(\frac{R}{2L}\right)^2 - \frac{1}{LC_{dec}}}. \qquad (18.16)$$

Note that (18.15) represents the solution of (18.13) as long as the system is overdamped. The damping factor is therefore greater than one, *i.e.*,

$$\left(\frac{R}{L}\right)^2 > \frac{4}{LC}. \qquad (18.17)$$

For a single line, from (18.17), the critical line length resulting in an overdamped system is

$$D > \frac{4l}{r^2 C_{dec}}, \qquad (18.18)$$

where C_{dec} is the on-chip decoupling capacitance, and l and r are the per length inductance and resistance, respectively. Inequality (18.18) determines the critical length of a line resulting in an overdamped system. Note that for typical values of r and l in a 90 nm CMOS technology, a power distribution system with a decoupling capacitor is overdamped for on-chip interconnects longer than several micrometers. Equation (18.15) is therefore a general solution of (18.13) for a scaled CMOS technology.

Initial conditions are applied to determine the arbitrary constants K_1 and K_2 in (18.15). The current charging the decoupling capacitor during the charging phase is

$$i_{ch}(t) = \frac{I_{max} (t_r + t_f)}{4 L C_{dec} \sqrt{\left(\frac{R}{2L}\right)^2 - \frac{1}{LC_{dec}}}} \qquad (18.19)$$

$$\times \left\{ \exp\left[\left(-\frac{R}{2L} + \sqrt{\left(\frac{R}{2L}\right)^2 - \frac{1}{LC_{dec}}}\right) t\right] \right.$$

$$\left. - \exp\left[\left(-\frac{R}{2L} - \sqrt{\left(\frac{R}{2L}\right)^2 - \frac{1}{LC_{dec}}}\right) t\right] \right\}.$$

The voltage across the decoupling capacitor during the charging phase can be determined by integrating (18.19) from zero to the charge time,

$$V_C(t) = \frac{1}{C_{\text{dec}}} \int_0^{t_{\text{ch}}} i_{\text{ch}}(t)\, dt, \tag{18.20}$$

where t_{ch} is the charge time, and $V_C(t)$ and $i_{\text{ch}}(t)$ are the voltage across the decoupling capacitor and the current flowing through the decoupling capacitor during the charging phase, respectively. Substituting (18.19) into (18.20) and integrating from zero to t_{ch}, the voltage across the decoupling capacitor during the charging phase is

$$V_{C_{\text{dec}}}(t_{\text{ch}}) = \frac{I_{\max}\,(t_r + t_f)}{4\,C_{\text{dec}}^2\,L\sqrt{\left(\dfrac{R}{2L}\right)^2 - \dfrac{1}{LC_{\text{dec}}}}} \tag{18.21}$$

$$\times \left\{ \frac{\exp\left[\left(-\dfrac{R}{2L} + \sqrt{\left(\dfrac{R}{2L}\right)^2 - \dfrac{1}{LC_{\text{dec}}}}\right)t_{\text{ch}}\right] - 1}{-\dfrac{R}{2L} + \sqrt{\left(\dfrac{R}{2L}\right)^2 - \dfrac{1}{LC_{\text{dec}}}}} \right.$$

$$\left. + \frac{1 - \exp\left[\left(-\dfrac{R}{2L} - \sqrt{\left(\dfrac{R}{2L}\right)^2 - \dfrac{1}{LC_{\text{dec}}}}\right)t_{\text{ch}}\right]}{-\dfrac{R}{2L} - \sqrt{\left(\dfrac{R}{2L}\right)^2 - \dfrac{1}{LC_{\text{dec}}}}} \right\}.$$

Observe that the criterion for estimating the maximum effective radius of an on-chip decoupling capacitor as determined by the charge time is transcendental. A closed-form expression is therefore not available for determining the maximum effective radius of an on-chip decoupling capacitor during the charging phase. Thus from (18.21), a design space can be graphically described in order to determine the maximum tolerable resistance and inductance that permit the decoupling capacitor to be recharged within a given t_{ch}, as shown in Fig. 18.9. The parasitic resistance and inductance should be maintained below the maximum tolerable values, permitting the decoupling capacitor to be charged during the relaxation time.

Note that as the parasitic resistance of the power delivery network decreases, the voltage across the decoupling capacitor increases exponentially. In contrast, the voltage across the decoupling capacitor during the charging phase is almost independent of the parasitic inductance, slightly increasing with inductance. This phenomenon is due to the behavior that an inductor resists sudden changes in the current. Alternatively, an inductor maintains the charging current at a particular level for a longer time. Thus, the decoupling capacitor is charged faster.

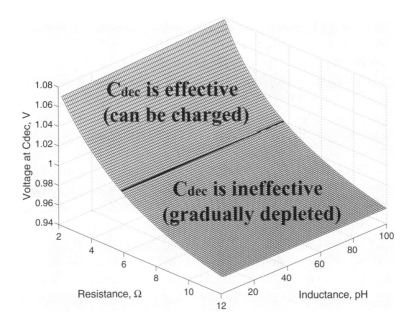

Fig. 18.9. Design space for determining the maximum tolerable parasitic resistance and inductance of a power distribution grid: $I_{\max} = 100\,\text{mA}$, $t_r = 100\,\text{ps}$, $t_f = 300\,\text{ps}$, $C_{\text{dec}} = 100\,\text{pF}$, $V_{\text{dd}} = 1\,\text{volt}$, and $t_{\text{ch}} = 400\,\text{ps}$. For a target charge time, the maximum resistance and inductance produce a voltage across the decoupling capacitor that is greater or equal to the power supply voltage (region above the dark line). Note that the maximum voltage across the decoupling capacitor is the power supply voltage. A design space that produces a voltage greater than the power supply means that the charge on the decoupling capacitor can be restored within t_{ch}.

18.5 Design methodology for placing on-chip decoupling capacitors

A design methodology for placing on-chip decoupling capacitors based on the maximum effective radii is illustrated in Fig. 18.10. The maximum effective radius based on the target impedance is determined from (18.2) for a particular current load (circuit block), power supply voltage, and allowable ripple. The minimum required on-chip decoupling capacitance is estimated to support the required current demand. If the resistive drop is larger than the inductive drop, (18.5) is used to determine the required on-chip decoupling capacitance. If $L\frac{dI}{dt}$ noise dominates, the on-chip decoupling capacitance is determined by (18.9). In the case of a single line connecting a decoupling capacitor to a current load, the critical wire length is determined by (18.11).

The maximum effective distance based on the charge time is determined from (18.21). Note that (18.21) results in a range of tolerable parasitic resistance and inductance of the metal lines connecting the decoupling capacitor to the power supply. Also note that the on-chip decoupling capacitor should be placed such that both the power supply and the current load are located inside the effective radius, as shown in Fig. 18.11. If this allocation is not possible, the current load (circuit block) should be partitioned into several blocks and the on-chip decoupling capacitors should be allocated for each block, satisfying both effective radii requirements. The effective radius as determined by the target impedance does not depend upon the decoupling capacitance. In contrast, the effective radius as determined by the charge time is inversely proportional to C_{dec}^2. The on-chip decoupling capacitors should be distributed across the circuit to provide sufficient charge for each functional unit.

18.6 Model of on-chip power distribution network

In order to determine the effective radii of an on-chip decoupling capacitor and the effect on the noise distribution, a model of a power distribution network is required. On-chip power distribution networks in high performance ICs are commonly modeled as a mesh. Early in the design process, minimal physical information characterizing the P/G structure is available. A simplified model of a power distribution system is therefore appropriate. For simplicity, equal segments within a mesh

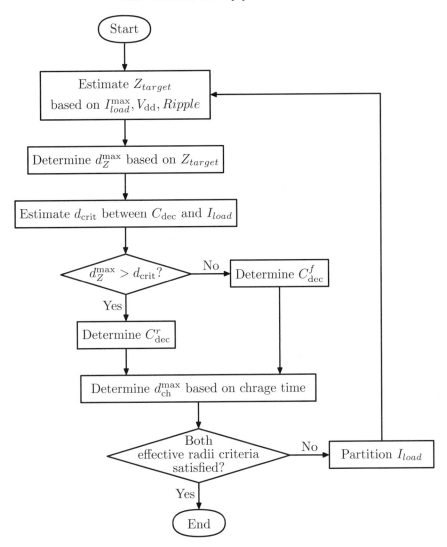

Fig. 18.10. Design flow for placing on-chip decoupling capacitors based on the maximum effective radii.

structure are assumed. The current demands of a particular module are modeled as current sources with equivalent magnitude and switching activities. The current load is located at the center of a circuit module which determines the connection point of the circuit module to the power grid. The parasitic resistance and inductance of the package are also included in the model as an equivalent series resistance R_p

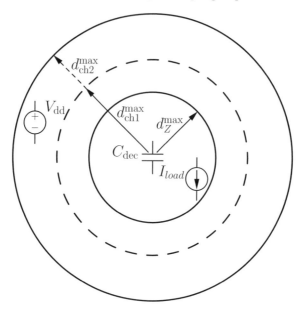

Fig. 18.11. The effective radii of an on-chip decoupling capacitor. The on-chip decoupling capacitor is placed such that both the current load and the power supply are located inside the effective radius. The maximum effective radius as determined by the target impedance d_Z^{max} does not depend on the decoupling capacitance. The maximum effective radius as determined by the charge time is inversely proportional to C_{dec}^2. If the power supply is located outside the effective radius $d_{\mathrm{ch1}}^{\mathrm{max}}$, the current load should be partitioned, resulting in a smaller decoupling capacitor and, therefore, an increased effective distance $d_{\mathrm{ch2}}^{\mathrm{max}}$.

and inductance L_p. Note that the parasitic capacitance of the power distribution grid provides a portion of the decoupling capacitance, providing additional charge to the current loads. The on-chip decoupling capacitance intentionally added to the IC is typically more than an order of magnitude greater than the parasitic capacitance of the on-chip power grid. The parasitic capacitance of the power delivery network is, therefore, neglected.

Typical effective radii of an on-chip decoupling capacitor is in the range of several hundreds micrometers. In order to determine the location of an on-chip decoupling capacitor, the size of each RL mesh segment should be much smaller than the effective radii. In modern high performance ICs such as microprocessors with die sizes approaching 1.5 inches by 1.5 inches, a fine mesh is infeasible to simulate. In the case of a coarse mesh, the effective radius is smaller than the size of each segment. The location of each on-chip decoupling capacitor, therefore, cannot be accurately determined. To resolve this dilemma, the accuracy of the capacitor location can be traded off with the complexity of the power distribution network. A hot spot (an area where the power supply voltage drops below the minimum tolerable level) is first determined based on a coarse mesh, as shown in Fig. 18.12. A finer mesh is used next within each hot spot to accurately estimate the effective radius of the on-chip decoupling capacitor. Note that in a mesh structure, the maximum effective radius is the Manhattan distance between two points. In disagreement with Fig. 18.11, the overall effective radius is actually shaped more like a diamond, as illustrated in Fig. 18.13.

In modern high performance ICs, up to 3000 I/O pins can be necessary [22]. Only half of the I/O pads are typically used to distribute power. The other half is dedicated to signaling. Assuming an equal distribution of power and ground pads, a quarter of the total number of pads is typically available for power *or* ground delivery. For high performance ICs with die sizes of 1.5 inches by 1.5 inches inside a flip-chip package, the distance between two adjacent power or ground pads is about $1300\,\mu$m. By modeling the flip chip area array by a six by six distributed RL mesh, the accuracy in determining the effective radii of an on-chip decoupling capacitor is traded off with the computational complexity required to analyze the power delivery network. In this chapter, an on-chip power distribution system composed of the four closest power pins is modeled as an RL mesh of forty by forty equal segments to accurately determine the maximum effective distance of an on-chip decoupling capacitor. Note that this approach of modeling a power distribution system is applicable to ICs with both conventional low cost and advanced high performance packaging.

18.7 Case study

The dependence of the effective radii of an on-chip decoupling capacitor on a power distribution system is described in this section to quantitatively illustrate these concepts. The load is modeled as a triangular current source with a 100 ps rise time and 300 ps fall time. The maximum tolerable ripple at the load is 10% of the power supply voltage. The relaxation time between two consecutive switching events (charge

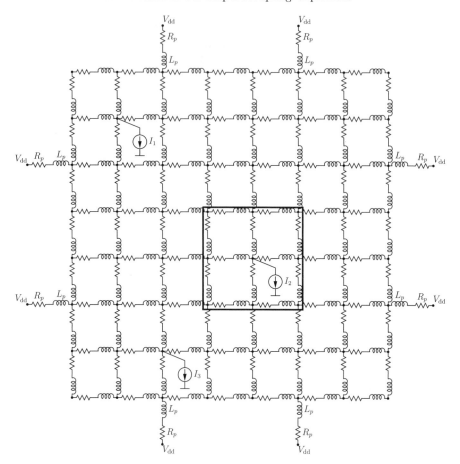

Fig. 18.12. Model of a power distribution network. The on-chip power delivery system is modeled as a distributed RL mesh with seven by seven equal segments. The current loads are modeled as current sources with equivalent magnitude and switching activities. R_p and L_p denote the parasitic resistance and inductance of the package, respectively. The rectangle denotes a "hot" spot – the area where the power supply voltage drops below the minimum tolerable level.

time) is 400 ps. Two scenarios are considered for determining the effective radii of an on-chip decoupling capacitor. In the first scenario, an on-chip decoupling capacitor is connected to the current load by a single line (local connectivity). In the second scenario, the on-chip decoupling capacitors are connected to the current loads by an on-chip

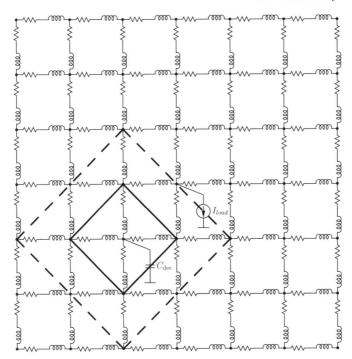

Fig. 18.13. Effective radii of an on-chip decoupling capacitor. For a power distribution system modeled as a distributed RL mesh, the maximum effective radius is the Manhattan distance between two points. The overall effective radius is therefore shaped like a diamond.

power distribution grid (global connectivity). A flip-chip package is assumed. An on-chip power distribution system with a flip-chip pitch (the area formed by the four closest pins) is modeled as an RL distributed mesh of forty by forty equal segments to accurately determine the maximum effective distance of an on-chip decoupling capacitor. The parasitic resistance and inductance of the package (the four closest pins of a flip-chip package) are also included in the model. The methodology for placing on-chip decoupling capacitors provides a highly accurate estimate of the magnitude and location of the on-chip decoupling capacitors. The maximum error of the resulting power noise is less than 0.1% as compared to SPICE.

For a single line, the maximum effective radii as determined by the target impedance and charge time for three sets of on-chip parasitic resistances and inductances are listed in Table 18.1. These three scenarios listed in Table 18.1 represent typical values of the parasitic resistance and inductance of the top, intermediate, and bottom layers of on-chip interconnects in a 90 nm CMOS technology [22]. In the case of the top metal layer, the maximum effective distance as determined by the target impedance is smaller than the critical distance as determined by (18.11). Hence, $IR \gg L\frac{dI}{dt}$, and the required on-chip decoupling capacitance is determined by (18.5). Note that the decoupling capacitance increases linearly with the current load. For a typical parasitic resistance and inductance of the intermediate and bottom layers of the on-chip interconnects, the effective radius as determined by the target impedance is longer than the critical distance d_{crit}. In this case, the overall voltage drop at the current load is determined by the inductive noise. The on-chip decoupling capacitance can therefore be estimated by (18.9).

Table 18.1. Maximum effective radii of an on-chip decoupling capacitor for a single line connecting a decoupling capacitor to a current load

Metal	Resistance	Inductance	I_{load}	C_{dec}	$d_{\max}$ (μm)	
Layer	(Ω/μm)	(pH/μm)	(A)	(pF)	Z	t_{ch}
	0.007	0.5	0.01	20	310.8	1166
Top	0.007	0.5	0.1	200	31.1	116
	0.007	0.5	1	2000	3.1	11.6
	0.04	0.3	0.01	183	226.2	24.2
Intermediate	0.04	0.3	0.1	1773	22.6	2.4
	0.04	0.3	1	45454	2.3	0.2
	0.1	0.1	0.01	50000	99.8	0
Bottom	0.1	0.1	0.1	∞	0	0
	0.1	0.1	1	∞	0	0
$V_{\mathrm{dd}} = 1$ V, $V_{ripple} = 100$ mV, $t_r = 100$ ps, $t_f = 300$ ps, $t_{\mathrm{ch}} = 400$ ps						

In the case of an RL mesh, the maximum effective radii as determined by the target impedance and charge time for three sets of on-chip parasitic resistances and inductances are listed in Table 18.2. From (18.11), for the parameters listed in Table 18.2, the critical voltage drop is 75 mV. If the voltage fluctuations at the current load do not exceed the critical voltage, $IR \gg L\frac{dI}{dt}$ and the required on-chip decoupling capacitance is determined by (18.5). Note that for the aforementioned three interconnect scenarios, assuming a 10 mA current load, the maximum effective radii of the on-chip decoupling capacitor based on the target impedance and charge time are larger than forty cells (the longest distance within the mesh from the center of the mesh to the corner). The maximum effective radii of the on-chip decoupling capacitor is therefore larger than the pitch size. The decoupling capacitor can therefore be placed anywhere inside the pitch. For a 100 mA current load, the voltage fluctuations at the current load exceed the critical voltage drop. The $L\frac{dI}{dt}$ noise dominates and the required on-chip decoupling capacitance is determined by (18.9).

Table 18.2. Maximum effective radii of an on-chip decoupling capacitor for an on-chip power distribution grid modeled as a distributed RL mesh

Metal	Resistance	Inductance	I_{load}	C_{dec}	d_{max} (cells)	
Layer	(Ω/μm)	(pH/μm)	(A)	(pF)	Z	t_{ch}
	0.007	0.5	0.01	20	>40	>40
Top	0.007	0.5	0.1	357	2	>40
	0.007	0.5	1	–	<1	–
	0.04	0.3	0.01	20	>40	>40
Intermediate	0.04	0.3	0.1	227	1	<1
	0.04	0.3	1	–	<1	–
	0.1	0.1	0.01	20	>40	>40
Bottom	0.1	0.1	0.1	–	<1	–
	0.1	0.1	1	–	<1	–
$V_{dd} = 1$ V, $V_{ripple} = 100$ mV, $t_r = 100$ ps,						
$t_f = 300$ ps, $t_{ch} = 400$ ps, cell size is $32.5\,\mu$m $\times$ $32.5\,\mu$m						

The effective radii of an on-chip decoupling capacitor decreases linearly with current load. The optimal size of an RL distributed mesh should therefore be determined for a particular current demand. If the magnitude of the current requirements is low, the mesh can be coarser, significantly decreasing the simulation time. For a 10 mA current load, the effective radii as determined from both the target impedance and charge time are longer than the pitch size. Thus, the distributed mesh is overly fine. For a current load of 1 A, the effective radii are shorter than one cell, meaning that the distributed RL mesh is overly coarse. A finer mesh should therefore be used to accurately estimate the maximum effective radii of the on-chip decoupling capacitor. In general, the cells within the mesh should be sized based on the current demand and the acceptable computational complexity (or simulation budget). As a rule of thumb, a coarser mesh should be used on the perimeter of each grid pitch. A finer mesh should be utilized around the current loads.

Note that in both cases, C_{dec}^r as determined by (18.9) increases rapidly with the effective radius based on the target impedance, becoming infinite at $d_Z^{\max}$. In this case study, the decoupling capacitor is allocated at almost the maximum effective distance $d_Z^{\max}$, simulating the worst case scenario. The resulting C_{dec} is therefore significantly large. As the decoupling capacitor is placed closer to the current load, the required on-chip decoupling capacitance as estimated by (18.9) can be reduced. A tradeoff therefore exists between the maximum effective distance as determined by the target impedance and the size of the minimum required on-chip decoupling capacitance (if the overall voltage drop at the current load is primarily caused by the inductive $L\frac{dI}{dt}$ drop).

The effective radii listed in Table 18.1 are determined for a single line between the current load or power supply and the decoupling capacitor. In the case of a power distribution grid modeled as a distributed RL mesh, multiple paths are connected in parallel, increasing the effective radii. For instance, comparing Table 18.1 to Table 18.2, note that the maximum effective radii as determined by the target impedance are increased about three times and two times for the top metal layers with a 10 mA and 100 mA current load, respectively. Note also that for typical values of the parasitic resistance and inductance of a power distribution grid, the effective radius as determined by the target impedance is longer than the radius based on the charge time for intermediate and bottom metal layers. For top metal layers, however, the effective radius as determined by the target impedance is typically shorter than the effective radius based on the charge time.

Also note that the maximum effective radius as determined by the charge time decreases quadratically with the decoupling capacitance. The maximum effective distance as determined by the charge time becomes impractically short for large decoupling capacitances. For the bottom metal layer, the maximum effective radius based on the charge time approaches zero. Note that the maximum effective radius during the charging phase has been evaluated for the case where the decoupling capacitor is charged to the power supply voltage. In practical applications, this constraint can be relaxed, assuming the voltage across the decoupling capacitor is several millivolts smaller than the power supply.

In this case, the effective radius of the on-chip decoupling capacitor as determined by the charge time can be significantly increased.

The maximum effective radius as determined by the charge time becomes impractically short for large decoupling capacitors, making the capacitors ineffective. In this case, the decoupling capacitor should be placed closer to the current load, permitting the decoupling capacitance to be decreased. Alternatively, the current load can be partitioned into several blocks, lowering the requirements on a specific local on-chip decoupling capacitance. The parasitic impedance between the decoupling capacitor and the current load and power supply should also be reduced, if possible, increasing the maximum effective radii of the on-chip decoupling capacitors.

18.8 Design implications

A larger on-chip decoupling capacitance is required to support increasing current demands. The maximum available on-chip decoupling capacitance, which can be placed in the vicinity of a particular circuit block, is limited however by the maximum capacitance density of a given technology, as described in Chapter 19. Large functional units (current loads) should therefore be partitioned into smaller blocks with local on-chip decoupling capacitors to enhance the likelihood of fault-free operation of the entire system. An important concept described in this chapter is that on-chip decoupling capacitors are a *local* phenomenon. Thus, the methodology for placing and sizing on-chip decoupling capacitors results in a greatly reduced budgeted on-chip decoupling capacitance as compared to a uniform (or blind) placement of on-chip decoupling capacitors into any available white space [268].

Typically, multiple current loads exist in an IC. An on-chip decoupling capacitor is placed in the vicinity of the current load such that both the current load and the power supply are within the maximum effective radius. Assuming a uniform distribution of the current loads, a schematic example placement of the on-chip decoupling capacitors is shown in Fig. 18.14. Each decoupling capacitor provides sufficient charge to the current load(s) within the maximum effective radius. Multiple on-chip decoupling capacitors are placed to provide charge to all of the circuit blocks. In general, the size and location of an on-chip decoupling capacitor are determined by the required charge (drawn by the local transient current loads) and certain system parameters (such as the per length resistance and inductance, power supply voltage, maximum tolerable ripple, and the switching characteristics of the current load).

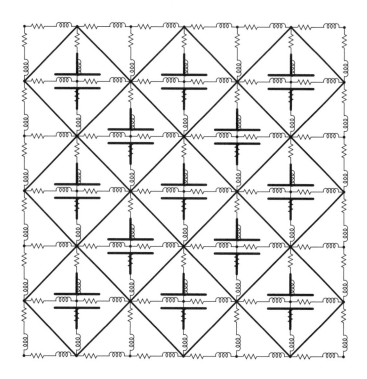

Fig. 18.14. A schematic example allocation of on-chip decoupling capacitors across an IC. Similar current loads are assumed to be uniformly distributed on the die. Each on-chip decoupling capacitor provides sufficient charge to the current load(s) within the maximum effective radius.

18.9 Summary

A design methodology for placing and sizing on-chip decoupling capacitors based on effective radii is presented in this chapter and can be summarized as follows:

- On-chip decoupling capacitors have traditionally been allocated into the available white space on a die, *i.e.*, using an unsystematic or *ad hoc* approach

- On-chip decoupling capacitors behave locally and should therefore be treated as a local phenomenon. The efficiency of on-chip decoupling capacitors depends upon the impedance of the power/ground lines connecting the capacitors to the current loads and power supplies

- Closed-form expressions for the maximum effective radii of an on-chip decoupling capacitor based on a target impedance (during discharge) and charge time (during charging phase) are described

- Depending upon the parasitic impedance of the power/ground lines, the maximum voltage drop is caused either by the dominant inductive $L\frac{dI}{dt}$ noise or by the dominant resistive IR noise

- Design expressions to estimate the minimum on-chip decoupling capacitance required to support expected current demands based on the dominant voltage drop are provided

- An expression for the critical length of the interconnect between the decoupling capacitor and the current load is described

- To be effective, an on-chip decoupling capacitor should be placed such that both the power supply and the current load are located inside the appropriate effective radius

- On-chip decoupling capacitors should be allocated within appropriate effective radii across an IC to satisfy local transient current demands

19

Efficient Placement of Distributed On-Chip Decoupling Capacitors

Decoupling capacitors are widely used to manage power supply noise [242] and are an effective way to reduce the impedance of power delivery systems operating at high frequencies [26], [27]. A decoupling capacitor acts as a local reservoir of charge, which is released when the power supply voltage at a particular current load drops below some tolerable level. Since the inductance scales slowly [207], the location of the decoupling capacitors significantly affects the design of the power/ground networks in high performance integrated circuits such as microprocessors. At higher frequencies, a distributed system of decoupling capacitors are placed on-chip to effectively manage the power supply noise [334].

The efficacy of decoupling capacitors depends upon the impedance of the conductors connecting the capacitors to the current loads and power sources. As described in [200], a maximum parasitic impedance between the decoupling capacitor and the current load (*or* power source) exists at which the decoupling capacitor is effective. Alternatively, to be effective, an on-chip decoupling capacitor should be placed such that both the power supply and the current load are located inside the appropriate effective radius [200]. The efficient placement of on-chip decoupling capacitors in nanoscale ICs is the subject of this chapter. Unlike the methodology for placing a single lumped on-chip decoupling capacitor presented in Chapter 18, a system of *distributed* on-chip decoupling capacitors is described in this chapter. A design methodology to estimate the parameters of the distributed system of on-chip decoupling capacitors is also presented, permitting the required on-chip decoupling capacitance to be allocated under existing technology constraints.

The chapter is organized as follows. Technology limitations in nanoscale integrated circuits are reviewed in Section 19.1. The problem of placing on-chip decoupling capacitors in nanoscale ICs while satisfying technology constraints is formulated in Section 19.2. The design of a distributed on-chip decoupling capacitor network is presented in Section 19.3. Various design tradeoffs are discussed in Section 19.4. A design methodology for placing distributed on-chip decoupling capacitors is presented in Section 19.5. Related simulation results for typical values of on-chip parasitic resistances are discussed in Section 19.6. Some specific conclusions are summarized in Section 19.7.

19.1 Technology constraints

On-chip decoupling capacitors have traditionally been designed as standard gate oxide CMOS capacitors [352]. As technology scales, leakage current through the gate oxide of an on-chip decoupling capacitor has greatly increased [299], [353], [354]. Moreover, in modern high performance ICs, a large portion (up to 40%) of the circuit area is occupied by the on-chip decoupling capacitance [355], [356]. Conventional gate oxide on-chip decoupling capacitors are therefore prohibitively expensive from an area and yield perspective, as well as greatly increasing the overall power dissipated on-chip [357].

To reduce the power consumed by an IC, MIM capacitors are frequently utilized as decoupling capacitors. The capacitance density of a MIM capacitor in a 90 nm CMOS technology is comparable to the maximum capacitance density of a CMOS capacitor and is typically $10\,\text{fF}/\mu\text{m}^2$ to $30\,\text{fF}/\mu\text{m}^2$ [179], [182], [193]. A maximum magnitude of an on-chip decoupling capacitor therefore exists for a specific distance between a current load and a decoupling capacitor (as constrained by the available on-chip metal resources). Alternatively, a minimum achievable impedance per unit length exists for a specified capacitance density of an on-chip decoupling capacitor placed at a specific distance from a circuit module, as illustrated in Fig. 19.1.

Observe from Fig. 19.1 that the available metal area for the second level of a distributed on-chip capacitance is greater than the fraction of metal resources dedicated to the first level of a distributed on-chip capacitance. Capacitor C_2 can therefore be larger than C_1. Note also that a larger capacitor can only be placed farther from the current load. Similarly, the metal resources required by the first level of interconnection (connecting C_1 to the current load) is smaller than the metal resources dedicated to the second level of interconnections. The impedance Z_2 is therefore smaller than Z_1.

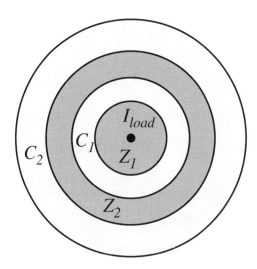

Fig. 19.1. Fundamental limits of on-chip interconnections. Two levels of a distributed on-chip decoupling capacitance are allocated around a current load. The interconnect impedance is inversely proportional to the fraction of metal area dedicated to the interconnect level, decreasing as the decoupling capacitor is farther from the current source ($Z_1 > Z_2$). The decoupling capacitance increases as the capacitor is farther from the current load due to the increased area ($C_1 < C_2$). The two levels of interconnection and distributed decoupling capacitance are shown in dark grey and light grey, respectively.

19.2 Placing on-chip decoupling capacitors in nanoscale ICs

Decoupling capacitors have traditionally been allocated into the white space (those areas not occupied by the circuit elements) available on the die based on an unsystematic or *ad hoc* approach [268], [272], as shown in Fig. 19.2. In this way, decoupling capacitors are often placed at a significant distance from the current load. Conventional approaches for placing on-chip decoupling capacitors result in oversized capacitors.

The conventional allocation strategy, therefore, results in increased power noise, compromising the signal integrity of an entire system, as illustrated in Fig. 19.3. This issue of power delivery cannot be alleviated by simply increasing the size of the on-chip decoupling capacitors. Furthermore, increasing the size of more distant on-chip decoupling capacitors results in wasted area, increased power, reduced reliability, and higher cost. A design methodology is therefore required to account for technology trends in nanoscale ICs, such as increasing frequencies, larger die sizes, higher current demands, and reduced noise margins.

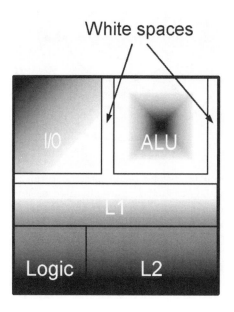

Fig. 19.2. Placement of on-chip decoupling capacitors using a conventional approach. Decoupling capacitors are allocated into the white space (those areas not occupied by the circuits elements) available on the die using an unsystematic or *ad hoc* approach. As a result, the power supply voltage drops below the minimum tolerable level for remote blocks (shown in dark grey). Low noise regions are light grey.

To be effective, a decoupling capacitor should be placed physically close to the current load. This requirement is naturally satisfied in board and package applications, since large capacitors are much smaller

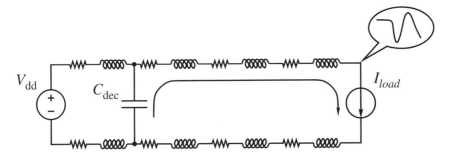

Fig. 19.3. A conventional on-chip decoupling capacitor. Typically, a large decoupling capacitor is placed farther from the current load due to physical limitations. Current flowing through the long power/ground lines results in large voltage fluctuations across the terminals of the current load.

than the dimensions of the circuit board (or package) [144]. In this case, a lumped model of a decoupling capacitor provides sufficient accuracy [358].

The size of an on-chip decoupling capacitor, however, is directly proportional to the area occupied by the capacitor and can require a significant portion of the on-chip area. The minimum impedance between an on-chip capacitor and the current load is fundamentally affected by the magnitude (and therefore the area) of the capacitor. Systematically partitioning the decoupling capacitor into smaller capacitors solves this issue. A system of distributed on-chip decoupling capacitors is illustrated in Fig. 19.4.

In a system of distributed on-chip decoupling capacitors, each decoupling capacitor is sized based on the impedance of the interconnect segment connecting the capacitor to the current load. A particular capacitor only provides charge to a current load during a short period. The rationale behind this scheme can be explained as follows. The capacitor closest to the current load is engaged immediately after the switching cycle is initiated. Once the first capacitor is depleted of charge, the next capacitor is activated, providing a large portion of the total current drawn by the load. This procedure is repeated until the last capacitor becomes active. Similar to the hierarchical placement of decoupling capacitors presented in [26], [107], this technique provides an efficient solution for providing the required on-chip decoupling capacitance based on specified capacitance density constraints. A system of distributed on-chip decoupling capacitors should therefore be utilized to provide a low impedance, cost effective power delivery network in nanoscale ICs.

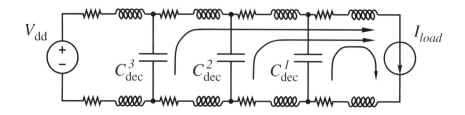

Fig. 19.4. A network of distributed on-chip decoupling capacitors. The magnitude of the decoupling capacitors is based on the impedance of the interconnect segment connecting a specific capacitor to a current load. Each decoupling capacitor is designed to only provide charge during a specific time interval.

19.3 Design of a distributed on-chip decoupling capacitor network

As described in Section 19.2, a system of distributed on-chip decoupling capacitors is an efficient solution for providing the required on-chip decoupling capacitance based on the maximum capacitance density available in a particular technology. A physical model of the technique is illustrated in Fig. 19.5. For simplicity, two decoupling capacitors are assumed to provide the required charge drawn by the current load. Note that as the capacitor is placed farther from the current load, the magnitude of an on-chip decoupling capacitor increases due to relaxed constraints. In the general case, the described methodology can be extended to any practical number of on-chip decoupling capacitors. Note that Z_1 is typically limited by a specific technology (determined by the impedance of a single metal wire) and the magnitude of C_1 (the area available in the vicinity of a circuit block).

A circuit model of a system of distributed on-chip decoupling capacitors is shown in Fig. 19.6. The impedance of the metal lines connecting the capacitors to the current load is modeled as resistors R_1 and R_2. A triangular current source is assumed to model the current load. The

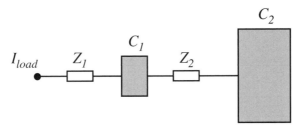

Fig. 19.5. A physical model of a system of distributed on-chip decoupling capacitors. Two capacitors are assumed to provide the required charge drawn by the load. Z_1 and Z_2 denote the impedance of the metal lines connecting C_1 to the current load and C_2 to C_1, respectively.

magnitude of the current source increases linearly, reaching the maximum current I_{max} at rise time t_r, i.e., $I_{load}(t) = I_{max}\frac{t}{t_r}$. The maximum tolerable ripple at the current load is 10% of the power supply voltage.

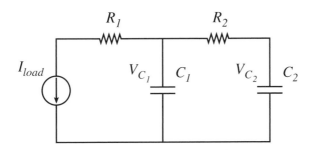

Fig. 19.6. A circuit model of an on-chip distributed decoupling capacitor network. The impedance of the metal lines is modeled as R_1 and R_2, respectively.

Note from Fig. 19.6 that since the charge drawn by the current load is provided by the on-chip decoupling capacitors, the voltage across the capacitors during discharge drops below the initial power supply voltage. The required charge during the entire switching event is thus determined by the voltage drop across C_1 and C_2.

The voltage across the decoupling capacitors at the end of the switching cycle ($t = t_r$) can be determined from Kirchhoff's laws [351]. Writing KVL and KCL equations for each of the loops (see Fig. 19.6), the system of differential equations describing the voltage across C_1 and C_2 at t_r is

$$\frac{dV_{C_1}}{dt} = \frac{V_{C_2} - V_{C_1}}{R_2 C_1} - \frac{I_{load}}{C_1}, \tag{19.1}$$

$$\frac{dV_{C_2}}{dt} = \frac{V_{C_1} - V_{C_2}}{R_2 C_2}. \tag{19.2}$$

Simultaneously solving (19.1) and (19.2) and applying the initial conditions, the voltage across C_1 and C_2 at the end of the switching activity is

$$V_{C_1}|_{t=t_r} = \frac{1}{2(C_1 + C_2)^3 t_r} \left[2C_1^3 t_r + C_1^2 t_r (6C_2 - I_{max} t_r) \right. \tag{19.3}$$
$$- C_2^2 t_r \left(2C_2 \left(I_{max} R_2 - 1 \right) + I_{max} t_r \right)$$
$$+ 2C_1 C_2 \left(C_2^2 \left(1 - e^{-\frac{(C_1 + C_2) t_r}{C_1 C_2 R_2}} \right) I_{max} R_2^2 \right.$$
$$\left. + C_2 \left(3 - I_{max} R_2 \right) t_r - I_{max} t_r^2 \right) \right],$$

$$V_{C_2}|_{t=t_r} = \frac{1}{2(C_1 + C_2)^3 t_r} \left[2C_1^3 t_r + C_2^2 t_r \left(2C_2 - I_{max} t_r \right) \right. \tag{19.4}$$
$$+ 2C_1 C_2 t_r \left(C_2 \left(3 + I_{max} R_2 \right) - I_{max} t_r \right)$$
$$+ C_1^2 \left(2C_2^2 \left(e^{-\frac{(C_1 + C_2) t_r}{C_1 C_2 R_2}} - 1 \right) I_{max} R_2^2 \right.$$
$$\left. + 2C_2 \left(3 + I_{max} R_2 \right) t_r - I_{max} t_r^2 \right) \right],$$

where I_{max} is the maximum magnitude of the current load and t_r is the rise time.

Note that the voltage across C_1 and C_2 after discharge is determined by the magnitude of the decoupling capacitors and the parasitic resistance of the metal line(s) between the capacitors. The voltage across C_1 after the switching cycle, however, depends upon the resistance of the P/G paths connecting C_1 to a current load and is

$$V_{C_1} = V_{load} + I_{max} R_1, \tag{19.5}$$

where V_{load} is the voltage across the terminals of a current load. Assuming $V_{load} \geq 0.9 V_{dd}$ and $V_{C_1}^{max} = V_{dd}$ (meaning that C_1 is infinitely large), the upper bound for R_1 is

$$R_1^{\max} = \frac{V_{dd}(1 - \alpha)}{I_{\max}}, \qquad (19.6)$$

where α is the ratio of the minimum tolerable voltage across the terminals of a current load to the power supply voltage ($\alpha = 0.9$ in this chapter). If $R_1 > R_1^{\max}$, no solution exists for providing sufficient charge drawn by the load. In this case, the circuit block should be partitioned, reducing the current demands ($I_{\max}$).

Note that expressions for determining the voltage across the decoupling capacitors are transcendental functions. No closed-form solution, therefore, exists. From (19.3) and (19.4), the design space can be graphically obtained for determining the maximum tolerable resistance R_2 and the minimum magnitude of the capacitors, maintaining the voltage across the load equal to or greater than the minimum allowable level. The voltage across C_1 after discharge as a function of C_1 and R_2 is depicted in Fig. 19.7.

Observe from Fig. 19.7 that the voltage across capacitor C_1 increases exponentially with capacitance, saturating for large C_1. The voltage across C_1, however, is almost independent of R_2, decreasing slightly with R_2 (see Fig. 19.7(a)). This behavior can be explained as follows. As a current load draws charge from the decoupling capacitors, the voltage across the capacitors drops below the initial level. The charge released by a capacitor is proportional to the capacitance and the change in voltage. A larger capacitance therefore results in a smaller voltage drop. From Fig. 19.6, note that as resistance R_2 increases, capacitor C_2 becomes less effective (a larger portion of the total current is provided by C_1). As a result, the magnitude of C_1 is increased to maintain the voltage across the load above the minimum tolerable level. Similarly, a larger C_2 results in a smaller C_1. As C_2 is increased, a larger portion of the total current is provided by C_2, reducing the magnitude of C_1. This phenomenon is well pronounced for small R_2, diminishing with larger R_2, as illustrated in Fig. 19.7(b).

In general, to determine the parameters of the system of distributed on-chip decoupling capacitors, the following assumptions are made. The parasitic resistance of the metal line(s) connecting capacitor C_1 to the current load is known. R_1 is determined by technology constraints (the sheet resistance) and by design constraints (the maximum available metal resources). The minimum voltage level at the load is $V_{load} = 0.9V_{dd}$. The maximum magnitude of the current load $I_{\max}$ is 0.01 A, the rise time t_r is 100 ps, and the power supply voltage V_{dd} is one volt.

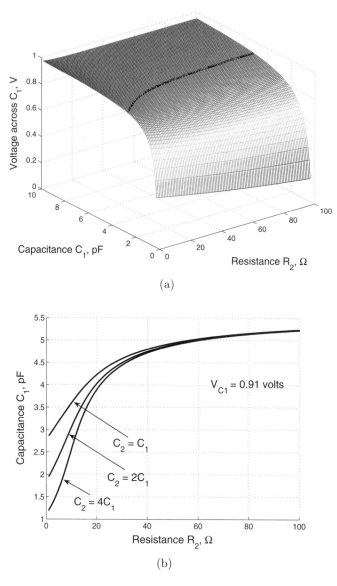

(a)

(b)

Fig. 19.7. Voltage across C_1 during discharge as a function of C_1 and R_2: $I_{max} = 0.01\,\text{mA}$, $V_{dd} = 1$ volt, and $t_r = 100\,\text{ps}$. a) Assuming $C_1 = C_2$ and $R_1 = 10\,\Omega$, the minimum tolerable voltage across C_1, resulting in $V_{load} \geq 0.9V_{dd}$, is 0.91 volts (shown as a black equipotential line). b) The design space for determining C_1 and R_2 resulting in the voltage across C_1 equal to 0.91 volts.

Note that the voltage across C_2 after discharge as determined by (19.4) is also treated as a design parameter. Since the capacitor C_2 is directly connected to the power supply (a shared power rail), the voltage drop across C_2 appears on the global power line, compromising the signal integrity of the overall system. The voltage across C_2 at t_r is therefore based on the maximum tolerable voltage fluctuations on the P/G line during discharge (the voltage across C_2 at the end of the switching cycle is set to 0.95 volts).

The system of equations to determine the parameters of an on-chip distributed decoupling capacitor network as depicted in Fig. 19.6 is

$$V_{load} = V_{C_1} - I_{max}R_1, \tag{19.7}$$

$$V_{C_1} = f(C_1, C_2, R_2), \tag{19.8}$$

$$V_{C_2} = f(C_1, C_2, R_2), \tag{19.9}$$

$$\frac{I_{max}t_r}{2} = C_1 (V_{dd} - V_{C_1}) + C_2 (V_{dd} - V_{C_2}), \tag{19.10}$$

where V_{C_1} and V_{C_2} are the voltage across C_1 and C_2 and determined by (19.3) and (19.4), respectively. Equation (19.10) states that the total charge drawn by the current load is provided by C_1 and C_2. Note that in the general case with the current load determined *a priori*, the total charge is the integral of $I_{load}(t)$ from zero to t_r. Solving (19.7) for V_{C_1} and substituting into (19.8), C_1, C_2, and R_2 are determined from (19.8), (19.9), and (19.10) for a specified $V_{C_2}(t_r)$, as discussed in the following section.

19.4 Design tradeoffs in a distributed on-chip decoupling capacitor network

To design a system of distributed on-chip decoupling capacitors, the parasitic resistances and capacitances should be determined based on design and technology constraints. As shown in Section 19.3, in a system composed of two decoupling capacitors (see Fig. 19.6) with known R_1; R_2, C_1, and C_2 are determined from the system of equations, (19.7)–(19.10). Note that since this system of equations involves transcendental functions, a closed-form solution cannot be determined. To determine the system parameters, the system of equations (19.7)–(19.10) is solved numerically [359].

Various tradeoff scenarios are discussed in this section. The dependence of the system parameters on R_1 is presented in Section 19.4.1.

The design of a distributed on-chip decoupling capacitor network with the minimum magnitude of C_1 is discussed in Section 19.4.2. The dependence of C_1 and C_2 on the parasitic resistance of the metal lines connecting the capacitors to the current load is presented in Section 19.4.3. The minimum total budgeted on-chip decoupling capacitance is also determined in this section.

19.4.1 Dependence of system parameters on R_1

The parameters of a distributed on-chip decoupling capacitor network for typical values of R_1 are listed in Table 19.1. Note that the minimum magnitude of R_2 exists for which the parameters of the system can be determined. If R_2 is sufficiently small, the distributed decoupling capacitor network degenerates to a system with a single capacitor (where C_1 and C_2 are combined). For the parameters listed in Table 19.1, the minimum magnitude of R_2 is four ohms, as determined from numerical simulations.

Table 19.1. Dependence of the parameters of a distributed on-chip decoupling capacitor network on R_1

R_1 (Ω)	$R_2 = 5\,(\Omega)$		$R_2 = 10\,(\Omega)$	
	C_1 (pF)	C_2 (pF)	C_1 (pF)	C_2 (pF)
1	1.35	7.57	3.64	3.44
2	2.81	5.50	4.63	2.60
3	4.54	3.64	5.88	1.77
4	6.78	1.87	7.56	0.92
5	10.00	0	10.00	0
	$V_{dd} = 1\,V$, $V_{load} = 0.9\,V$,			
	$t_r = 100\,ps$, and $I_{max} = 0.01\,A$			

Note that the parameters of a distributed on-chip decoupling capacitor network are determined by the parasitic resistance of the P/G line(s) connecting C_1 to the current load. As R_1 increases, the capacitor C_1 increases substantially (see Table 19.1). This increase in C_1 is due to R_1 becoming comparable to R_2, and C_1 providing a greater portion of the total current. Alternatively, the system of distributed on-chip decoupling capacitors degenerates to a single oversized capacitor. The system of distributed on-chip decoupling capacitors should therefore be carefully designed. Since the distributed on-chip decoupling capacitor network is strongly dependent upon the first level of interconnection (R_1), C_1 should be placed as physically close as possible to the current load, reducing R_1. If such an allocation is not practically possible, the current load should be partitioned, permitting an efficient allocation of the distributed on-chip decoupling capacitors under specific technology constraints.

19.4.2 Minimum C_1

In practical applications, the size of C_1 (the capacitor closest to the current load) is typically limited by technology constraints, such as the maximum capacitance density and available area. The magnitude of the first capacitor in the distributed system is therefore typically small. In this section, the dependence of the distributed on-chip decoupling capacitor network on R_1 is determined for minimum C_1. A target magnitude of 1 pF is assumed for C_1. The parameters of a system of distributed on-chip decoupling capacitors as a function of R_1 under the constraint of a minimum C_1 are listed in Table 19.2. Note that V_{C_2} denotes the voltage across C_2 after discharge.

Table 19.2. Distributed on-chip decoupling capacitor network as a function of R_1 under the constraint of a minimum C_1

R_1	$V_{C_2} \neq$ const				$V_{C_2} = 0.95$ volt	
(Ω)	$R_2\,(\Omega)$	$C_2\,(\mathrm{pF})$	$R_2\,(\Omega)$	$C_2\,(\mathrm{pF})$	$R_2\,(\Omega)$	$C_2\,(\mathrm{pF})$
1	2	5.59	5	8.69	4.68	8.20
2	2	6.68	5	11.64	3.46	8.40
3	2	8.19	5	17.22	2.28	8.60
4	2	10.46	5	31.70	1.13	8.80
5	2	14.21	5	162.10	–	–
	$V_{\mathrm{dd}} = 1$ V, $V_{load} = 0.9$ V, $t_r = 100$ ps,					
	$I_{\max} = 0.01$ A, and $C_1 = 1$ pF					

Note that two scenarios are considered in Table 19.2 to evaluate the dependence of a distributed system of on-chip decoupling capacitors on R_1 and R_2. In the first scenario, the distributed on-chip decoupling capacitor network is designed to maintain the minimum tolerable voltage across the terminals of a current load. In this case, the magnitude of C_2 increases with R_1, becoming impractically large for large R_2. In the second scenario, an additional constraint (the voltage across C_2) is applied to reduce the voltage fluctuations on the shared P/G lines. In this case, as R_1 increases, C_2 slightly increases. In order to satisfy the constraint for V_{C_2}, R_2 should be significantly reduced for large values of R_1, meaning that the second capacitor should be placed close to the first capacitor. As R_1 is further increased, R_2 becomes negligible, implying that capacitors C_1 and C_2 should be merged to provide the required charge to the distant current load. Alternatively, the system of distributed on-chip decoupling capacitors degenerates to a conventional scheme with a single oversized capacitor [360].

Note that simultaneously satisfying both the voltage across the terminals of the current load and the voltage across the last decoupling capacitor is not easy. The system of on-chip distributed decoupling capacitors in this case depends upon the parameters of the first decoupling stage (R_1 and C_1). If C_1 is too small, no solution exists to satisfy V_{load}^{min} and $V_{C_2}^{min}$. Sufficient circuit area should therefore be allocated for C_1 early in the design process to provide the required on-chip decoupling capacitance in order to satisfy specific design and technology constraints.

19.4.3 Minimum total budgeted on-chip decoupling capacitance

As discussed in Sections 19.4.1 and 19.4.2, the design of a system of distributed on-chip decoupling capacitors is greatly determined by the parasitic resistance of the metal lines connecting C_1 to the current load and by the magnitude of C_1. Another important design constraint is the total budgeted on-chip decoupling capacitance. Excessive on-chip decoupling capacitance results in increased circuit area and greater leakage currents. Large on-chip decoupling capacitors can also compromise the reliability of the overall system, creating a short circuit between the plates of a capacitor [357]. It is therefore important to reduce the required on-chip decoupling capacitance while providing sufficient charge to support expected current demands.

To estimate the total required on-chip decoupling capacitance, $C_{total} = C_1 + C_2$ is plotted as a function of R_1 and R_2, as depicted in Fig. 19.8. Note that if R_2 is large, C_2 is ineffective and the system of distributed on-chip decoupling capacitors behaves as a single capacitor. Observe from Fig. 19.8 that C_{total} increases with R_1 for large R_2. In this case, C_1 is oversized, providing most of the required charge. C_1 should therefore be placed close to the current load to reduce the total required on-chip decoupling capacitance.

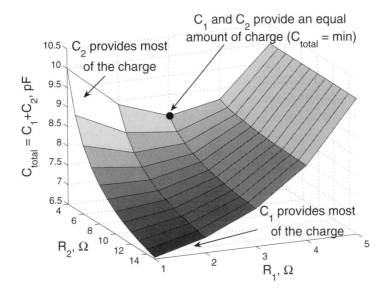

Fig. 19.8. The total budgeted on-chip decoupling capacitance as a function of the parasitic resistance of the metal lines, R_1 and R_2: $I_{max} = 10\,\mathrm{mA}$, $V_{dd} = 1\,\mathrm{volt}$, $V_{load} = 0.9\,\mathrm{volt}$, and $t_r = 100\,\mathrm{ps}$. In the system of distributed on-chip decoupling capacitors, an optimal ratio $\frac{R_2}{R_1}$ exists, resulting in the minimum total budgeted on-chip decoupling capacitance.

Similarly, if R_2 is reduced with small R_1, C_2 provides most of the charge drawn by the current load. The distributed on-chip decoupling capacitor network degenerates to a conventional system with a single capacitor. As R_1 increases, however, the total required on-chip decoupling capacitance decreases, reaching the minimum (see Fig. 19.8 for $R_1 = 3\,\Omega$ and $R_2 = 4\,\Omega$). In this case, C_1 and C_2 each provide an equal amount of the total charge. As R_1 is further increased (C_1 is placed farther from the current load), C_1 and C_2 increase substantially to compensate for the increased voltage drop across R_1. In the system of distributed on-chip decoupling capacitors, an optimal ratio $\frac{R_2}{R_1}$

exists which requires the minimum total budgeted on-chip decoupling capacitance.

Note that in the previous scenario, the magnitude of the on-chip decoupling capacitors has not been constrained. In practical applications, however, the magnitude of the first decoupling capacitor (placed close to the current load) is limited. To determine the dependence of the total required on-chip decoupling capacitance under the magnitude constraint of C_1, C_1 is fixed and set to $1\,\mathrm{pF}$. $C_{total} = C_1 + C_2$ is plotted as a function of R_1 and R_2, as shown in Fig. 19.9. In contrast to the results depicted in Fig. 19.8, the total budgeted on-chip decoupling capacitance required to support expected current demands increases with R_1 and R_2. Alternatively, C_2 provides the major portion of the total charge. Thus, the system behaves as a single distant on-chip decoupling capacitor. In this case, C_1 is too small. A larger area should therefore be allocated for C_1, resulting in a balanced system with a reduced total on-chip decoupling capacitance. Also note that as R_1 and R_2 further increase (beyond $4\,\Omega$, see Fig. 19.9), the total budgeted on-chip decoupling capacitance increases rapidly, becoming impractically large.

Comparing Fig. 19.8 to Fig. 19.9, note that if C_1 is constrained, a larger total decoupling capacitance is required to provide the charge drawn by the current load. Alternatively, the system of distributed on-chip decoupling capacitors under a magnitude constraint of C_1 behaves as a single distant decoupling capacitor. As a result, the magnitude of a single decoupling capacitor is significantly increased to compensate for the IR voltage drop across R_1 and R_2. The system of distributed on-chip decoupling capacitors should therefore be carefully designed to reduce the total budgeted on-chip decoupling capacitance. If the magnitude of C_1 is limited, C_2 should be placed close to the current load to be effective, reducing the total required on-chip decoupling capacitance. Alternatively, the parasitic impedance of the P/G lines connecting C_1 and C_2 should be reduced (e.g., utilizing wider lines and/or multiple lines in parallel) [70].

19.5 Design methodology for a system of distributed on-chip decoupling capacitors

An overall methodology for designing a distributed system of on-chip decoupling capacitors is illustrated in Fig. 19.10. General differential

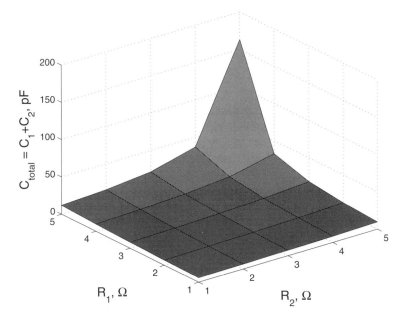

Fig. 19.9. The total budgeted on-chip decoupling capacitance as a function of the parasitic resistance of the metal lines, R_1 and R_2: $I_{\max} = 10\,\text{mA}$, $V_{\text{dd}} = 1\,\text{volt}$, $V_{load} = 0.9\,\text{volt}$, and $t_r = 100\,\text{ps}$. C_1 is fixed and set to $1\,\text{pF}$. The total budgeted on-chip decoupling capacitance increases with R_1 and R_2. As the parasitic resistance of the metal lines is further increased beyond $4\,\Omega$, C_{total} increases substantially, becoming impractically large.

equations for voltages $V_{C_1}(t)$ and $V_{C_2}(t)$ across capacitors C_1 and C_2 are derived based on Kirchhoff's laws. The maximum parasitic resistance $R_1^{\max}$ between C_1 and the current load is determined from (19.6) for specific parameters of the system, such as the power supply voltage V_{dd}, the minimum voltage across the terminals of the current load V_{load}, the maximum magnitude of the current load $I_{\max}$, and the rise time t_r. If $R_1 > R_1^{\max}$, no solution exists for the system of distributed on-chip decoupling capacitors. Alternatively, the voltage across the terminals of a current load always drops below the minimum acceptable level. In this case, the current load should be partitioned to reduce $I_{\max}$, resulting in $R_1 < R_1^{\max}$.

Simultaneously solving (19.1) and (19.2), the voltage across C_1 and C_2 is estimated at the end of a switching cycle ($t = t_r$), as determined by (19.3) and (19.4). The parameters of the distributed on-chip decoupling

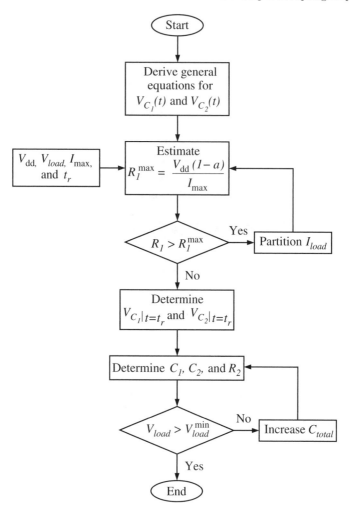

Fig. 19.10. Design flow for determining the parameters of a system of distributed on-chip decoupling capacitors.

capacitor network C_1, C_2, and R_2, are determined from (19.7)–(19.10). Note that different tradeoffs exist in a system of distributed on-chip decoupling capacitors, as discussed in Section 19.4. If the voltage across the terminals of a current load drops below the minimum tolerable level, the total budgeted on-chip decoupling capacitance should be increased. The system of equations, (19.7) to (19.10), is solved for an increased total on-chip decoupling capacitance, resulting in different C_1, C_2, and

R_2 until the criterion for the maximum tolerable power noise $V_{load} > V_{load}^{min}$ is satisfied, as shown in Fig. 19.10.

Note that the system of distributed on-chip decoupling capacitors permits the design of an effective power distribution system under specified technology constraints. The techniques presented in this chapter are also applicable to future technology generations. The methodology also provides a computationally efficient way to determine the required on-chip decoupling capacitance to support expected current demands. In the worst case example presented in this chapter, the simulation time to determine the parameters of the system of on-chip distributed decoupling capacitors is under one second on a Pentium III PC with one gigabyte of RAM. A methodology for efficiently placing on-chip decoupling capacitors can also be integrated into a standard IC design flow. In this way, the circuit area required to allocate on-chip decoupling capacitors is estimated early in the design process, significantly reducing the number of iterations and the eventual time to market.

19.6 Case study

The dependence of the system of distributed on-chip decoupling capacitors on the current load and the parasitic impedance of the power delivery system is described in this section to quantitatively illustrate the previously presented concepts. Resistive power and ground lines are assumed to connect the decoupling capacitors to the current load and are modeled as resistors (see Fig. 19.6). The load is modeled as a ramp current source with a 100 ps rise time. The minimum tolerable voltage across the load terminals is 90% of the power supply. The magnitude of the on-chip decoupling capacitors for various parasitic resistances of the metal lines connecting the capacitors to the current load is listed in Table 19.3. The parameters of the distributed on-chip decoupling capacitor network listed in Table 19.3 are determined for two amplitudes of the current load. Note that the values of R_1 and R_2 are typical parasitic resistances of an on-chip power distribution grid for a 90 nm CMOS technology.

The parameters of the system of distributed on-chip decoupling capacitors are analytically determined from (19.7)–(19.10). The resulting power supply noise is estimated using SPICE and compared to the maximum tolerable level (the minimum voltage across the load terminals V_{load}^{min}). The maximum voltage drop across C_2 at the end of the switching activity is also estimated and compared to $V_{C_2}^{\min}$. Note that the analytic solution produces an accurate estimate of the on-chip decoupling capacitors for typical parasitic resistances of a power distribution grid. The maximum error in this case study is 0.003%.

From Table 19.3, note that in the case of a large R_2, the distributed decoupling capacitor network degenerates into a system with a single capacitor. Capacitor C_1 is therefore excessively large. Conversely, if C_2 is placed close to C_1 (R_2 is small), C_2 is excessively large and the system again behaves as a single capacitor. An optimal ratio $\frac{R_2}{R_1}$ therefore exists for specific characteristics of the current load that results in a minimum required on-chip decoupling capacitance. Alternatively, in this case, both capacitors provide an equal portion of the total charge (see Table 19.3 for $R_1 = 0.5\,\Omega$ and $R_2 = 10\,\Omega$). Also note that as the magnitude of the current load increases, larger on-chip decoupling capacitors are required to provide the expected current demands.

The parameters of a distributed on-chip decoupling capacitor network listed in Table 19.3 have been determined for the case where the magnitude of the decoupling capacitors is not limited. In most practical systems, however, the magnitude of the on-chip decoupling capacitor placed closest to the current load is limited by technology and design constraints. A case study of a system of distributed on-chip decoupling capacitors with a limited value of C_1 is listed in Table 19.4. Note that in contrast to Table 19.3, where both R_1 and R_2 are design parameters, in the system with a limit on C_1, R_2 and C_2 are determined by R_1. Alternatively, both the magnitude and location of the second capacitor are determined from the magnitude and location of the first capacitor.

The parameters of the distributed on-chip decoupling capacitor network listed in Table 19.4 are determined for two amplitudes of the current load with R_1 representing a typical parasitic resistance of the metal line connecting C_1 to the current load. The resulting power supply noise at the current load and across the last decoupling stage is estimated using SPICE and compared to the maximum tolerable levels $V_{load}^{\min}$ and $V_{C_2}^{\min}$, respectively. Note that the analytic solution accurately estimates the parameters of the distributed on-chip decoupling capacitor network, producing a worst case error of 0.0001%.

Comparing results from Table 19.4 for two different magnitudes of C_1, note that a larger C_1 results in a smaller C_2. A larger C_1 also relaxes the constraints for the second decoupling stage, permitting C_2 to be placed farther from C_1. The first stage of a system of distributed

Table 19.3. The magnitude of the on-chip decoupling capacitors as a function of the parasitic resistance of the power/ground lines connecting the capacitors to the current load

R_1 (Ω)	R_2 (Ω)	I_{max} (A)	C_1 (pF)	C_2 (pF)	V_{load} (mV)		Error (%)	V_{C_2} (mV)		Error (%)
					V_{load}^{min}	SPICE		$V_{C_2}^{min}$	SPICE	
0.5	4.5	0.01	0	9.99999	900	899.999	0.0001	950	949.999	0.0001
0.5	6	0.01	1.59747	6.96215	900	899.986	0.002	950	949.983	0.002
0.5	8	0.01	2.64645	4.97091	900	899.995	0.0006	950	949.993	0.0004
0.5	10	0.01	3.22455	3.87297	900	899.997	0.0003	950	949.996	0.0004
0.5	12	0.01	3.59188	3.17521	900	899.998	0.0002	950	949.997	0.0003
0.5	14	0.01	3.84641	2.69168	900	899.998	0.0002	950	949.997	0.0003
0.5	16	0.01	4.03337	2.33650	900	899.999	0.0001	950	949.998	0.0002
0.5	18	0.01	4.17658	2.06440	900	899.998	0.0002	950	949.998	0.0002
0.5	20	0.01	4.28984	1.84922	900	899.999	0.0001	950	949.998	0.0002
0.5	1.5	0.025	0	24.99930	900	899.998	0.0002	950	949.998	0.0002
0.5	2	0.025	4.25092	17.56070	900	899.999	0.0001	950	949.999	0.0001
0.5	3	0.025	7.97609	11.04180	900	899.999	0.0001	950	949.999	0.0001
0.5	4	0.025	9.67473	8.06921	900	899.999	0.0001	950	949.999	0.0001
0.5	5	0.025	10.65000	6.36246	900	899.999	0.0001	950	949.999	0.0001
0.5	6	0.025	11.2838	5.25330	900	899.999	0.0001	950	949.999	0.0001
0.5	7	0.025	11.72910	4.47412	900	899.999	0.0001	950	949.999	0.0001
0.5	8	0.025	12.05910	3.89653	900	899.999	0.0001	950	949.999	0.0001
0.5	9	0.025	12.31110	3.44905	900	899.980	0.002	950	949.973	0.003
1	4	0.01	0	9.99999	900	899.999	0.0001	950	949.999	0.0001
1	6	0.01	2.16958	6.09294	900	899.990	0.001	950	949.988	0.001
1	8	0.01	3.11418	4.39381	900	899.996	0.0004	950	949.994	0.0006
1	10	0.01	3.64403	3.44040	900	899.997	0.0003	950	949.996	0.0004
1	12	0.01	3.98393	2.82871	900	899.998	0.0002	950	949.997	0.0003
1	14	0.01	4.22079	2.40240	900	899.998	0.0002	950	949.997	0.0003
1	16	0.01	4.39543	2.08809	900	899.998	0.0002	950	949.997	0.0003
1	18	0.01	4.52955	1.84668	900	899.998	0.0002	950	949.998	0.0002
1	20	0.01	4.63582	1.65540	900	899.998	0.0002	950	949.998	0.0002
1	1	0.025	0	24.99940	900	899.998	0.0002	950	949.998	0.0002
1	2	0.025	9.08053	11.37910	900	899.999	0.0001	950	949.999	0.0001
1	3	0.025	11.74820	7.37767	900	899.999	0.0001	950	949.999	0.0001
1	4	0.025	13.02600	5.46100	900	899.999	0.0001	950	949.999	0.0001
1	5	0.025	13.77630	4.33559	900	899.999	0.0001	950	949.999	0.0001
1	6	0.025	14.27000	3.59504	900	899.999	0.0001	950	949.999	0.0001
1	7	0.025	14.61950	3.07068	900	899.999	0.0001	950	949.999	0.0001
1	8	0.025	14.88010	2.67987	900	899.999	0.0001	950	949.999	0.0001
1	9	0.025	15.08180	2.37733	900	899.999	0.0001	950	949.999	0.0001

$V_{\mathrm{dd}} = 1\,\mathrm{V}$ and $t_r = 100\,\mathrm{ps}$

Table 19.4. The magnitude of the on-chip decoupling capacitors as a function of the parasitic resistance of the power/ground lines connecting the capacitors to the current load for a limit on C_1

R_1 (Ω)	I_{max} (A)	R_2 (Ω)	C_2 (pF)	V_{load}^{min}	V_{load} (mV) SPICE	Error (%)	$V_{C_2}^{min}$	V_{C_2} (mV) SPICE	Error (%)
					$C_1 = 0.5\,\mathrm{pF}$				
1	0.005	10.6123	4.05	900	899.999	0.0001	950	949.999	0.0001
2	0.005	9.3666	4.10	900	899.999	0.0001	950	949.999	0.0001
3	0.005	8.1390	4.15	900	899.999	0.0001	950	949.999	0.0001
4	0.005	6.9290	4.20	900	899.999	0.0001	950	949.999	0.0001
5	0.005	5.7354	4.25	900	899.999	0.0001	950	949.999	0.0001
0.5	0.01	4.8606	9.05	900	899.999	0.0001	950	949.999	0.0001
1	0.01	4.3077	9.10	900	900.000	0.0000	950	949.999	0.0001
2	0.01	3.2120	9.20	900	899.999	0.0001	950	949.999	0.0001
3	0.01	2.1290	9.30	900	899.999	0.0001	950	949.999	0.0001
4	0.01	1.0585	9.40	900	899.999	0.0001	950	949.999	0.0001
					$C_1 = 1\,\mathrm{pF}$				
1	0.005	13.2257	3.1	900	899.999	0.0001	950	949.999	0.0001
2	0.005	11.5092	3.2	900	899.999	0.0001	950	949.999	0.0001
3	0.005	9.8686	3.3	900	899.999	0.0001	950	949.999	0.0001
4	0.005	8.2966	3.4	900	899.999	0.0001	950	949.999	0.0001
5	0.005	6.7868	3.5	900	899.999	0.0001	950	949.999	0.0001
0.5	0.01	5.3062	8.1	900	899.999	0.0001	950	949.999	0.0001
1	0.01	4.6833	8.2	900	899.999	0.0001	950	949.999	0.0001
2	0.01	3.4644	8.4	900	899.999	0.0001	950	949.999	0.0001
3	0.01	2.2791	8.6	900	899.999	0.0001	950	949.999	0.0001
4	0.01	1.1250	8.8	900	899.999	0.0001	950	949.999	0.0001
				$V_{dd} = 1$ V and $t_r = 100$ ps					

on-chip decoupling capacitors should therefore be carefully designed to provide a balanced distributed decoupling capacitor network with a minimum total required capacitance, as discussed in Section 19.4.3.

On-chip decoupling capacitors have traditionally been allocated during a post-layout iteration (after the initial allocation of the standard cells). The on-chip decoupling capacitors are typically inserted into the available white space. If significant area is required for an on-chip decoupling capacitor, the circuit blocks are iteratively rearranged until the timing and signal integrity constraints are satisfied. Traditional strategies for placing on-chip decoupling capacitors therefore result in increased time to market, design effort, and cost.

The methodology for placing on-chip decoupling capacitors presented in this chapter permits simultaneous allocation of the on-chip decoupling capacitors and the circuit blocks. In this methodology, a current profile of a specific circuit block is initially estimated [361]. The magnitude and location of the distributed on-chip decoupling capacitors are determined based on expected current demands and technology constraints, such as the maximum capacitance density and parasitic resistance of the metal lines connecting the decoupling capacitors to the current load. Note that the magnitude of the decoupling capacitor closest to the current load should be determined for each circuit block, resulting in a balanced system and the minimum required total on-chip decoupling capacitance. As the number of decoupling capacitors increases, the parameters of a distributed on-chip decoupling capacitor network are relaxed, permitting the decoupling capacitors to be placed farther from the optimal location (permitting the parasitic resistance of the metal lines connecting the decoupling capacitors to vary over a larger range). In this way, the maximum effective radii of a distant on-chip decoupling capacitor is significantly increased [200]. A trade-off therefore exists between the magnitude and location of the on-chip decoupling capacitors comprising the distributed decoupling capacitor network.

19.7 Summary

A design methodology for placing distributed on-chip decoupling capacitors in nanoscale ICs can be summarized as follows:

- On-chip decoupling capacitors have traditionally been allocated into the available white space using an unsystematic approach. In this way, the on-chip decoupling capacitors are often placed far from the current load

- Existing allocation strategies result in increased power noise, compromising the signal integrity of an entire system

- Increasing the size of the on-chip decoupling capacitors allocated with conventional techniques does not enhance power delivery

- An on-chip decoupling capacitor should be placed physically close to the current load to be effective

- Since the area occupied by the on-chip decoupling capacitor is directly proportional to the magnitude of the capacitor, the minimum impedance between the on-chip decoupling capacitor and the current load is fundamentally affected by the magnitude of the capacitor

- A system of distributed on-chip decoupling capacitors has been described in this chapter to resolve this dilemma. A distributed on-chip decoupling capacitor network is an efficient solution for providing sufficient on-chip decoupling capacitance while satisfying existing technology constraints

- An optimal ratio of the parasitic resistance of the metal lines connecting the capacitors exists, permitting the total budgeted on-chip decoupling capacitance to be significantly reduced

- Simulation results for typical values of the on-chip parasitic resistances are also presented, demonstrating high accuracy of the analytic solution. In the worst case, the maximum error is 0.003% as compared to SPICE

- A distributed on-chip decoupling capacitor network permits the on-chip decoupling capacitors and the circuit blocks to be simultaneously placed within a single design step

Impedance/Noise Issues
in On-Chip Power Distribution Networks

The high frequency response of a power distribution system is the focus of this chapter. The impedance of the power distribution system at high frequencies is determined by the characteristics of the on-chip power distribution network. The impedance of a power system at a specific on-chip location is determined by the local resistive, inductive, and capacitive characteristics of the on-chip network. The resistive and inductive characteristics of the on-chip power distribution grids are characterized in Chapters 9 and 13. In this chapter, the impedance characteristics of both the on-chip power interconnect and the decoupling capacitors are combined to evaluate the noise characteristics of a power network. The inductance of an on-chip power distribution network is shown under specific conditions to be a significant design issue in high speed integrated circuits.

As discussed in Chapter 5, the inductance of the on-chip power and ground interconnect affects the impedance characteristics at relatively high frequencies; specifically, from the chip-package resonance to the highest frequencies of interest. The on-chip interconnect is a part of the current loop from the on-chip decoupling capacitors to the package decoupling capacitors. Typically, the inductance of this current loop is dominated by other parts of the loop — the bonding solder bumps, package conductors, and package decoupling capacitors. This situation is changing with technology scaling, as discussed in Section 20.1. The propagation of the power supply noise through the on-chip power distribution network is discussed in Section 20.2. The on-chip interconnect also provides a current path between the on-chip decoupling capacitors and the load. As the switching speed of the load increases, the inductance of the on-chip power lines can degrade the effectiveness of the on-chip capacitors, as discussed in Section 20.3. The chapter concludes with a summary.

20.1 Scaling effects in chip-package resonance

The continuous improvement in the performance characteristics of integrated circuits is primarily due to decreasing feature sizes, as discussed in Chapter 1. Technology scaling, however, has highly unfavorable implications for the impedance characteristics of a power distribution system. The manner in which these scaling trends affect the impedance characteristics of a power distribution system are described in this section. Specifically, the impedance characteristics near the chip-package resonance is the topic of primary concern.

Ideal scaling theory is briefly reviewed in Section 12.1. The current density increases as S and the supply voltage decreases as $1/S$ in the ideal scaling scenario, as discussed in Chapter 12. To maintain the power noise margin at the same fraction of the power supply voltage, the impedance of the power distribution system will decrease as

$$Z_{\text{pds}} \propto \frac{V}{I} = \frac{1}{S^2 S_C^2}, \tag{20.1}$$

assuming that the circuit area increases by a factor S_C.

Power supply scaling is impeded in sub-100 nanometer technologies by the difficulties in reducing the transistor threshold voltages. A decrease in the power supply voltage therefore significantly deviates from the ideal scaling scenario [10]. Consider a scenario where the voltage levels are scaled by a factor S_V ($S_V < S$). The power supply V_{dd} decreases as $1/S_V$, while the transistor current I_{tr} scales as S/S_V^2. The current per circuit area I_a increases by a factor of $S^2 \cdot I_{\text{tr}} = S^3/S_V^2$. The impedance of the power supply system therefore decreases as

$$Z_{\text{pds}} \propto \frac{V}{I} \propto \frac{1}{S_V} \frac{1}{S_C^2 S^3 / S_V^2} \propto \frac{S_V}{S} \frac{1}{S^2 S_C^2}. \tag{20.2}$$

This rate of decrease in the impedance is greater by a factor of S/S_V as compared to the ideal scaling scenario represented by (20.1).

The evolution of the impedance of a power distribution system in microprocessors is illustrated in Fig. 20.1. The rate of decrease in the impedance is approximately 2.7 times per technology generation (there

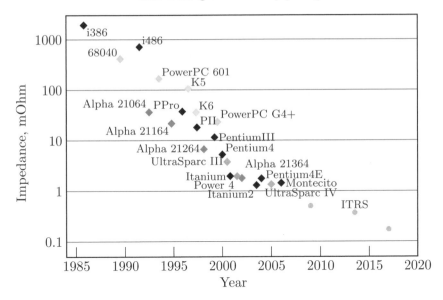

Fig. 20.1. Evolution of the impedance of a power distribution system in micro-processors. Several families of microprocessors and ITRS predictions [10] are shown in different shades of gray. The power supply noise margin is assumed to be 10% of the power supply voltage.

are approximately four technology generations per decade). This rate is significantly greater than the dimension scaling factor $\sqrt{2}$, in good agreement with the scaling analysis characterized by (20.1) and (20.2). As described in Chapter 1, the rate of decrease in the target imped-ance has recently saturated (1.25 times per computer generation [22]). This decrease is due to the limited power dissipation capabilities of traditional air cooled packaging.

Consider the magnitude of the impedance at the frequency of the chip-package resonance, where the impedance is typically the greatest. The minimum impedance at the resonant frequency is the character-istic impedance of the tank circuit, $Z_0 = \sqrt{\frac{L}{C}}$, where C is the on-chip decoupling capacitance and L is the inductance of the current loop from the on-chip load to the package decoupling capacitance, as discussed in Section 5.6. The decoupling capacitance per circuit area increases by a factor of S, and the overall capacitance of a circuit increases as $S_C^2 S$. The flip-chip contact density increases by a factor of S, assuming a contact pitch scaling factor of $\sqrt{S}$, as discussed in Chapter 12. Assum-ing a proportional decrease in the other components of the resonant inductance L, such as the inductance of the package conductors and the series inductance of the package capacitors, the overall loop inductance L decreases by a factor of S. Including an increase in the chip area as a factor of S_C^2, the inductance decreases by a factor of $1/S_C^2 S$. In this scenario, the resonant impedance only decreases by a factor of $\sqrt{\frac{1/S_C^2 S}{S_C^2 S}} = 1/S_C^2 S$, which is smaller than the requirements determined by (20.1) and (20.2). This reduction in impedance is therefore insufficient to satisfy a target noise margin. Further improvements in the circuit characteristics are necessary to approach the target specifications. The on-chip decoupling capacitance C is expensive to increase. When the available on-chip area is filled with decoupling capacitors, any additional decoupling capacitors increase die area and, consequently, the overall cost. Furthermore, in sub-100 nanometer technologies, the on-chip decoupling capacitors increase the static power consumption due to gate tunneling leakage current [362], [363].

The inductance of the current loop between the on-chip circuits and the package capacitors should be decreased to achieve the target impedance. The required decrease in inductance is particularly significant due to the square root dependence of the impedance on inductance. The inductance is reduced in advanced packaging technologies through improvements in the structure of the package and package decoupling capacitors. Application of finely spaced metal layers and replacing the solder bump connections with denser microvia contacts have been proposed to achieve this objective [114].

Due to the aggressive reduction in the package inductance, the inductance of the off-chip portion of the current loop becomes comparable to the inductance of the on-chip power interconnect structures. This trend is in agreement with the general approach of using a system of hierarchical decoupling capacitors to achieve a low impedance power distribution system. The upper frequency limit of the low impedance characteristics of a power distribution system should be extended as the switching speed of the on-chip circuitry increases with technology scaling. Exclusively using on-chip capacitors to improve the high frequency impedance characteristics is a relatively expensive solution. Reducing the package inductance typically offers a more economical alternative by extending the frequency range of the package decoupling capacitors, thereby relaxing the requirements placed on the on-chip decoupling capacitors.

20.2 Propagation of power distribution noise

The one-dimensional model described in Chapter 5 is inadequate to accurately describe the high frequency operation of a distributed circuit. As discussed in Chapter 7, the high frequency behavior of the on-chip power distribution network cannot be adequately described by a lumped model. A disturbance in the power supply voltage due to switching a local load propagates relatively slowly through the on-chip power distribution network. The power supply voltage is consequently non-uniform across the circuit die. These important effects are absent in a one-dimensional model. A two-dimensional model of the power distribution network is essential to accurately capture the high frequency impedance characteristics. The propagation of the power distribution noise through the on-chip power distribution network based on a simplified circuit model is discussed in this section.

The speed of noise propagation is an important characteristic of a power distribution network. The propagation speed of an undamped signal can be estimated using an idealized model of an on-chip power distribution network, where the decoupling capacitance is assumed to be uniformly distributed across a uniform power distribution grid. Assuming one-dimensional signal propagation, the power distribution grid is analogous to a capacitively loaded transmission line. The corresponding velocity of the signal propagation is

$$v = \frac{1}{\sqrt{L_\square C_a}}, \tag{20.3}$$

where $L_\square$ is the sheet inductance of the power distribution grid (as described in Chapter 9) and C_a is the area density of the on-chip decoupling capacitance.

As a practical example, assume a global power distribution grid consists of two layers with mutually perpendicular lines. The width and pitch of the grid lines are 50 μm and 100 μm, respectively, similar to the characteristics of the upper layer of the two layer grid considered in Section 13.2. The corresponding sheet inductance is approximately 0.2 $\frac{nH}{\square}$. A typical decoupling capacitance density C_a in high performance digital circuits manufactured in a 130 nm CMOS process is approximately 2 nF/mm^2. The velocity of the signal propagation based on these characteristics is approximately 1.6 mm/ns. This velocity is two orders of magnitude smaller than the speed of light in the circuit dielectric — approximately 150 mm/ns in silicon dioxide. The low velocity of the signal propagation is due to the high capacitance across the power and ground interconnect. This estimate is the upper bound on the signal velocity, as the resistance of the grid is neglected in (20.3). The resistance of the power lines further reduces the velocity of the signal. In overdamped power distribution networks, the signal propagation is determined by an RC rather than an LC time constant and approaches a diffusive RC-like signal behavior.

The relatively slow propagation of the power distribution noise has important circuit implications. From the perspective of a switching circuit, the low velocity noise propagation means that only the decoupling capacitance in the immediate proximity of the switching load is effective in limiting power supply variations at the load terminals. No charge sharing occurs during the switching transient between the load and the decoupling capacitors located farther than the propagation velocity times the switching time of the load, as described in Chapter 18. Alternatively, from the perspective of a quiescent circuit, the low propagation velocity means that the power supply level of the circuit is only affected by the switching loads that are located in close proximity to the circuit.

The idealized uniform model can also be used to estimate the inductive behavior of the on-chip power distribution grid. For one-dimensional signal propagation, the metric of inductive behavior for transmission lines described by (2.40) can be applied, yielding

$$\frac{t_r}{2\sqrt{L_\square C_a}} < l < \frac{2}{R_\square}\sqrt{\frac{L_\square}{C_a}}, \tag{20.4}$$

where $R_\square$ is the sheet resistance of the grid. Assuming a sheet resistance of $0.2\,\Omega/\square$ and a signal rise time t_r of the load current of $100\,\text{ps}$,

$$0.1\,\text{mm} < l < 3\,\text{mm}, \tag{20.5}$$

the power supply noise exhibits a significant inductive component only within a limited distance from the switching load. In other terms, the damping factor ζ of the current path within a power distribution grid,

$$\zeta = \frac{R_\square l}{2}\sqrt{\frac{C_a}{L_\square}}, \tag{20.6}$$

is smaller than unity if the path length is smaller than $3\,\text{mm}$. Within this distance from the load, the response of a power distribution network is underdamped and the power supply noise can exhibit significant ringing. At greater distances from the load, the propagation of the power supply noise approaches a diffusive RC-like behavior.

The inductance of a pair of wide global power and ground lines is comparable to that of an on-chip signal line. The capacitive load of the power lines, however, is approximately three orders of magnitude greater, while the resistance is ten to a hundred times lower. The range of length where the power interconnect exhibits inductive behavior, as indicated by (20.5), is similar to that of an on-chip signal line.

Note, however, that the characteristic impedance of a power-ground line pair is approximately two orders of magnitude lower than the characteristic impedance of an on-chip signal path. The magnitude of the power distribution noise induced by switching an on-chip signal line is therefore two orders of magnitude smaller than the swing of a signal line transition.

20.3 Local inductive behavior

The idealized model used in the preceding section provides a reasonable approximation of the noise propagation at a relatively large geometric scale, *i.e.*, where the wavelength of the signal is significantly larger than the pitch of the power lines. At smaller scales (and, consequently, shorter propagation times), the discrete nature of both the power load and the decoupling capacitors may be significant under certain conditions. Particularly, the high frequency characteristics of the power distribution interconnect become crucial. These local effects are discussed in this section.

A low impedance power distribution system during and immediately after the switching of an on-chip load is maintained using on-chip decoupling capacitors. The on-chip decoupling capacitance limits the variation of the power supply until the package decoupling capacitors become effective. As discussed in Section 6.3, the intrinsic parasitic capacitance of the load circuit typically provides a small fraction of the required decoupling capacitance. The intrinsic capacitance is embedded in the circuit structure. Consequently, the impedance between the intrinsic capacitance and the switching load capacitance is small. The intentional decoupling capacitors augment the intrinsic capacitance of the circuit to reach the required level of capacitance. The intentional capacitance, however, is typically added at the final stages of the circuit design process and is often physically located at a significant distance from the switching load. As the switching time of the load decreases, the impedance of the power interconnect becomes increasingly important. The significance of the power line impedance on the efficacy of the decoupling capacitors is demonstrated in the following example.

Consider an integrated circuit manufactured in a sub-100 nm CMOS technology. A high power local circuit macro, $200\,\mu m \times 200\,\mu m$ in size, switches a 20 pF load capacitance C_{load} within a 100 ps time period t_r. Assuming a 1 volt power supply, the maximum power current of the circuit is estimated as

$$I_{max} \approx \frac{C_{load}V_{dd}}{t_r/2} = \frac{20\,\text{pF} \times 1\,\text{volt}}{100\,\text{ps}/2} = 400\,\text{mA}, \qquad (20.7)$$

and the maximum current transient as

$$\left(\frac{dI}{dt}\right)_{max} \approx \frac{I_{max}}{t_r/2} = \frac{400\,\text{mA}}{100\,\text{ps}/2} = 8 \times 10^9\,\frac{\text{A}}{\text{s}}. \qquad (20.8)$$

The decoupling capacitance embedded within the circuit is assumed to be insufficient, supplying only half of the required current. The rest of the current is supplied by the nearest on-chip decoupling capacitor. To limit the resistive and inductive voltage drops to below 100 millivolts, 10% of the power supply in this 1 volt system, the resistance and inductance of the current path between the load circuit and the decoupling capacitor should be smaller than

$$R_{max} = \frac{0.1V_{dd}}{0.5I_{max}} = \frac{0.1\,\text{V}}{0.2\,\text{A}} = 0.5\,\Omega \qquad (20.9)$$

and

$$L_{max} = \frac{0.1V_{dd}}{0.5\,(dI/dt)_{max}} = \frac{0.1\,\text{volts}}{4 \times 10^9\,\text{A/s}} = 25\,\text{pH}, \qquad (20.10)$$

respectively. These impedance specifications are demanding. Assume that the physical distance between the load and the capacitor is $100\,\mu m$. Consider a scenario where the load and capacitor are connected by two global power and ground lines that are $50\,\mu m$ wide, $1\,\mu m$ thick, and are placed on a $100\,\mu m$ pitch, as illustrated in Fig. 20.2(a). The resistance of the current path is approximately 0.08 ohms, well below the limit set by (20.9). The inductance of the path, however, is approximately 80 pH, exceeding the limit set by (20.10).

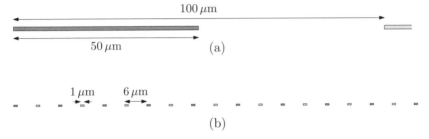

Fig. 20.2. Cross section of a current path connecting the load and decoupling capacitance. The power lines are shown in a darker gray, while the ground lines are shown in a lighter gray. The connection between the load and decoupling capacitance can be made using either thick and wide global power lines (a) or finer local power lines (b). The dimensions are drawn to scale.

Alternatively, if the load and decoupling capacitors are connected by fine lines of a local distribution network, for example, by 32 interdigitated power and ground lines with a $0.4\,\mu\text{m} \times 1\,\mu\text{m}$ cross section and a $6\,\mu\text{m}$ line pitch, as illustrated in Fig. 20.2(b), the inductance of the current path is reduced to 7 pH, well below the limit set by (20.10). The resistance in this case, however, is approximately 0.63 ohms, exceeding the limit set by (20.9). The high current density in these fine lines is also likely to violate existing electromigration reliability constraints.

The target impedance characteristics are therefore more readily achieved if both wide and thick lines in the upper metal layers and fine lines in the lower metal layers are used. The superior impedance characteristics of such interconnect structures are described in Chapter 13. Alternatively, the width of the global power and ground lines in the upper metal layers should be greatly decreased. This approach will decrease the inductance of the grid, at the expense of a moderate increase in resistance, as discussed in Chapter 11.

Due to the limitations of the power interconnect, the on-chip decoupling capacitance should be placed in the immediate vicinity of the switching load in order to be effective. This requirement necessitates novel approaches to allocating the on-chip decoupling capacitance, as described in Chapter 18. A common design approach, where the bulk of the decoupling capacitance is placed among the circuit blocks after the initial design of the blocks has been completed, as shown in Fig. 6.18, does not permit placing the decoupling capacitors sufficiently close to the circuits far from the block boundary. As the feature size of the on-chip circuits decreases, the capacitance allocation process should be performed at a commensurately finer scale, as schematically illustrated in Fig. 20.3.

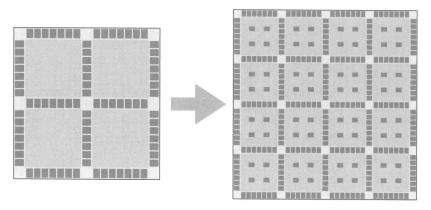

Fig. 20.3. The effect of circuit scaling on allocation of the on-chip decoupling capacitance. The circuit blocks are shown in darker gray, the decoupling capacitors are shown in lighter gray. As the circuit feature sizes decrease, the allocation of the on-chip decoupling capacitance should be performed at a commensurately finer scale. If the size of the circuit blocks does not decrease in proportion to the feature size, the decoupling capacitors are placed within the circuit blocks, as shown on the right.

The high frequency characteristics of the on-chip power interconnect are particularly important when the power load is non-uniformly distributed. Those circuits with the greatest peak power consumption require a significant decoupling capacitance in close proximity while the surrounding lower power circuits tend to require a relatively low decoupling capacitance.

The power consumption is particularly non-uniform in high performance digital circuits with a highly irregular structure, such as microprocessors, which are comprised of both low and high power circuit blocks. More than half of the die area in state-of-the-art microprocessors is occupied by memory circuit structures, which are characterized by a low switching activity and, consequently, low power consumption. The power density of a microprocessor core can be an order of magnitude higher as compared to the memory arrays; the core and synchronization circuits dissipate the dominant share of the overall circuit power. The distribution of the power consumption within the high power blocks is also typically non-uniform. The peak power demand circuitry with high load capacitances and switching activities, such as the arithmetic units and bus drivers, can exceed severalfold the worst case requirements of the surrounding circuits.

Contemporary trends in circuit design exacerbate the uneven distribution of the dissipated power. Similar to microprocessors, system-on-chip circuits integrate diverse circuit structures and also tend to exhibit a highly non-uniform power distribution pattern. Since the power consumption of integrated circuits has become a primary design priority, as described in Chapter 1, aggressive power saving techniques have become mandatory. Clock and power gating have gained wider use in order to decrease dynamic and leakage power consumption, respectively, in idle circuit blocks. While the power consumed by the circuit blocks is greatly reduced in the gated mode, abrupt transients in the power current are induced when a circuit block transitions from a power saving mode to active operation or vice versa. Power gating presents particular challenges to the analysis and verification of power distribution networks. A significant share of the decoupling capacitance is often disconnected from the global network during a power-down mode. This change in the decoupling capacitance can potentially cause power integrity problems in the surrounding circuits.

20.4 Summary

The effect of the decoupling capacitance and the inductance of on-chip interconnect on the high frequency impedance characteristics of a power distribution system is discussed in this chapter. The primary conclusions are summarized as follows.

- The inductance of the on-chip interconnect becomes more significant as the inductance of the package conductors is reduced

- The power noise propagates through the on-chip power distribution network at a relatively low velocity

- The response of the on-chip power distribution network is underdamped in close proximity to the load

- The impedance of the current path between the on-chip load and the on-chip decoupling capacitors becomes a critical design parameter as the power supply and circuit switching times decrease

- Allocating the on-chip decoupling capacitance should be performed at a finer scale as the feature size of the on-chip circuits decreases

21

Conclusions

The operation of an integrated circuit relies on the supply of power to the on-chip circuitry. Power distribution systems serve the purpose of supplying an integrated circuit with current while maintaining specific voltage levels. The power current is distributed across an integrated circuit through an on-chip power distribution network, an integral part of the overall power distribution system. The on-chip power distribution network delivers current to hundreds of millions of high speed transistors comprising a high complexity integrated circuit. Tens of amperes must be efficiently distributed to supply power to the on-chip circuits. Due to the high currents and high frequencies, the impedance of a power distribution system should be maintained sufficiently low over a wide range of frequencies in order to limit the voltage variations at the power load — the millions of on-chip transistors.

Maintaining a low impedance over a wide range of frequencies is a complex task. Decoupling capacitors effectively reduce the impedance of a power distribution system near the capacitor resonant frequency by allowing high frequency currents to bypass high inductance interconnect structures. The decoupling capacitance and interconnect inductance, however, create tank resonant modes within a power distribution system, increasing the impedance near the tank resonant frequency. The magnitude of the tank resonance is controlled by maintaining appropriate damping characteristics within the system. The design of a low impedance power distribution system therefore requires a careful balance among the resistive, inductive, and capacitive impedances of the comprising elements.

This balance should be maintained throughout the hierarchical structure of the system — at the board, package, and integrated circuit levels. The low impedance characteristics of an entire power distribution system, from the voltage regulator through the printed circuit board and package onto the integrated circuit to the power terminals of the on-chip circuitry, are maintained using a hierarchy of decoupling capacitors. In a hierarchical decoupling scheme, the power current loop is terminated progressively closer to the load as the frequency increases. The capacitance at each decoupling stage is constrained by the inductive and resistive characteristics of the capacitors and the power distribution network.

To maintain the impedance of a power distribution system below a specified level (the target impedance), multiple decoupling capacitors are placed in parallel at different levels of the power grid hierarchy. Two capacitors with different magnitudes connected in parallel result in antiresonance — an increase in the impedance of the power distribution system. If not properly controlled, the antiresonant peak may exceed the target impedance, jeopardizing the signal integrity of the system. The frequency of the antiresonant spike depends upon the ratio of the effective series inductance of the decoupling capacitors. As the parasitic inductance of the decoupling capacitors is reduced, the antiresonant spike is shifted to a higher frequency. A power distribution system with decoupling capacitors should therefore be carefully designed to control the effective series inductance of the capacitors. Alternatively, multiple decoupling capacitors with progressively decreasing magnitude should be allocated to cancel the antiresonance, shifting the antiresonant spikes to a frequency greater than the maximum operating frequency of the system.

Maintaining balanced impedance characteristics at the integrated circuit level is particularly challenging. The power current requirements and impedance characteristics can vary significantly across the die area. Electromigration reliability considerations place additional constraints on the design of the power distribution network. The design of the on-chip power distribution interconnect, placement of the on-chip circuits, allocation of the on-chip decoupling capacitors, and the analysis of the chip-package interface characteristics should all be carefully choreographed to achieve the target power noise characteristics.

Controlling the inductive characteristics of the interconnect comprising a power distribution network in a complex on-chip environment has become of significant importance in high speed circuits. The inductance is an essential parameter in determining the high frequency response of a power distribution grid. The significant inductive behavior of an on-chip power distribution network makes the power supply noise difficult to predict, exacerbating the analysis and verification process. The grid inductance can be effectively reduced with a moderate penalty in either the area or resistance of the grid. The impedance characteristics of multi-layer grids can result in the efficient distribution of power in conventional high speed circuits with relatively thick and wide lines in the upper layers and fine lines in the lower layers. The upper layers provide a low impedance current path at low frequencies, while the lower layers serve as a low impedance path at high frequencies. As a result, a low impedance power distribution network is maintained over a wide range of frequencies.

Despite recent advancements in integrated circuit technologies and packaging solutions, on-chip decoupling capacitors remain an attractive cost effective solution for providing a low impedance power distribution network supplying current over a wide range of frequencies. A decoupling capacitor acts as a local reservoir of charge, which is released when the power supply voltage across a particular current load drops below some tolerable level. MOS transistors have historically been used as on-chip decoupling capacitors, exploiting the relatively high gate capacitance of these structures. In sub-100 nanometer technologies, however, the application of on-chip MOS decoupling capacitors has become undesirable due to prohibitively high leakage currents. Occupying up to 40% of the circuit area, on-chip MOS decoupling capacitors contribute more than half of the total leakage power in modern high speed, high complexity ICs. New types of on-chip decoupling capacitors, such as MIM and lateral flux capacitors, have recently emerged as better candidates for on-chip decoupling capacitors.

On-chip decoupling capacitors have traditionally been allocated into the white space available on the die based on an unsystematic *ad hoc* approach. Conventional approaches for placing on-chip decoupling capacitors result in oversized capacitors often placed at a significant physical distance from the current loads. As a result, the power noise increases, compromising the signal integrity of the entire system. The efficacy of the decoupling capacitors depends upon the impedance of the conductors connecting the capacitors to the current loads and power supplies. To be effective, an on-chip decoupling capacitor should be placed such that both the power supply and the current load are located inside the appropriate effective radii of each decoupling capacitor.

The size of an on-chip decoupling capacitor, however, is directly proportional to the area occupied by the capacitor and can require a significant portion of the on-chip area. A system of distributed on-chip decoupling capacitors should therefore be utilized in nanoscale ICs to satisfy technology and performance constraints. The methodologies for placing on-chip decoupling capacitors presented in this book provide a computationally efficient way for determining the required on-chip decoupling capacitance to support expected current demands. In this way, the circuit area required for the on-chip decoupling capacitors is estimated at an early stage of the design process, significantly reducing the number of design iterations and the eventual time to market.

The topics presented in this book on the design of power distribution networks with decoupling capacitors are intended to provide insight into the electrical behavior and design principles of these high performance systems. A thorough understanding of the electrical phenomena in complex multi-layer power distribution networks with on-chip decoupling capacitors is therefore essential for developing effective methodologies and computer-aided tools for designing the next generation of nanoscale CMOS integrated systems.

Appendices

A

Mutual Loop Inductance in Fully Interdigitated Power Distribution Grids with DSDG

Assuming $d_I^i = s_I^i = d$, from (15.3), the mutual inductance between the power and ground paths of the different voltage domains for a fully interdigitated power distribution grid with DSDG is

$$L_{V\mathrm{dd1}-V\mathrm{dd2}} = 0.2l \left(\ln\frac{2l}{2d} - 1 + \frac{2d}{l} - \ln\gamma + \ln k \right), \qquad (A.1)$$

$$L_{V\mathrm{dd1}-G\mathrm{nd2}} = 0.2l \left(\ln\frac{2l}{3d} - 1 + \frac{3d}{l} - \ln\gamma + \ln k \right), \qquad (A.2)$$

$$L_{G\mathrm{nd1}-G\mathrm{nd2}} = 0.2l \left(\ln\frac{2l}{2d} - 1 + \frac{2d}{l} - \ln\gamma + \ln k \right), \qquad (A.3)$$

$$L_{V\mathrm{dd2}-G\mathrm{nd1}} = 0.2l \left(\ln\frac{2l}{d} - 1 + \frac{d}{l} - \ln\gamma + \ln k \right). \qquad (A.4)$$

Substituting (A.1) — (A.4) into (15.8), the mutual inductive coupling M_{loop}^{intI} between the two current loops in a fully interdigitated power distribution grid with DSDG is

$$M_{loop}^{\mathrm{intI}} = 0.2l \left(\ln\frac{2l}{2d} - 1 + \frac{2d}{l} - \ln\gamma + \ln k - \ln\frac{2l}{3d} + 1 - \frac{3d}{l} + \right.$$
$$\ln\gamma - \ln k + \ln\frac{2l}{2d} - 1 + \frac{2d}{l} - \ln\gamma + \ln k - \ln\frac{2l}{d} + 1 -$$
$$\left. \frac{d}{l} + \ln\gamma - \ln k \right). \qquad (A.5)$$

Simplifying (A.5) and considering that $\ln\gamma$ and $\ln k$ are approximately the same for different distances between the lines, M_{loop}^{intI} is

$$
\begin{aligned}
M_{loop}^{intI} &= 0.2l \left(\ln\frac{2l}{2d} - \ln\frac{2l}{3d} + \ln\frac{2l}{2d} - \ln\frac{2l}{d} \right) \\
&= 0.2l \ln\frac{2l \times 3d \times 2l \times d}{2d \times 2l \times 2d \times 2l} \\
&= 0.2l \ln\frac{3}{4} < 0.
\end{aligned}
\tag{A.6}
$$

B

Mutual Loop Inductance in Pseudo-Interdigitated Power Distribution Grids with DSDG

Assuming $d_{II}^i = 2d$ and $s_{II}^i = d$, from (15.3), the mutual inductance between the power and ground paths of the different voltage domains for a pseudo-interdigitated power distribution grid with DSDG is

$$L_{Vdd1-Vdd2} = 0.2l \left(\ln\frac{2l}{d} - 1 + \frac{d}{l} - \ln\gamma + \ln k \right), \qquad (B.1)$$

$$L_{Vdd1-Gnd2} = 0.2l \left(\ln\frac{2l}{3d} - 1 + \frac{3d}{l} - \ln\gamma + \ln k \right), \qquad (B.2)$$

$$L_{Gnd1-Gnd2} = 0.2l \left(\ln\frac{2l}{d} - 1 + \frac{d}{l} - \ln\gamma + \ln k \right), \qquad (B.3)$$

$$L_{Vdd2-Gnd1} = 0.2l \left(\ln\frac{2l}{d} - 1 + \frac{d}{l} - \ln\gamma + \ln k \right). \qquad (B.4)$$

Substituting (B.1) — (B.4) into (15.8), the mutual inductive coupling M_{loop}^{intII} between the two current loops in a pseudo-interdigitated power distribution grid with DSDG is

$$
M_{loop}^{\text{intII}} = 0.2l \left(\ln\frac{2l}{d} - 1 + \frac{d}{l} - \ln\gamma + \ln k - \ln\frac{2l}{3d} + 1 - \frac{3d}{l} + \right.
$$
$$
\ln\gamma - \ln k + \ln\frac{2l}{d} - 1 + \frac{d}{l} - \ln\gamma + \ln k - \ln\frac{2l}{d} + 1 -
$$
$$
\left. \frac{d}{l} + \ln\gamma - \ln k \right). \qquad (B.5)
$$

Simplifying (B.5) and considering that $\ln\gamma$ and $\ln k$ are approximately the same for different distances between the lines, M_{loop}^{intII} is

$$
\begin{aligned}
M_{loop}^{\mathrm{intII}} &= 0.2l \left(\ln\frac{2l}{d} - \ln\frac{2l}{3d} + \ln\frac{2l}{d} - \ln\frac{2l}{d} - \frac{2d}{l} \right) \\
&= 0.2l \left(\ln\frac{2l \times 3d \times 2l \times d}{d \times 2l \times d \times 2l} - \frac{2d}{l} \right) \\
&= 0.2l \left(\ln 3 - \frac{2d}{l} \right) > 0.
\end{aligned}
\tag{B.6}
$$

C

Mutual Loop Inductance in Fully Paired Power Distribution Grids with DSDG

Assuming the separation between the pairs is n times larger than the distance between the power and ground lines inside each pair d (see Fig. 15.8), from (15.3), the mutual inductance between the power and ground paths of the different voltage domains for a fully paired power distribution grid with DSDG is

$$L_{Vdd1-Vdd2} = 0.2l \left[\ln \frac{2l}{(n+1)d} - 1 + \frac{(n+1)d}{l} - \ln\gamma + \ln k \right], \quad (C.1)$$

$$L_{Vdd1-Gnd2} = 0.2l \left[\ln \frac{2l}{(n+2)d} - 1 + \frac{(n+2)d}{l} - \ln\gamma + \ln k \right], \quad (C.2)$$

$$L_{Gnd1-Gnd2} = 0.2l \left[\ln \frac{2l}{(n+1)d} - 1 + \frac{(n+1)d}{l} - \ln\gamma + \ln k \right], \quad (C.3)$$

$$L_{Vdd2-Gnd1} = 0.2l \left(\ln \frac{2l}{nd} - 1 + \frac{nd}{l} - \ln\gamma + \ln k \right). \quad (C.4)$$

Substituting (C.1) — (C.4) into (15.8), the mutual inductive coupling M_{loop}^{prdl} between the two current loops in a fully paired power distribution grid with DSDG is

$$
\begin{aligned}
M_{loop}^{prdl} = 0.2l \Big[&\ln \frac{2l}{(n+1)d} - 1 + \frac{(n+1)d}{l} - \ln\gamma + \\
&\ln k - \ln \frac{2l}{(n+2)d} + 1 - \frac{(n+2)d}{l} + \ln\gamma - \\
&\ln k + \ln \frac{2l}{(n+1)d} - 1 + \frac{(n+1)d}{l} - \ln\gamma + \\
&\ln k - \ln \frac{2l}{nd} + 1 - \frac{nd}{l} + \ln\gamma - \ln k \Big].
\end{aligned}
\quad (C.5)
$$

Simplifying (C.5) and considering that $\ln \gamma$ and $\ln k$ are approximately the same for different distances between the lines, M_{loop}^{prdI} is

$$
\begin{aligned}
M_{loop}^{prdI} &= 0.2l \left[\ln\frac{2l}{(n+1)d} + \frac{(n+1)d}{l} - \ln\frac{2l}{(n+2)d} - \right. \\
&\quad \left. \frac{(n+2)d}{l} + \ln\frac{2l}{(n+1)d} + \frac{(n+1)d}{l} - \ln\frac{2l}{nd} - \frac{nd}{l} \right] \\
&= 0.2l \left[\ln\frac{2l \times (n+2)d \times 2l \times nd}{(n+1)d \times 2l \times (n+1)d \times 2l} + \right. \\
&\quad \left. \frac{(n+1)d - (n+2)d + (n+1)d - nd}{l} \right] \\
&= 0.2l \ln \left[\frac{(n+2)n}{(n+1)^2} \right] < 0 \ \text{ for } n \geq 1.
\end{aligned}
\tag{C.6}
$$

D

Mutual Loop Inductance in Pseudo-Paired Power Distribution Grids with DSDG

Observing that the effective distance between the power and ground lines in a specific power delivery network is $n+1$ times greater than the separation d between the lines making up the pair (see Fig. 15.9), from (15.3), the mutual inductance between the power and ground paths of the different voltage domains for a pseudo-paired power distribution grid with DSDG is

$$L_{V\mathrm{dd1}-V\mathrm{dd2}} = 0.2l \left(\ln\frac{2l}{d} - 1 + \frac{d}{l} - \ln\gamma + \ln k \right), \qquad (D.1)$$

$$L_{V\mathrm{dd1}-G\mathrm{nd2}} = 0.2l \left[\ln\frac{2l}{(n+2)d} - 1 + \frac{(n+2)d}{l} - \ln\gamma + \ln k \right], \quad (D.2)$$

$$L_{G\mathrm{nd1}-G\mathrm{nd2}} = 0.2l \left(\ln\frac{2l}{d} - 1 + \frac{d}{l} - \ln\gamma + \ln k \right), \qquad (D.3)$$

$$L_{V\mathrm{dd2}-G\mathrm{nd1}} = 0.2l \left(\ln\frac{2l}{nd} - 1 + \frac{nd}{l} - \ln\gamma + \ln k \right). \qquad (D.4)$$

Substituting (D.1) — (D.4) into (15.8), the mutual inductive coupling $M_{loop}^{\mathrm{prdII}}$ between the two current loops in a pseudo-paired power distribution grid with DSDG is

$$
\begin{aligned}
M_{loop}^{\mathrm{prdII}} = 0.2l \Bigg[& \ln\frac{2l}{d} - 1 + \frac{d}{l} - \ln\gamma + \ln k - \ln\frac{2l}{(n+2)d} + 1 - \\
& \frac{(n+2)d}{l} + \ln\gamma - \ln k + \ln\frac{2l}{d} - 1 + \frac{d}{l} - \ln\gamma + \ln k - \\
& \ln\frac{2l}{nd} + 1 - \frac{nd}{l} + \ln\gamma - \ln k \Bigg].
\end{aligned}
\qquad (D.5)
$$

Simplifying (D.5) and considering that $\ln \gamma$ and $\ln k$ are approximately the same for different distances between the lines, M_{loop}^{prdII} is

$$
\begin{aligned}
M_{loop}^{prdII} &= 0.2l \left[\ln \frac{2l}{d} + \frac{d}{l} - \ln \frac{2l}{(n+2)d} - \frac{(n+2)d}{l} + \ln \frac{2l}{d} + \frac{d}{l} - \right. \\
&\quad \left. \ln \frac{2l}{nd} - \frac{nd}{l} \right] \\
&= 0.2l \left[\ln \frac{2l \times (n+2)d \times 2l \times nd}{d \times 2l \times d \times 2l} + \frac{2d - (n+2)d - nd}{l} \right] \\
&= 0.2l \left[\ln \left(n^2 + 2n \right) - \frac{2nd}{l} \right] > 0 \ \text{ for } n \geq 1. \quad\quad (D.6)
\end{aligned}
$$

References

1. C. Pirtle, *Engineering the World: Stories from the First 75 Years of Texas Instruments*, Southern Methodist University Press, Dallas, Texas, 2005.
2. T. R. Reid, *The Chip: How Two Americans Invented the Microchip and Launched a Revolution*, Random House, Inc., New York, New York, 2001.
3. L. Berlin, *Man Behind the Microchip: Robert Noyce and the Invention of Silicon Valley*, Oxford University Press, New York, New York, 2005.
4. J. S. Kilby, "Miniaturized Electronic Circuits," U.S. Patent # 3,138,743, June 23, 1964.
5. J. A. Hoerni, "Planar Silicon Transistors and Diodes," *IRE Transactions on Electron Devices*, Vol. 8, No. 2, pp. 178, March 1961.
6. D. Kahng, "A Historical Perspective on the Development of MOS Transistors and Related Devices," *IEEE Transactions on Electron Devices*, Vol. 23, No. 7, pp. 655–657, July 1976.
7. R. E. Kerwin, D. L. Klein, and J. C. Sarace, "Method for Making MIS Structures," U.S. Patent # 3,475,234, October 28, 1969.
8. G. E. Moore, "Cramming More Components onto Integrated Circuits," *Electronics*, pp. 114–117, April 19, 1965.
9. G. E. Moore, "Progress in Digital Integrated Electronics," *Proceedings of the IEEE International Electron Devices Meeting*, pp. 11–13, December 1975.
10. *International Technology Roadmap for Semiconductors, 2001 Edition*, Semiconductor Industry Association, http://public.itrs.net, 2001.
11. B. T. Murphy, D. E. Haggan, and W. W. Troutman, "From Circuit Miniaturization to the Scalable IC," *Proceedings of the IEEE*, Vol. 88, No. 5, pp. 691–703, May 2000.
12. J. Millman, *Microelectronics*, McGraw-Hill, New York, New York, 1979.
13. F. Faggin and M. E. Hoff, "Standard Parts and Custom Design Merge in Four-Chip Processor Kit," *Electronics*, pp. 112–116, April 24, 1972.
14. C. McNairy and R. Bhatia, "Montecito: a Dual-Core, Dual-Thread Itanium Processor," *IEEE Micro*, Vol. 25, No. 2, pp. 10–20, March/April 2005.
15. S. Rusu *et al.*, "A 65-nm Dual-Core Multithreaded Xeon Processor with 16-MB L3 Cache," *IEEE Journal of Solid-State Circuits*, Vol. 42, No. 1, pp. 17–25, January 2007.
16. J. Chang *et al.*, "The 65-nm 16-MB Shared On-Die L3 Cache for the Dual-Core Intel Xeon Processor 7100 Series," *IEEE Journal of Solid-State Circuits*, Vol. 42, No. 4, pp. 846–852, April 2007.

17. J. M. Hart *et al.*, "Implementation of a Forth-Generation 1.8-GHz Dual-Core SPARC V9 Microprocessor," *IEEE Journal of Solid-State Circuits*, Vol. 41, No. 1, pp. 210–217, January 2006.

18. R. Kalla, B. Sinharoy, and J. M. Tendler, "IBM Power5 Chip: a Dual-Core Multithreaded Processor," *IEEE Micro*, Vol. 24, No. 2, pp. 40–47, March/April 2004.

19. K. T. Tang and E. G. Friedman, "Estimation of Transient Voltage Fluctuations in the CMOS-Based Power Distribution Networks," *Proceedings of the IEEE International Symposium on Circuits and Systems*, Vol. 5, pp. 463–466, May 2001.

20. K. T. Tang and E. G. Friedman, "On-Chip ΔI Noise in the Power Distribution Networks of High Speed CMOS Integrated Circuits," *Proceedings of the IEEE International ASIC/SOC Conference*, pp. 53–57, September 2000.

21. K. T. Tang and E. G. Friedman, "Simultaneous Switching Noise in On-Chip CMOS Power Distribution Networks," *IEEE Transactions on Very Large Scale Integration (VLSI) Systems*, Vol. 10, No. 4, pp. 487–493, August 2002.

22. *International Technology Roadmap for Semiconductors, 2005 Edition*, Semiconductor Industry Association, http://public.itrs.net, 2005.

23. M. Benoit, S. Taylor, D. Overhauser, and S. Rochel, "Power Distribution in High-Performance Design," *Proceedings of the IEEE International Symposium on Low Power Electronics and Design*, pp. 274–278, August 1998.

24. L. C. Tsai, "A 1 GHz PA-RISC Processor," *Proceedings of the IEEE International Solid-State Circuits Conference*, pp. 322–323, February 2001.

25. C. J. Anderson *et al.*, "Physical Design of a Fourth-Generation POWER GHz Microprocessor," *Proceedings of the IEEE International Solid-State Circuits Conference*, pp. 232–233, February 2001.

26. M. Popovich and E. G. Friedman, "Decoupling Capacitors for Multi-Voltage Power Distribution Systems," *IEEE Transactions on Very Large Scale Integration (VLSI) Systems*, Vol. 14, No. 3, pp. 217–228, March 2006.

27. M. Popovich and E. G. Friedman, "Impedance Characteristics of Decoupling Capacitors in Multi-Power Distribution Systems," *Proceedings of the IEEE International Conference on Electronics, Circuits and Systems*, pp. 160–163, December 2004.

28. A. V. Mezhiba and E. G. Friedman, *Power Distribution Networks in High Speed Integrated Circuits*, Kluwer Academic Publishers, Norwell, Massachusetts, 2004.

29. K. T. Tang and E. G. Friedman, "Delay Uncertainty Due to On-Chip Simultaneous Switching Noise in High Performance CMOS Integrated Circuits," *Proceedings of the IEEE Workshop on Signal Processing Systems*, pp. 633–642, October 2000.

30. K. T. Tang and E. G. Friedman, "Incorporating Voltage Fluctuations of the Power Distribution Network into the Transient Analysis of CMOS Logic Gates," *Analog Integrated Circuits and Signal Processing*, Vol. 31, No. 3, pp. 249–259, June 2002.

31. M. Saint-Laurent and M. Swaminathan, "Impact of Power Supply Noise on Timing in High-Frequency Microprocessors," *Proceedings of the IEEE Topical Meeting on Electrical Performance of Electronic Packaging*, pp. 261–264, October 2002.

32. M. Saint-Laurent and M. Swaminathan, "Impact of Power-Supply Noise on Timing in High-Frequency Microprocessors," *IEEE Transactions on Advanced Packaging*, Vol. 27, No. 1, pp. 135–144, February 2004.

33. A. Waizman and C.-Y. Chung, "Package Capacitor Impact on Microprocessor Maximum Operating Frequency," *Proceedings of the IEEE Electronic Components and Technology Conference*, pp. 118–122, June 2001.

34. E. G. Friedman, *Clock Distribution Networks in VLSI Circuits and Systems*, IEEE Press, Piscataway, New Jersey, 1995.

35. E. G. Friedman, *High Performance Clock Distribution Networks*, Kluwer Academic Publishers, Norwell, Massachusetts, 1997.

36. I. S. Kourtev and E. G. Friedman, *Timing Optimization Through Clock Skew Scheduling*, Kluwer Academic Publishers, Norwell, Massachusetts, 2000.

37. J. P. Eckhardt and K. A. Jenkins, "PLL Phase Error and Power Supply Noise," *Proceedings of the IEEE Conference on Electrical Performance of Electronic Packaging*, pp. 73–76, October 1998.

38. A. W. Strong *et al.*, *Reliability Wearout Mechanisms in Advanced CMOS Technologies*, John Wiley & Sons, Inc., New York, New York, 2006.

39. L. Smith, "Reliability and Performance Tradeoffs in the Design of On-Chip Power Delivery and Interconnects," *Proceedings of the IEEE Conference on Electrical Performance of Electronic Packaging*, pp. 49–52, November 1999.

40. J. D. Jackson, *Classical Electrodynamics*, John Wiley & Sons, Inc., New York, New York, 1975.

41. F. W. Grover, *Inductance Calculations: Working Formulas and Tables*, D. Van Nostrand Company, Inc., New York, New York, 1946.

42. A. E. Ruehli, "Inductance Calculations in a Complex Integrated Circuit Environment," *IBM Journal of Research and Development*, Vol. 16, No. 5, pp. 470–481, September 1972.

43. E. B. Rosa, "The Self and Mutual Inductance of Linear Conductors," *Bulletin of the National Bureau of Standards*, Vol. 4, No. 2, pp. 301–344. Government Printing Office, Washington, D.C., January 1908.

44. E. B. Rosa and L. Cohen, "Formulæ and Tables for the Calculation of Mutual and Self-Inductance," *Bulletin of the National Bureau of Standards*, Vol. 5, No. 1, pp. 1–132. Government Printing Office, Washington, August 1908.

45. E. B. Rosa and F. W. Grover, "Formulæ and Tables for the Calculation of Mutual and Self-Inductance," *Bulletin of the National Bureau of Standards*, Vol. 8, No. 1, pp. 1–237. Government Printing Office, Washington, August 1912.

46. R. F. German, H. W. Ott, and C. R. Paul, "Effect of an Image Plane on Printed Circuit Board Radiation," *Proceedings of the IEEE International Symposium on Electromagnetic Compatibility*, pp. 284–291, August 1990.

47. T. S. Smith and C. R. Paul, "Effect of Grid Spacing on the Inductance of Ground Grids," *Proceedings of the IEEE International Symposium on Electromagnetic Compatibility*, pp. 72–77, August 1991.

48. C. R. Paul, *Introduction to Electromagnetic Compatibility*, John Wiley & Sons, Inc., New York, New York, 1992.

49. R. E. Matick, *Transmission Lines for Digital and Communication Networks*, McGraw-Hill, New York, New York, 1969.

50. D. W. Bailey and B. J. Benschneider, "Clocking Design and Analysis for a 600-MHz Alpha Microprocessor," *IEEE Journal of Solid-State Circuits*, Vol. 33, No. 11, pp. 1627–1633, November 1998.

51. R. M. Averill III *et al.*, "Chip Integration Methodology for the IBM S/390 G5 and G6 Custom Microprocessors," *IBM Journal of Research and Development*, Vol. 43, No. 5/6, pp. 681–706, September/November 1999.

52. H. A. Wheeler, "Formulas for the Skin Effect," *Proceedings of the IRE*, pp. 412–424, September 1942.

53. C.-S. Yen, Z. Fazarinc, and R. L. Wheeler, "Time-Domain Skin-Effect Model for Transient Analysis of Lossy Transmission Lines," *Proceedings of the IEEE*, Vol. 70, No. 7, pp. 750–757, July 1982.

54. T. V. Dinh, B. Cabon, and J. Chilo, "New Skin-Effect Equivalent Circuit," *Electronic Letters*, Vol. 26, No. 19, pp. 1582–1584, September 13, 1990.

55. S. Kim and D. P. Neikirk, "Compact Equivalent Circuit Model for the Skin Effect," *Proceedings of the IEEE International Microwave Symposium*, pp. 1815–1818, June 1996.

56. B. Krauter and S. Mehrotra, "Layout Based Frequency Dependent Inductance and Resistance Extraction for On-Chip Interconnect Timing Analysis," *Proceedings of the IEEE/ACM Design Automation Conference*, pp. 303–308, June 1998.

57. G. V. Kopcsay *et al.*, "A Comprehensive 2-D Inductance Modeling Approach for VLSI Interconnects: Frequency-Dependent Extraction and Compact Model Synthesis," *IEEE Transactions on Very Large Scale Integration (VLSI) Systems*, Vol. 10, No. 6, pp. 695–711, December 2002.

58. S. Mei, C. Amin, and Y. I. Ismail, "Efficient Model Order Reduction Including Skin Effect," *Proceedings of the IEEE/ACM Design Automation Conference*, pp. 232–237, June 2003.

59. A. Deutsch *et al.*, "When Are Transmission-Line Effects Important for On-Chip Interconnections?," *IEEE Transactions on Microwave Theory and Techniques*, Vol. 45, No. 10, pp. 1836–1846, October 1997.

60. Y. I. Ismail, E. G. Friedman, and J. L. Neves, "Figures of Merit to Characterize the Importance of On-Chip Inductance," *IEEE Transactions on Very Large Scale Integration (VLSI) Systems*, Vol. 7, No. 4, pp. 442–449, December 1999.

61. Y. I. Ismail and E. G. Friedman, *On-Chip Inductance in High Speed Integrated Circuits*, Kluwer Academic Publishers, Norwell, Massachusetts, 2001.

62. R. Schmitt, *Electromagnetics Explained*, Newnes—Elsevier Science, Boston, 2002.

63. B. Krauter and L. T. Pileggi, "Generating Sparse Partial Inductance Matrices with Guaranteed Stability," *Proceedings of the IEEE/ACM Design Automation Conference*, pp. 45–52, November 1995.

64. Z. He, M. Celik, and L. T. Pileggi, "SPIE: Sparse Partial Inductance Extraction," *Proceedings of the IEEE/ACM Design Automation Conference*, pp. 137–140, June 1997.

65. A. J. Dammers and N. P. van der Meijs, "Virtual Screening: A Step Towards a Sparse Partial Inductance Matrix," *Proceedings of the IEEE/ACM International Conference on Computer-Aided Design*, pp. 445–452, June 1999.

66. M. Beattie and L. Pileggi, "Efficient Inductance Extraction via Windowing," *Proceedings of the IEEE/ACM Design Automation and Test in Europe Conference*, pp. 430–436, March 2001.

67. M. Kamon, M. J. Tsuk, and J. White, "FastHenry: A Multipole-Accelerated 3-D Inductance Extraction Program," *IEEE Transactions on Microwave Theory and Techniques*, Vol. 42, No. 9, pp. 1750–1758, September 1994.

68. A. V. Mezhiba and E. G. Friedman, "Properties of On-Chip Inductive Current Loops," *Proceedings of the ACM Great Lakes Symposium on Very Large Scale Integration*, pp. 12–17, April 2002.

69. A. V. Mezhiba and E. G. Friedman, "Inductive Characteristics of Power Distribution Grids in High Speed Integrated Circuits," *Proceedings of the IEEE International Symposium on Quality Electronic Design*, pp. 316–321, March 2002.

70. A. V. Mezhiba and E. G. Friedman, "Inductive Properties of High-Performance Power Distribution Grids," *IEEE Transactions on Very Large Scale Integration (VLSI) Systems*, Vol. 10, No. 6, pp. 762–776, December 2002.

71. J. M. Maxwell, *A Treatise on Electricity and Magnetism*, Vol. 2, Part IV, Chapter XIII, The Clarendon Press, Oxford, United Kingdom, 2nd Edition, 1881.

72. J. J. Clement, "Electromigration Reliability," *Design of High-Performance Microprocessor Circuits*, Chandrakasan, Bowhill, and Fox, (Eds.), Chapter 20, pp. 429–448, IEEE Press, New York, New York, 2001.

73. I. A. Blech and H. Sello, *Mass Transport of Aluminum by Moment Exchange with Conducting Electrons*, Vol. 5, pp. 496–505, USAF-RADC Series, 1966.

74. J. R. Black, "Mass Transport of Aluminum by Moment Exchange with Conducting Electrons," *Proceedings of the IEEE International Reliability Physics Symposium*, pp. 148–159, April 1967.

75. F. M. D'Heurle, "Electromigration and Failure in Electronics: an Introduction," *Proceedings of the IEEE*, Vol. 59, No. 10, pp. 1409–1417, October 1971.

76. C.-K. Hu *et al.*, "Electromigration and Stress-Induced Voiding in Fine Al and Al-alloy Thin-Film Lines," *IBM Journal of Research and Development*, Vol. 39, No. 4, pp. 465–497, July 1995.

77. M. J. Attardo and R. Rosenberg, "Electromigration Damage in Aluminum Film Conductors," *Journal of Applied Physics*, Vol. 41, No. 5, pp. 2381–2386, May 1970.

78. F. G. Yost, D. E. Amos, and A. D. Romig, Jr., "Stress-Driven Diffusive Voiding of Aluminum Conductor Lines," *Proceedings of the IEEE International Reliability Physics Symposium*, pp. 193–201, April 1989.

79. I. A. Blech and K. L. Tai, "Measurements of Stress Gradients Generated by Electromigration," *Applied Physics Letters*, Vol. 30, No. 8, pp. 387–389, April 1977.

80. I. A. Blech, "Electromigration in Thin Aluminum Films on Titanium Nitride," *Journal of Applied Physics*, Vol. 47, No. 4, pp. 1203–1208, April 1976.

81. R. G. Filippi *et al.*, "The Effect of Current Density and Stripe Length on Resistance Saturation During Electromigration Testing," *Applied Physics Letters*, Vol. 69, No. 16, pp. 2350–2352, October 1996.

82. P. Børgesen *et al.*, "Stress Evolution During Stress Migration and Electromigration in Passivated Interconnect Lines," *Proceedings of the American Institute of Physics Conference*, Vol. 305, pp. 231–253, 1994.

83. J. J. Clement, J. R. Lloyd, and C. V. Thompson, "Failure in Tungsten-Filled Via Structures," *Proceedings of the Materials Research Society*, Vol. 391, pp. 423–428, 1995.

84. B. N. Argarwala, M. J. Attardo, and A. J. Ingraham, "Dependence of Electromigration-Induced Failure Time on Length and Width of Aluminum Thin Film Conductors," *Journal of Applied Physics*, Vol. 41, pp. 3954–3960, September 1970.

85. J. Cho and C. V. Thompson, "Grain Size Dependence of Electromigration-Induced Failures in Narrow Interconnects," *Applied Physics Letters*, Vol. 54, No. 25, pp. 2577–2579, June 19, 1989.

86. J. R. Black, "Electromigration — A Brief Survey and Some Recent Results," *IEEE Transactions on Electron Devices*, Vol. 4, No. 16, pp. 338–347, April 1969.

87. J. J. Clement, "Electromigration Modeling for Integrated Circuit Interconnect Relibility Analysis," *IEEE Transactions on Device and Materials Reliability*, Vol. 1, No. 1, pp. 33–42, March 2001.

88. J. M. Towner and E. P. van de Ven, "Aluminum Electromigration under Pulsed DC Conditions," *Proceedings of the IEEE International Reliability Physics Symposium*, pp. 36–39, April 1983.

89. J. A. Maiz, "Characterization of Electromigration under Bidirectional (BC) and Pulsed Unidirectional (PDC) Currents," *Proceedings of the IEEE International Reliability Physics Symposium*, pp. 220–228, April 1989.

90. L. M. Ting, J. S. May, W. R. Hunter, and J. W. McPherson, "AC Electromigration Characterization and Modeling of Multilayered Interconnects," *Proceedings of the IEEE International Reliability Physics Symposium*, pp. 311–316, March 1993.

91. E. T. Ogawa, K.-D. Lee, V. A. Blaschke, and P. S. Ho, "Electromigration Reliability Issues in Dual-Damascene Cu Interconnections," *IEEE Transactions on Reliability*, Vol. 51, No. 4, pp. 403–419, December 2002.

92. S. Thrasher *et al.*, "Blech Effect in Single-Inlaid Cu Interconnects," *Proceedings of the IEEE International Interconnect Technology Conference*, pp. 177–179, June 2001.

93. P. C. Wang, R. G. Filippi, and L. M. Gignac, "Electromigration Threshold in Single-Damascene Copper Interconnects with SiO_2 Dielectrics," *Proceedings of the IEEE International Interconnect Technology Conference*, pp. 263–265, June 2001.

94. E. T. Ogawa, "Direct Observation of a Critical Length Effect in Dual-Damascene Cu/Oxide Interconnects," *Applied Physics Letters*, Vol. 78, No. 18, pp. 2652–2654, April 2001.

95. S. P. Hau-Riege, "Probabilistic Immortality of Cu Damascene Interconnects," *Journal of Applied Physics*, Vol. 91, No. 4, pp. 2014–2022, February 2002.

96. C.-K. Hu *et al.*, "Bimodal Electromigration Mechanisms in Dual-Damascene Cu Line/Via on W," *Proceedings of the IEEE International Interconnect Technology Conference*, pp. 133–135, June 2002.

97. B. Li, T. D. Sullivan, and T. C. Lee, "Line Depletion Electromigration Characteristics of Cu Interconnects," *Proceedings of the IEEE International Reliability Physics Symposium*, pp. 140–145, March 2003.

98. P. Justison *et al.*, "Electromigration in Multi-Level Interconnects with Polymeric Low-k Interlevel Dielectrics," *Proceedings of the IEEE International Interconnect Technology Conference*, pp. 202–204, June 2000.

99. K.-D. Lee *et al.*, "Electromigration Study of Cu/Low k Dual Damascene Interconnects," *Proceedings of the IEEE International Reliability Physics Symposium*, pp. 322–326, March 2002.

100. C. S. Hau-Riege, A. P. Marathe, and V. Pham, "The Effect of Low-k ILD on the Electromigration Reliability of Cu Interconnects with Different Line Lengths," *Proceedings of the IEEE International Reliability Physics Symposium*, pp. 173–177, March 2003.

101. E. T. Ogawa *et al.*, "Statistics of Electromigration Early Failures in Cu/Oxide Dual Damascene Interconnects," *Proceedings of the IEEE International Reliability Physics Symposium*, pp. 341–349, March 2001.

102. F. G. Yost, D. E. Amos, and A. D. Romig, Jr., "Statistical Electromigration Budgeting for Reliable Design and Verification in a 300-MHz Microprocessor," *Proceedings of the IEEE Symposium on VLSI Circuits*, pp. 115–116, June 1995.

103. R. R. Tummala, E. J. Rymaszewski, and A. G. Klopfenstein, (Eds.), *Microelectronics Packaging Handbook*, Chapman & Hall, New York, New York, 1997.

104. R. Evans and M. Tsuk, "Modeling and Measurement of a High-Performance Computer Power Distribution System," *IEEE Transactions on Components, Packaging, and Manufacturing Technology, Part B: Advanced Packaging*, Vol. 17, No. 4, pp. 467–471, November 1994.

105. D. J. Herrell and B. Beker, "Modeling of Power Distribution Systems for High-Performance Processors," *IEEE Transactions on Advanced Packaging*, Vol. 22, No. 3, pp. 240–248, August 1999.

106. T. R. Arabi *et al.*, "Design and Validation of the Core and IOs Decoupling of the Pentium III and Pentium 4 Processors," *Proceedings of the IEEE Topical Meeting on Electrical Performance of Electronic Packaging*, pp. 249–252, October 2002.

107. L. D. Smith *et al.*, "Power Distribution System Design Methodology and Capacitor Selection for Modern CMOS Technology," *IEEE Transactions on Advanced Packaging*, Vol. 22, No. 3, pp. 284–291, August 1999.

108. A. Waizman and C.-Y. Chung, "Resonant Free Power Network Design Using Extended Adaptive Voltage Positioning (EAVP) Methodology," *IEEE Transactions on Advanced Packaging*, Vol. 24, No. 3, pp. 236–244, August 2001.

109. I. Novak *et al.*, "Distributed Matched Bypassing for Board-Level Power Distribution Networks," *IEEE Transactions on Advanced Packaging*, Vol. 25, No. 2, pp. 230–242, May 2002.

110. G. F. Taylor, C. Deutschle, T. Arabi, and B. Owens, "An Approach to Measuring Power Supply Impedance of Microprocessors," *Proceedings of the IEEE Topical Meeting on Electrical Performance of Electronic Packaging*, pp. 211–214, October 2001.

111. R. Panda *et al.*, "Model and Analysis of Combined Package and On-Chip Power Grid Simulation," *Proceedings of the IEEE International Symposium on Low Power Electronics and Design*, pp. 179–184, August 2000.

112. A. Hasan, A. Sarangi, A. Sathe, and G. Ji, "High Performance Mobile Pentium III Package Development and Design," *Proceedings of the IEEE Electronic Components and Technology Conference*, pp. 378–385, June 2002.

113. I. Novak, "Lossy Power Distribution Networks with Thin Dielectric Layers and/or Thin Conductive Layers," *IEEE Transactions on Advanced Packaging*, Vol. 23, No. 3, pp. 353–360, August 2000.

114. H. Braunisch *et al.*, "Electrical Performance of Bumpless Build-Up Layer Packaging," *Proceedings of the IEEE Electronic Components and Technology Conference*, pp. 353–358, June 2002.

115. B. W. Amick, C. R. Gauthier, and D. Liu, "Macro-Modeling Concepts for the Chip Electrical Interface," *Proceedings of the IEEE/ACM Design Automation Conference*, pp. 391–394, June 2002.

116. G. E. R. Lloyd, *Early Greek Science: Thales to Aristotle*, W W Norton & Co., Inc., New York, New York, 1974.

117. A. D. Moore, *Electrostatics and Its Applications*, John Wiley & Sons, Inc., New York, New York, 1973.
118. J. L. Heilborn, *Electricity in the 17th & 18th Centuaries: A Study in Early Modern Physics*, Dover Publications, Mineola, New York, 1999.
119. A. Guillemin, *Electricity and Magnetism*, Macmillam, London, 1891.
120. M. Faraday, *Experimental Researches in Electricity*, Dover Publications, Mineola, New York, 2004.
121. J. D. Cutnell and K. W. Johnson, *Physics, 6th Edition*, John Wiley & Sons, Inc., New York, New York, 2003.
122. T. H. Lee, *The Design of CMOS Radio-Frequency Integrated Circuits, 2nd Edition*, Cambridge University Press, New York, New York, 2004.
123. C. P. Yuan and T. N. Trick, "A Simple Formula for the Estimation of the Capacitance of Two-Dimensional Interconnects in VLSI Circuits," *IEEE Electron Device Letters*, Vol. 3, No. 12, pp. 391–393, December 1982.
124. T. Sakurai and K. Tamaru, "Simple Formulas for Two- and Three-Dimensional Capacitance," *IEEE Transactions on Electron Devices*, Vol. 30, No. 2, pp. 183–185, February 1983.
125. J.-H. Chern *et al.*, "Multilevel Metal Capacitance Models for CAD Design Synthesis Systems," *IEEE Electron Device Letters*, Vol. 13, No. 1, pp. 32–34, January 1992.
126. S.-C. Wong, G.-Y. Lee, and D.-J. Ma, "Modeling of Interconnect Capacitance, Delay, and Crosstalk in VLSI," *IEEE Transactions on Semiconductor Manufacturing*, Vol. 13, No. 1, pp. 108–111, January 2000.
127. E. Barke, "Line-to-Ground Capacitance Calculation for VLSI: A Comparison," *IEEE Transactions on Computer-Aided Design of Integrated Circuits and Systems*, Vol. 7, No. 2, pp. 295–298, February 1988.
128. N. v.d. Meijs and J. T. Fokkema, "VLSI Circuit Reconstruction from Mask Topology," *Integration*, Vol. 2, No. 2, pp. 85–119, February 1984.
129. T. Roy, L. Smith, and J. Prymak, "ESR and ESL of Ceramic Capacitor Applied to Decoupling Applications," *Proceedings of the IEEE Conference on Electrical Performance of Electronic Packaging*, pp. 213–216, October 1998.
130. D. A. Neamen, *Semiconductor Physics and Devices: Basic Principles, 3rd Edition*, McGraw-Hill, New York, New York, 2002.
131. "Power Distribution System (PDS) Design: Using Bypass/Decoupling Capacitors," http://direct.xilinx.com/bvdocs/appnotes/xapp623.pdf.
132. H. B. Bakoglu, *Circuits, Interconnections, and Packaging for VLSI*, Addison-Wesley, Reading, Massachusetts, 1990.
133. W. Becker *et al.*, "Mid-Frequency Simultaneous Switching Noise in Computer Systems," *Proceedings of the IEEE Electronic Components and Technology Conference*, pp. 676–681, May 1997.
134. W. D. Becker *et al.*, "Modeling, Simulation, and Measurement of Mid-Frequency Simultaneous Switching Noise in Computer Systems," *IEEE Transactions on Components, Packaging, and Manufacturing Technology, Part B: Advanced Packaging*, Vol. 21, No. 2, pp. 157–163, May 1998.
135. T. Zhou, T. Strach, and W. D. Becker, "On Chip Circuit Model for Accurate Mid-Frequency Simultaneous Switching Noise Prediction," *Proceedings of the IEEE Conference on Electrical Performance of Electronic Packaging*, pp. 275–278, October 2005.

136. S. Bobba, T. Thorp, K. Aingaran, and D. Liu, "IC Power Distribution Challenges," *Proceedings of the IEEE/ACM International Conference on Computer-Aided Design*, pp. 643–650, November 2001.

137. S. Chun *et al.*, "Physics Based Modeling of Simultaneous Switching Noise in High Speed Systems," *Proceedings of the IEEE Electronic Components and Technology Conference*, pp. 760–768, May 2000.

138. S. Chun *et al.*, "Modeling of Simultaneous Switching Noise in High Speed Systems," *IEEE Transactions on Advanced Packaging*, Vol. 24, No. 2, pp. 132–142, May 2001.

139. L. Smith, "Simultaneous Switching Noise and Power Plane Bounce for CMOS Technology," *Proceedings of the IEEE Conference on Electrical Performance of Electronic Packaging*, pp. 163–166, October 1999.

140. F. Y. Yuan, "Electromagnetic Modeling and Signal Integrity Simulations of Power/Ground Networks in High Speed Digital Packages and Printed Circuit Boards," *Proceedings of the IEEE/ACM Design Automation Conference*, pp. 421–426, June 1998.

141. Z. Mu, "Simulation and Modeling of Power and Ground Planes in High Speed Printed Circuit Boards," *Proceedings of the IEEE International Symposium on Circuits and Systems*, pp. 459–462, May 2001.

142. N. Na *et al.*, "Modeling and Transient Simulation of Planes in Electronic Packages," *IEEE Transactions on Advanced Packaging*, Vol. 23, No. 3, pp. 340–352, August 2000.

143. T.-G. Yew, Y.-L. Li, C.-Y. Chung, and D. G. Figueroa, "Design and Performance Evaluation of Chip Capacitors on Microprocessor Packaging," *Proceedings of the IEEE Conference on Electrical Performance of Electronic Packaging*, pp. 175–178, October 1999.

144. J. Kim *et al.*, "Separated Role of On-Chip and On-PCB Decoupling Capacitors for Reduction of Radiated Emission on Printed Circuit Board," *Proceedings of the IEEE International Symposium on Electromagnetic Compatibility*, pp. 531–536, August 2001.

145. B. Garben, G. A. Katopis, and W. D. Becker, "Package and Chip Design Optimization for Mid-Frequency Power Distribution Decoupling," *Proceedings of the IEEE Conference on Electrical Performance of Electronic Packaging*, pp. 245–248, October 2002.

146. M. Xu *et al.*, "Power-Bus Decoupling with Embedded Capacitance in Printed Circuit Board Design," *IEEE Transactions on Electromagnetic Compatibility*, Vol. 45, No. 1, pp. 22–30, February 2003.

147. M. I. Montrose, "Analysis on Loop Area Trace Radiated Emissions from Decoupling Capacitor Placement on Printed Circuit Boards," *Proceedings of the IEEE International Symposium on Electromagnetic Compatibility*, pp. 423–428, August 1999.

148. P. Muthana *et al.*, "Mid Frequency Decoupling Using Embedded Decoupling Capacitors," *Proceedings of the IEEE Conference on Electrical Performance of Electronic Packaging*, pp. 271–274, October 2005.

149. O. P. Mandhana, "Design Oriented Analysis of Package Power Distribution System Considering Target Impedance for High Performance Microprocessors," *Proceedings of the IEEE Conference on Electrical Performance of Electronic Packaging*, pp. 273–276, October 2001.

150. L. D. Smith and D. Hockanson, "Distributed SPICE Circuit Model for Ceramic Capacitors," *Proceedings of the IEEE Electronic Components and Technology Conference*, pp. 523–528, May/June 2001.

151. P. Larsson, "Resonance and Damping in CMOS Circuits with On-Chip Decoupling Capacitance," *IEEE Transactions on Circuits and Systems—I: Fundamental Theory and Applications*, Vol. 45, No. 8, pp. 849–858, August 1998.

152. L. D. Smith, R. E. Anderson, and T. Roy, "Chip-Package Resonance in Core Power Supply Structures for a High Power Microprocessor," *Proceedings of the ASME International Electronic Packaging Technical Conference and Exhibition*, July 2001.

153. C. R. Paul, "Effectiveness of Multiple Decoupling Capacitors," *IEEE Transactions on Electromagnetic Compatibility*, Vol. 34, No. 2, pp. 130–133, May 1992.

154. M. Popovich and E. G. Friedman, "Decoupling Capacitors for Power Distribution Systems with Multiple Power Supplies," *Proceedings of the IEEE EDS/CAS Activities in Western New York Conference*, p. 9, November 2004.

155. A. Waizman and C.-Y. Chung, "Extended Adaptive Voltage Positioning (EAVP)," *Proceedings of the IEEE Conference on Electrical Performance of Electronic Packaging*, pp. 65–68, October 2000.

156. M. Sotman, M. Popovich, A. Kolodny, and E. G. Friedman, "Leveraging Symbiotic On-Die Decoupling Capacitance," *Proceedings of the IEEE Conference on Electrical Performance of Electronic Packaging*, pp. 111–114, October 2005.

157. H. H. Chen and J. S. Neely, "Interconnect and Circuit Modeling Techniques for Full-Chip Power Supply Noise Analysis," *IEEE Transactions on Components, Packaging, and Manufacturing Technology, Part B: Advanced Packaging*, Vol. 21, No. 3, pp. 209–215, August 1998.

158. S. Bobba and I. Hajj, "Input Vector Generation for Maximum Intrinsic Decoupling Capacitance of VLSI Circuits," *Proceedings of the IEEE International Symposium on Circuits and Systems*, pp. 195–198, May 2001.

159. R. Panda, S. Sundareswaran, and D. Blaauw, "On the Interaction of Power Distribution Network with Substrate," *Proceedings of the IEEE International Symposium on Low Power Electronics and Design*, pp. 388–393, August 2001.

160. R. Panda, S. Sundareswaran, and D. Blaauw, "Impact of Low-Impedance Substrate on Power Supply Integrity," *IEEE Transactions on Design & Test of Computers*, Vol. 20, No. 3, pp. 16–22, May/June 2003.

161. G. Steele, D. Overhauser, S. Rochel, and S. Z. Hussain, "Full-Chip Verification Methods for DSM Power Distribution Systems," *Proceedings of the IEEE/ACM Design Automation Conference*, pp. 744–749, June 1998.

162. N. H. Pham, "On-Chip Capacitor Measurement for High Performance Microprocessor," *Proceedings of the IEEE Conference on Electrical Performance of Electronic Packaging*, pp. 65–68, October 1998.

163. H. Seidl *et al.*, "A Fully Integrated Al$_2$O$_3$ Trench Capacitor DRAM for Sub-100 nm Technology," *Proceedings of the IEEE International Electron Devices Meeting*, pp. 839–842, December 2002.

164. K. V. Rao *et al.*, "Trench Capacitor Design Issues in VLSI DRAM Cells," *Proceedings of the IEEE International Electron Devices Meeting*, pp. 140–143, December 1986.

165. P. E. Gronowski *et al.*, "High-Performance Microprocessor Design," *IEEE Journal of Solid-State Circuits*, Vol. 33, No. 5, pp. 676–686, May 1998.

166. B. J. Bowhill *et al.*, "Circuit Implementation of a 300 MHz 64-bit Second Generation CMOS Alpha CPU," *Digital Technical Journal*, Vol. 7, No. 1, pp. 100–118, 1995.

167. P. Larsson, "Parasitic Resistance in an MOS Transistor Used as On-Chip Decoupling Capacitance," *IEEE Journal of Solid-State Circuits*, Vol. 32, No. 4, pp. 574–576, April 1997.

168. A. Hastings, *The Art of Analog Layout*, Prentice Hall, Upper Saddle River, New Jersey, 2001.

169. J. L. McCreary, "Matching Properties, and Voltage and Temperature Dependence of MOS Capacitors," *IEEE Journal of Solid-State Circuits*, Vol. 16, No. 6, pp. 608–616, December 1981.

170. R. T. Howe and C. G. Sodini, *Microelectronics: An Integrated Approach*, Prentice Hall, Upper Saddle River, New Jersey, 1996.

171. C. T. Black *et al.*, "High-Capacity, Self-Assembled Metal-Oxide-Semiconductor Decoupling Capacitors," *IEEE Electron Device Letters*, Vol. 25, No. 9, pp. 622–624, September 2004.

172. A. R. Alvarez, *BiCMOS Technology and Applications*, Kluwer Academic Publishers, Norwell, Massachusetts, 1993.

173. A. Behr, M. Schneider, S. Filho, and C. Montoro, "Harmonic Distortion Caused by Capacitors Implemented with MOSFET Gates," *IEEE Journal of Solid-State Circuits*, Vol. 27, No. 10, pp. 1470–1475, October 1992.

174. S. Rusu *et al.*, "A 1.5-GHz 130-nm Itanium 2 Processor with 6-MB On-Die L3 Cache," *IEEE Journal of Solid-State Circuits*, Vol. 38, No. 11, pp. 1887–1895, November 2003.

175. "Optimization of Metal-Metal Comb-Capacitors for RF Applications," http://www.oea.com/document/OptimizMetal.pdf.

176. M. J. Deen and T. A. Fjeldly, *CMOS RF Modeling, Characterization and Applications*, World Scientific, River Edge, New Jersey, 2004.

177. B. Razavi, *RF Microelectronics*, Prentice Hall, Upper Saddle River, New Jersey, 1998.

178. R. K. Ulrich and L. W. Schaper, *Integrated Passive Component Technology*, Wiley-IEEE Press, New York, New York, 2003.

179. S. B. Chen *et al.*, "High-Density MIM Capacitors Using Al_2O_3 and $AlTiO_x$ Dielectrics," *IEEE Electron Device Letters*, Vol. 23, No. 4, pp. 185–187, April 2002.

180. M. Y. Yang *et al.*, "High-Density MIM Capacitors Using $AlTaO_x$ Dielectrics," *IEEE Electron Device Letters*, Vol. 24, No. 5, pp. 306–308, May 2003.

181. X. Yu *et al.*, "A High-Density MIM Capacitor (13 fF/μm^2) Using ALD HfO_2 Dielectrics," *IEEE Electron Device Letters*, Vol. 24, No. 2, pp. 63–65, February 2003.

182. H. Hu *et al.*, "High Performance ALD HfO_2–Al_2O_3 Laminate MIM Capacitors for RF and Mixed Signal IC Applications," *Proceedings of the IEEE International Electron Devices Meeting*, pp. 15.6.1–15.6.4, December 2003.

183. S.-J. Ding *et al.*, "High-Density MIM Capacitor Using ALD High-k HfO_2 Laminate Dielectrics," *IEEE Electron Device Letters*, Vol. 24, No. 12, pp. 730–732, December 2003.

184. S.-J. Kim *et al.*, "Metal-Insulator-Metal RF Bypass Capacitor Using Niobium Oxide (Nb_2O_5) With HfO_2–Al_2O_3 Barriers," *IEEE Electron Device Letters*, Vol. 26, No. 9, pp. 625–627, September 2005.

185. Y. H. Wu *et al.*, "The Fabrication of Very High Resistivity Si with Low Loss and Cross Talk," *IEEE Electron Device Letters*, Vol. 21, No. 9, pp. 442–444, September 2000.

186. A. Kar-Roy *et al.*, "High Density Metal Insulator Metal Capacitors Using PECVD Nitride for Mixed Signal and RF Circuits," *Proceedings of the IEEE International Conference on Interconnect Technology*, pp. 245–247, May 1999.

187. J. A. Babcock *et al.*, "Analog Characteristics of Metal-Insulator-Metal Capacitors Using PECVD Nitride Dielectrics," *IEEE Electron Device Letters*, Vol. 22, No. 5, pp. 230–232, May 2001.

188. P. Zurcher *et al.*, "Integration of Thin Film MIM Capacitors and Resistors into Copper Metallization Based RF-CMOS and Bi-CMOS Technologies," *Proceedings of the IEEE International Electron Devices Meeting*, pp. 153–156, December 2000.

189. M. Armacost *et al.*, "A High Reliability Metal Insulator Metal Capacitor for $0.18\,\mu$m Copper Technology," *Proceedings of the IEEE International Electron Devices Meeting*, pp. 157–160, December 2000.

190. C. H. Ng *et al.*, "Characterization and Comparison of Two Metal-Insulator-Metal Capacitor Schemes in $0.13\,\mu$m Copper Dual Damascene Metallization Process for Mixed-Mode and RF Applications," *Proceedings of the IEEE International Electron Devices Meeting*, pp. 241–244, December 2002.

191. N. Inoue *et al.*, "High Performance High-k MIM Capacitor with Plug-in Plate (PiP) for Power Delivery Line of High-Speed MPUs," *Proceedings of the IEEE International Interconnect Technology Conference*, pp. 63–65, June 2006.

192. T. Soorapanth, *CMOS RF Filtering at GHz Frequency*, Ph.D. Thesis, Stanford University, Stanford, California, 2002.

193. "Applications of Metal-Insulator-Metal (MIM) Capacitors," *International SEMATECH*, Technology Transfer # 00083985A-ENG, October 2000.

194. O. E. Akcasu, "High Capacitance Structures in a Semiconductor Device," U.S. Patent # 5,208,725, May 4, 1993.

195. B. B. Mandelbrot, *The Fractal Geometry of Nature*, Freeman, New York, New York, 1983.

196. H. Samavati *et al.*, "Fractal Capacitors," *IEEE Journal of Solid-State Circuits*, Vol. 33, No. 12, pp. 2035–2041, December 1998.

197. R. Aparicio and A. Hajimiri, "Capacity Limits and Matching Properties of Integrated Capacitors," *IEEE Journal of Solid-State Circuits*, Vol. 37, No. 3, pp. 384–393, March 2002.

198. A. C. C. Ng and M. Saran, "Capacitor Structure for an Integrated Circuit," U.S. Patent # 5,583,359, December 10, 1996.

199. M. Popovich and E. G. Friedman, "Decoupling Capacitors for Power Distribution Systems with Multiple Power Supply Voltages," *Proceedings of the IEEE International SOC Conference*, pp. 331–334, September 2004.

200. M. Popovich, E. G. Friedman, M. Sotman, A. Kolodny, and R. M. Secareanu, "Maximum Effective Distance of On-Chip Decoupling Capacitors in Power Distribution Grids," *Proceedings of the ACM/IEEE Great Lakes Symposium on VLSI*, pp. 173–179, March 2006.

201. M. Ang, R. Salem, and A. Taylor, "An On-Chip Voltage Regulator Using Switched Decoupling Capacitors," *Proceedings of the IEEE International Solid-State Circuits Conference*, pp. 438–439, February 2000.

202. D. Blaauw, R. Panda, and R. Chaudhry, "Design and Analysis of Power Distribution Networks," *Design of High-Performance Microprocessor Circuits*, Chandrakasan, Bowhill, and Fox, (Eds.), Chapter 24, pp. 499–522, IEEE Press, New York, 2001.

203. A. Dalal, L. Lev, and S. Mitra, "Design of an Efficient Power Distribution Network for the UltraSPARC-I Microprocessor," *Proceedings of the IEEE International Conference on Computer Design*, pp. 118–123, October 1995.

204. M. K. Gowan, L. L. Biro, and D. Jackson, "Power Considerations in the Design of the Alpha 21264 Microprocessor," *Proceedings of the IEEE/ACM Design Automation Conference*, pp. 726–731, June 1998.

205. L. Cao and J. P. Krusius, "A New Power Distribution Strategy for Area Array Bonded ICs and Packages of Future Deep Sub-Micron ULSI," *Proceedings of the IEEE Electronic Components and Technology Conference*, pp. 915–920, June 1999.

206. A. V. Mezhiba and E. G. Friedman, "Scaling Trends of On-Chip Power Distribution Noise," *Proceedings of the ACM International Workshop on System Level Interconnect Prediction*, pp. 47–53, April 2002.

207. A. V. Mezhiba and E. G. Friedman, "Scaling Trends of On-Chip Power Distribution Noise," *IEEE Transactions on Very Large Scale Integration (VLSI) Systems*, Vol. 12, No. 4, pp. 386–394, April 2004.

208. D. A. Priore, "Inductance on Silicon for Sub-Micron CMOS VLSI," *Proceedings of the IEEE Symposium on VLSI Circuits*, pp. 17–18, May 1993.

209. L.-R. Zheng and H. Tenhunen, "Effective Power and Ground Distribution Scheme for Deep Submicron High Speed VLSI Circuits," *Proceedings of the IEEE International Symposium on Circuits and Systems*, Vol. I, pp. 537–540, May 1999.

210. B. J. Benschneider et al., "A 1 GHz Alpha Microprocessor," *Proceedings of the IEEE International Solid-State Circuits Conference*, pp. 86–87, February 2000.

211. A. Dharchoudhury et al., "Design and Analysis of Power Distribution Networks in PowerPC Microprocessors," *Proceedings of the IEEE/ACM Design Automation Conference*, pp. 738–743, June 1998.

212. R. M. Averill III et al., "Chip Integration Methodology for the IBM S/390 G5 and G6 Custom Microprocessors," *IBM Journal of Research and Development*, Vol. 43, No. 5/6, pp. 681–706, September/November 1999.

213. Y.-L. Li, T.-G. Yew, C.-Y. Chung, and D. G. Figueroa, "Design and Performance Evaluation of Microprocessor Packaging Capacitors Using Integrated Capacitor-Via-Plane Model," *IEEE Transactions on Advanced Packaging*, Vol. 23, No. 3, pp. 361–367, August 2000.

214. S. H. Hashemi, P. A. Sandborn, D. Disko, and R. Evans, "The Close Attached Capacitor: A Solution to Switching Noise Problems," *IEEE Transactions on Advanced Packaging*, Vol. 15, No. 6, pp. 1056–1063, December 1992.

215. B. Gieseke et al., "A 600 MHz Superscalar RISC Microprocessor with Out-of-Order Execution," *Proceedings of the IEEE International Solid-State Circuits Conference*, pp. 176–177, February 1997.

216. S. H. Hall, G. W. Hall, and J. A. McCall, *High-Speed Digital System Design: A Handbook of Interconnect Theory and Design Practices*, John Wiley & Sons, Inc., New York, 2000.

217. A. Jain *et al.*, "A 1.2 GHz Alpha Microprocessor with 44.8 GB/s Chip Pin Bandwidth," *Proceedings of the IEEE International Solid-State Circuits Conference*, pp. 240–241, February 2001.

218. R. Heald *et al.*, "Implementation of a 3rd Generation SPARC V9 64b Microprocessor," *Proceedings of the IEEE International Solid-State Circuits Conference*, pp. 412–413, February 2000.

219. J. D. Warnock *et al.*, "The Circuit and Physical Design of the POWER4 Microprocessor," *IBM Journal of Research and Development*, Vol. 46, No. 1, pp. 27–51, January 2002.

220. L. A. Arledge Jr. and W. T. Lynch, "Scaling and Performance Implications for Power Supply and Other Signal Routing Constraints Imposed by I/O Limitations," *Proceedings of the IEEE Symposium on IC/Package Design Integration*, pp. 45–50, February 1998.

221. W. T. Lynch and L. A. Arledge Jr., "Power Supply Distribution and Other Wiring Issues for Deep-Submicron ICs," *Proceedings of the Material Research Society Symposia*, Vol. 514, pp. 11–27, April 1998.

222. L. Zu *et al.*, "Improving Microprocessor Performance with Flip Chip Package Design," *Proceedings of the IEEE Symposium on IC/Package Design Integration*, pp. 82–87, February 1998.

223. S. Lipa, J. T. Schaffer, A. W. Glaser, and P. D. Franzon, "Flip-Chip Power Distribution," *Proceedings of the IEEE Topical Meeting on Electrical Performance of Electronic Packaging*, pp. 39–41, October 1998.

224. P. D. Franzon, J. T. Schaffer, S. Lipa, and A. W. Glaser, "Issues in Chip-Package Codesign with MCM-D/Flip-Chip Technology," *Proceedings of the IEEE Topical Meeting on Electrical Performance of Electronic Packaging*, pp. 88–92, October 1998.

225. A. Sarangi, G. Ji, T. Arabi, and G. F. Taylor, "Design and Performance Evaluation of Pentium III Microprocessor Packaging," *Proceedings of the IEEE Topical Meeting on Electrical Performance of Electronic Packaging*, pp. 291–294, October 2001.

226. R. Mahajan *et al.*, "High Performance Package Design for a 1 GHz Microprocessor," *Intel Technology Journal*, Vol. 6, No. 2, pp. 62–75, May 2002.

227. T. Kawahara, "SuperCSP^{TM}," *IEEE Transactions on Advanced Packaging*, Vol. 23, No. 2, pp. 215–219, May 2000.

228. S. N. Towle *et al.*, "Bumpless Build-Up Layer Packaging," *Proceedings of ASME International Mechanical Engineering Congress and Exposition*, November 2001.

229. A. Hasan *et al.*, "High Performance Package Design for a 1 GHz Microprocessor," *IEEE Transactions on Advanced Packaging*, Vol. 24, No. 4, pp. 470–476, November 2001.

230. H. Xie *et al.*, "Modeling and Measurement of the Alpha 21364 Package," *Proceedings of the IEEE Electronic Components and Technology Conference*, pp. 583–589, June 2002.

231. P. Saxena and S. Gupta, "Shield Count Minimization in Congested Regions," *Proceedings of the ACM International Symposium on Physical Design*, pp. 78–83, April 2002.

232. H. Su, J. Hu, S. S. Sapatnekar, and S. R. Nassif, "Congestion-Driven Codesign of Power and Signal Networks," *Proceedings of the IEEE/ACM Design Automation Conference*, pp. 64–69, June 2002.

233. P. Saxena and S. Gupta, "On Integrating Power and Signal Routing for Shield Count Minimization in Congested Regions," *IEEE Transactions on Computer-Aided Design of Integrated Circuits and Systems*, Vol. 22, No. 4, pp. 437–445, April 2003.

234. P. Restle, A. Ruehli, and S. G. Walker, "Dealing with Inductance in High-Speed Chip Design," *Proceedings of the IEEE/ACM Design Automation Conference*, pp. 904–909, June 1999.

235. P. J. Restle, A. E. Ruehli, and S. G. Walker, "Multi-GHz Interconnect Effects in Microprocessors," *Proceedings of the ACM International Symposium on Physical Design*, pp. 93–97, April 2001.

236. R. R. Troutman, *Latch-Up in CMOS Technology: The Problem and Its Cure*, Kluwer Academic Publishers, Boston, 1986.

237. R. M. Secareanu *et al.*, "Placement of Substrate Contacts to Minimize Substrate Noise in Mixed-Signal Integrated Circuits," *Analog Integrated Circuits and Signal Processing*, Vol. 28, No. 3, pp. 253–264, September 2001.

238. N. H. E. Weste and K. Eshraghian, *Principles of CMOS VLSI Design*, Addison-Wesley Publishing Company, Boston, 1992.

239. H. H. Chen and D. D. Ling, "Power Supply Noise Analysis Methodology for Deep-Submicron VLSI Chip Design," *Proceedings of the IEEE/ACM Design Automation Conference*, pp. 638–643, June 1997.

240. M. Horowitz and R. W. Dutton, "Resistance Extraction from Mask Layout Data," *IEEE Transactions on Computer-Aided Design of Integrated Circuits and Systems*, Vol. 2, No. 3, pp. 145–150, July 1983.

241. L. Ladage and R. Leupers, "Resistance Extraction Using a Routing Algorithm," *Proceedings of the IEEE/ACM Design Automation Conference*, pp. 38–42, June 1993.

242. H. H. Chen and S. E. Schuster, "On-Chip Decoupling Capacitor Optimization for High-Performance VLSI Design," *Proceedings of the IEEE International Symposium on VLSI Technology, Systems, and Applications*, pp. 99–103, May 1995.

243. E. Macii, M. Pedram, and F. Somenzi, "High-Level Power Modeling, Estimation, and Optimization," *IEEE Transactions on Computer-Aided Design of Integrated Circuits and Systems*, Vol. 17, No. 11, pp. 1061–1079, November 1998.

244. D. Brooks, T. Tiwari, and M. Martonosi, "Wattch: A Framework for Architectural-Level Analysis and Optimization," *Proceedings of the ACM International Symposium on Computer Architecture*, pp. 83–94, June 2000.

245. F. N. Najm, "A Survey of Power Estimation Techniques in VLSI Circuits," *IEEE Transactions on Very Large Scale Integration (VLSI) Systems*, Vol. 2, No. 4, pp. 446–455, December 1994.

246. H. Kriplani, F. N. Najm, and I. N. Hajj, "Pattern Independent Maximum Current Estimation in Power and Ground Buses of CMOS VLSI Circuits: Algorithms, Signal Correlations, and Their Resolution," *IEEE Transactions on Computer-Aided Design of Integrated Circuits and Systems*, Vol. 14, No. 8, pp. 998–1012, August 1995.

247. S. Bobba and I. N. Hajj, "Estimation of Maximum Current Envelope for Power Bus Analysis and Design," *Proceedings of the ACM International Symposium on Physical Design*, pp. 141–146, April 1998.

248. S. Bobba and I. N. Hajj, "Maximum Voltage Variation in the Power Distribution Network of VLSI Circuits with *RLC* Models," *Proceedings of the IEEE International Symposium on Low Power Electronics and Design*, pp. 376–381, August 2001.

249. G. Bai and I. N. Hajj, "Simultaneous Switching Noise and Resonance Analysis of On-Chip Power Distribution Network," *Proceedings of the IEEE International Symposium on Quality Electronic Design*, pp. 163–168, March 2002.

250. L. T. Pillage, R. A. Rohrer, and C. Visweswariah, *Electronic Circuit and System Simulation Methods*, McGraw-Hill, Inc., New York, 1994.

251. G. Golub and C. Van Loan, *Matrix Computations*, Johns Hopkins University Press, Baltimore, 1989.

252. L.-R. Zheng and H. Tenhunen, "Design and Analysis of Power Integrity in Deep Submicron System-on-Chip Circuits," *Analog Integrated Circuits and Signal Processing*, Vol. 30, No. 1, pp. 15–29, January 2002.

253. L.-R. Zheng, *Design, Analysis, and Integration of Mixed-Signal Systems for Signal and Power Integrity*, Ph.D. Thesis, Royal Institute of Technology, Stockholm, Sweden, 2001.

254. M. Zhao, R. V. Panda, S. S. Sapatnekar, and D. Blaauw, "Hierarchical Analysis of Power Distribution Networks," *IEEE Transactions on Computer-Aided Design of Integrated Circuits and Systems*, Vol. 21, No. 2, pp. 159–168, February 2002.

255. S. R. Nassif and J. Kozhaya, "Multi-Grid Methods for Power Grid Simulation," *Proceedings of the IEEE International Symposium on Circuits and Systems*, Vol. V, pp. 457–460, May 2000.

256. S. R. Nassif and J. Kozhaya, "Fast Power Grid Simulation," *Proceedings of the IEEE/ACM Design Automation Conference*, pp. 156–161, June 2000.

257. J. N. Kozhaya, S. R. Nassif, and F. N. Najm, "A Multigrid-like Technique for Power Grid Analysis," *IEEE Transactions on Computer-Aided Design of Integrated Circuits and Systems*, Vol. 21, No. 10, pp. 1148–1160, October 2002.

258. H. Su, K. H. Gala, and S. S. Sapatnekar, "Fast Analysis and Optimization of Power/Ground Networks," *Proceedings of the IEEE/ACM International Conference on Computer-Aided Design*, pp. 477–480, November 2000.

259. A. Odabasioglu, M. Celik, and L. T. Pileggi, "PRIMA: Passive Reduced-Order Interconnect Macromodeling Algorithm," *IEEE Transactions on Computer-Aided Design of Integrated Circuits and Systems*, Vol. 17, No. 8, pp. 645–654, August 1998.

260. S. Zhao, K. Roy, and C.-K. Koh, "Estimation of Inductive and Resistive Switching Noise on Power Supply Network in Deep Submicron CMOS Circuits," *Proceedings of the IEEE International Conference on Computer Design*, pp. 65–72, October 2000.

261. S. Zhao, K. Roy, and C.-K. Koh, "Frequency Domain Analysis of Switching Noise on Power Supply Network," *Proceedings of the IEEE/ACM International Conference on Computer-Aided Design*, pp. 487–492, November 2000.

262. L. Smith, "Decoupling Capacitor Calculations for CMOS Circuits," *Proceedings of the IEEE Conference on Electrical Performance of Electronic Packaging*, pp. 101–105, November 1994.

263. M. D. Pant, P. Pant, and D. S. Wills, "On-Chip Decoupling Capacitor Optimization Using Architectural Level Current Signature Prediction," *Proceedings of the IEEE International ASIC/SOC Conference*, pp. 288–292, September 2000.

264. M. D. Pant, P. Pant, and D. S. Wills, "On-Chip Decoupling Capacitor Optimization Using Architectural Level Prediction," *Proceedings of the IEEE Midwest Symposium on Circuit and Systems*, pp. 772–775, August 2000.

265. M. D. Pant, P. Pant, and D. S. Wills, "On-Chip Decoupling Capacitor Optimization Using Architectural Level Prediction," *IEEE Transactions on Very Large Scale Integration (VLSI) Systems*, Vol. 10, No. 3, pp. 319–326, June 2002.

266. S. Zhao, K. Roy, and C.-K. Koh, "Decoupling Capacitance Allocation for Power Supply Noise Suppression," *Proceedings of the ACM International Symposium on Physical Design*, pp. 66–71, April 2001.

267. S. Zhao, K. Roy, and C.-K. Koh, "Power Supply Noise Aware Floorplanning and Decoupling Capacitance Placement," *Proceedings of the IEEE International Conference on VLSI Design*, pp. 489–495, January 2002.

268. S. Zhao, K. Roy, and C.-K. Koh, "Decoupling Capacitance Allocation and Its Application to Power-Supply Noise-Aware Floorplanning," *IEEE Transactions on Computer-Aided Design of Integrated Circuits and Systems*, Vol. 21, No. 1, pp. 81–92, January 2002.

269. A. R. Conn, R. A. Haring, and C. Visweswariah, "Noise Considerations in Circuit Optimization," *Proceedings of the IEEE/ACM International Conference on Computer-Aided Design*, pp. 220–227, November 1998.

270. C. Viswesvariah, R. A. Haring, and A. R. Conn, "Noise Considerations in Circuits Optimization," *IEEE Transactions on Computer-Aided Design of Integrated Circuits and Systems*, Vol. 19, No. 6, pp. 679–690, June 2000.

271. H. Su, S. S. Sapatnekar, and S. R. Nassif, "An Algorithm for Optimal Decoupling Capacitor Sizing and Placement for Standard Cell Layouts," *Proceedings of the ACM International Symposium on Physical Design*, pp. 68–73, April 2002.

272. H. Su, S. S. Sapatnekar, and S. R. Nassif, "Optimal Decoupling Capacitor Sizing and Placement for Standard Cell Layout Designs," *IEEE Transactions on Computer-Aided Design of Integrated Circuits and Systems*, Vol. 22, No. 4, pp. 428–436, April 2003.

273. D. Edelstein *et al.*, "Full Copper Wiring in a Sub-0.25 μm CMOS ULSI Technology," *Proceedings of the IEEE International Electron Device Meeting*, pp. 773–776, December 1997.

274. S. Venkatesan *et al.*, "A High Performance 1.8 V 0.20 μm CMOS Technology with Copper Metalization," *Proceedings of the IEEE International Electron Device Meeting*, pp. 769–772, December 1997.

275. *International Technology Roadmap for Semiconductors, 2000 Update*, Semiconductor Industry Association, http://public.itrs.net, 2000.

276. A. V. Mezhiba and E. G. Friedman, "Inductance/Area/Resistance Tradeoffs in High Performance Power Distribution Grids," *Proceedings of the IEEE International Symposium on Circuits and Systems*, Vol. I, pp. 101–104, May 2002.

277. A. V. Mezhiba and E. G. Friedman, "Variation of Inductance with Frequency in High Performance Power Distribution Grids," *Proceedings of the IEEE International ASIC/SOC Conference*, pp. 421–425, September 2002.

278. A. V. Mezhiba and E. G. Friedman, "Frequency Characteristics of High Speed Power Distribution Grids," *Analog Integrated Circuits and Signal Processing*, Vol. 35, No. 2/3, pp. 207–214, May/June 2003.

279. *International Technology Roadmap for Semiconductors, 2006 Update*, Semiconductor Industry Association, http://public.itrs.net, 2006.

280. R. H. Dennard *et al.*, "Design of Ion-Implanted MOSFET's with Very Small Physical Dimensions," *IEEE Journal of Solid-State Circuits*, Vol. SC–33, No. 5, pp. 256–268, October 1974.

281. K. C. Saraswat and E. Mohammadi, "Effect of Scaling of Interconnections on the Time Delay of VLSI Circuits," *IEEE Transactions on Electron Devices*, Vol. ED–29, No. 4, pp. 645–650, April 1982.

282. W. S. Song and L. A. Glasser, "Power Distribution Techniques for VLSI Circuits," *IEEE Journal of Solid-State Circuits*, Vol. SC–21, No. 1, pp. 150–156, February 1986.

283. P. Larsson, "*di/dt* Noise in CMOS Integrated Circuits," *Analog Integrated Circuits and Signal Processing*, Vol. 14, No. 1/2, pp. 113–129, September 1997.

284. G. A. Katopis, "Delta-I Noise Specification for a High-Performance Computing Machine," *Proceedings of the IEEE*, Vol. 73, No. 9, pp. 1405–1415, September 1985.

285. B. D. McCredie and W. D. Becker, "Modeling, Measurement, and Simulation of Simultaneous Switching Noise," *IEEE Transactions on Components, Packaging, and Manufacturing Technology, Part B: Advanced Packaging*, Vol. 19, No. 3, pp. 461–472, August 1996.

286. S. R. Nassif and O. Fakhouri, "Technology Trends in Power-Grid-Induced Noise," *Proceedings of the ACM International Workshop on System Level Interconnect Prediction*, pp. 55–59, April 2002.

287. *International Technology Roadmap for Semiconductors, 1999 Edition*, Semiconductor Industry Association, http://public.itrs.net, 1999.

288. *International Technology Roadmap for Semiconductors, 1997 Edition*, Semiconductor Industry Association, http://public.itrs.net, 1997.

289. *International Technology Roadmap for Semiconductors, 1998 Update*, Semiconductor Industry Association, http://public.itrs.net, 1998.

290. D. Sylvester and H. Kaul, "Future Performance Challenges in Nanometer Design," *Proceedings of the IEEE/ACM Design Automation Conference*, pp. 3–8, June 2001.

291. A. V. Mezhiba and E. G. Friedman, "Electrical Characteristics of Multi-Layer Power Distribution Grids," *Proceedings of the IEEE International Symposium on Circuits and Systems*, Vol. 5, pp. 473–476, May 2003.

292. A. Deutsch *et al.*, "The Importance of Inductance and Inductive Coupling for On-Chip Wiring," *Proceedings of the IEEE Topical Meeting on Electrical Performance of Electronic Packaging*, pp. 53–56, October 1997.

293. A. Chandrakasan and R. W. Brodersen, *Low-Power CMOS Design*, Wiley-IEEE Press, New York, New York, 1998.

294. C. Piguet, *Low-Power Processors and Systems on Chips*, CRC Press, Boca Raton, Florida, 2005.

295. A. Chandrakasan, M. Potkonjak, J. Rabaey, and R. W. Brodersen, "HYPER-LP: A System for Power Minimization Using Architectural Transformations," *Proceedings of the IEEE/ACM International Conference on Computer-Aided Design*, pp. 300–303, November 1992.

296. A. P. Chandrakasan *et al.*, "Optimizing Power Using Transformations," *IEEE Transactions on Computer-Aided Design of Integrated Circuits and Systems*, Vol. 14, No. 1, pp. 12–31, January 1995.

297. A. P. Chandrakasan, S. Sheng, and R. W. Brodersen, "Low-Power CMOS Digital Design," *IEEE Journal of Solid-State Circuits*, Vol. 27, No. 4, pp. 473–484, April 1992.

298. S. Mutoh *et al.*, "1-V Power Supply High-Speed Digital Circuit Technology with Multithreshold-Voltage CMOS," *IEEE Journal of Solid-State Circuits*, Vol. 30, No. 8, pp. 847–854, August 1995.

299. V. Kursun and E. G. Friedman, *Multi-Voltage CMOS Circuit Design*, John Wiley & Sons, Inc., New York, New York, 2006.

300. T. Kuroda *et al.*, "A High-Speed Low-Power 0.3 μm CMOS Gate Array with Variable Threshold Voltage (VT) Scheme," *Proceedings of the IEEE Custom Integrated Circuits Conference*, pp. 53–56, May 1996.

301. V. Kursun and E. G. Friedman, "Domino Logic with Variable Threshold Keeper," *IEEE Transactions on Very Large Scale Integration (VLSI) Systems*, Vol. 11, No. 6, pp. 1080–1093, December 2003.

302. V. Kursun and E. G. Friedman, "Sleep Switch Dual Threshold Voltage Domino Logic with Reduced Standby Leakage Current," *IEEE Transactions on Very Large Scale Integration (VLSI) Systems*, Vol. 12, No. 5, pp. 485–496, May 2004.

303. K. Usami, T. Ishikawa, M. Kanazawa, and H. Kotani, "Low-Power Design Technique for ASIC's by Partially Reducing Supply Voltage," *Proceedings of the IEEE International ASIC Conference*, pp. 301–304, September 1996.

304. D. Marculescu, "Power Efficient Processors Using Multiple Supply Voltages," *Proceedings of Workshop on Compilers and Operating Systems for Low Power*, October 2000.

305. J.-M. Chang and M. Pedram, "Energy Minimization Using Multiple Supply Voltages," *Proceedings of the IEEE International Symposium on Low Power Electronics and Design*, pp. 157–162, August 1996.

306. J.-M. Chang and M. Pedram, "Energy Minimization Using Multiple Supply Voltages," *IEEE Transactions on Very Large Scale Integration (VLSI) Systems*, Vol. 5, No. 4, pp. 436–443, December 1997.

307. R. I. Bahar *et al.*, "An Application of ADD-Based Timing Analysis to Combinational Low Power Re-synthesis," *Proceedings of the ACM/IEEE International Workshop on Low Power Design*, pp. 39–44, April 1994.

308. V. Kursun, R. M. Secareanu, and E. G. Friedman, "CMOS Voltage Interface Circuit for Low Power Systems," *Proceedings of the IEEE International Symposium on Circuits and Systems*, pp. 3667–3670, May 2002.

309. K. Usami *et al.*, "Automated Low-Power Technique Exploiting Multiple Supply Voltages Applied to a Media Processor," *IEEE Journal of Solid-State Circuits*, Vol. 33, No. 3, pp. 463–472, March 1998.

310. K. Usami and M. Horowitz, "Clustered Voltage Scaling Technique for Low-Power Design," *Proceedings of the IEEE International Symposium on Low Power Electronics and Design*, pp. 3–8, April 1995.

311. K. Usami *et al.*, "Automated Low-Power Technique Exploiting Multiple Supply Voltages Applied to a Media Processor," *Proceedings of the IEEE Custom Integrated Circuit Conference*, pp. 131–134, May 1997.

312. V. Kursun, S. G. Narendra, V. K. De, and E. G. Friedman, "Low-Voltage-Swing Monolithic DC–DC Conversion," *IEEE Transactions on Circuits and Systems II: Express Briefs*, Vol. 51, No. 5, pp. 241–248, May 2004.

313. V. Kursun, V. K. De, E. G. Friedman, and S. G. Narendra, "Monolithic Voltage Conversion in Low-Voltage CMOS Technologies," *Microelectronics Journal*, Vol. 36, No. 9, pp. 863–867, September 2005.

314. R. K. Krishnamurthy, A. Alvandpour, V. De, and S. Borkar, "High-Performance and Low-Power Challenges for Sub-70 nm Microprocessor

Circuits," *Proceedings of the IEEE Custom Integrated Circuit Conference*, pp. 125–128, May 2002.

315. S. H. Kulkarni and D. Sylvester, "High Performance Level Conversion for Dual V_{dd} Design," *IEEE Transactions on Very Large Scale Integration (VLSI) Systems*, Vol. 12, No. 9, pp. 926–936, September 2004.

316. V. Kursun, S. G. Narendra, V. K. De, and E. G. Friedman, "Analysis of Buck Converters for On-Chip Integration with a Dual Supply Voltage Microprocessor," *IEEE Transactions on Very Large Scale Integration (VLSI) Systems*, Vol. 11, No. 3, pp. 514–522, June 2003.

317. P. Hazucha *et al.*, "A 233-MHz 80%–87% Efficient Four-Phase DC–DC Converter Utilizing Air-Core Inductors on Package," *IEEE Journal of Solid-State Circuits*, Vol. 40, No. 4, pp. 838–845, April 2005.

318. M. Igarashi *et al.*, "A Low-Power Design Method Using Multiple Supply Voltages," *Proceedings of the IEEE International Symposium on Low Power Electronics and Design*, pp. 36–41, August 1997.

319. J.-S. Wang, S.-J. Shieh, J.-C. Wang, and C.-W. Yeh, "Design of Standard Cells Used in Low-Power ASIC's Exploiting the Multiple-Supply-Voltage Scheme," *Proceedings of the IEEE International ASIC Conference*, pp. 119–123, September 1998.

320. M. Hamada, Y. Ootaguro, and T. Kuroda, "Utilizing Surplus Timing for Power Reduction," *Proceedings of the IEEE Conference on Custom Integrated Circuits*, pp. 89–92, May 2001.

321. T. Sakurai and A. R. Newton, "Alpha-Power Law MOSFET Model and Its Application to CMOS Inverter Delay and Other Formulas," *IEEE Journal of Solid-State Circuits*, Vol. 25, No. 2, pp. 584–594, April 1990.

322. W. Hung *et al.*, "Total Power Optimization through Simultaneously Multiple-V_{dd} Multiple-V_{TH} Assignment and Device Sizing with Stack Forcing," *Proceedings of the IEEE International Symposium on Low Power Electronics and Design*, pp. 144–149, August 2004.

323. S. K. Mathew *et al.*, "A 4-GHz 300-mW 64-bit Integer Execution ALU with Dual Supply Voltages in 90-nm CMOS," *IEEE Journal of Solid-State Circuits*, Vol. 40, No. 1, pp. 44–51, January 2005.

324. D. Nguyen *et al.*, "Minimization of Dynamic and Static Power Through Joint Assignment of Threshold Voltages and Sizing Optimization," *Proceedings of the IEEE International Symposium on Low Power Electronics and Design*, pp. 158–163, August 2003.

325. M. Takahashi *et al.*, "A 60-mW MPEG4 Video Codec Using Clustered Voltage Scaling with Variable Supply-Voltage Scheme," *IEEE Journal of Solid-State Circuits*, Vol. 33, No. 11, pp. 1772–1780, November 1998.

326. K. Zhang *et al.*, "A 3-GHz 70-Mb SRAM in 65-nm CMOS Technology With Integrated Column-Based Dynamic Power Supply," *IEEE Journal of Solid-State Circuits*, Vol. 41, No. 1, pp. 146–151, January 2006.

327. S. Raje and M. Sarrafzadeh, "Variable Voltage Scheduling," *Proceedings of the ACM International Symposium on Low Power Design*, pp. 9–14, April 1995.

328. A. V. Mezhiba and E. G. Friedman, "Impedance Characteristics of Power Distribution Grids in Nanoscale Integrated Circuits," *IEEE Transactions on Very Large Scale Integration (VLSI) Systems*, Vol. 12, No. 11, pp. 1148–1155, November 2004.

329. M. Popovich, E. G. Friedman, M. Sotman, and A. Kolodny, "On-Chip Power Distribution Grids with Multiple Supply Voltages for High Performance Integrated Circuits," *Proceedings of the ACM/IEEE Great Lakes Symposium on VLSI*, pp. 2–7, April 2005.

330. J. Mermet and W. Nebel, *Low Power Design in Deep Submicron Electronics*, Kluwer Academic Publishers, Norwell, Massachusetts, 1997.

331. A. Chandrakasan, W. J. Bowhill, and F. Fox, *Design of High-Performance Microprocessor Circuits*, Wiley-IEEE Press, New York, New York, 2000.

332. L. D. Smith, "Packaging and Power Distribution Design Considerations for a Sun Microsystems Desktop Workstation," *Proceedings of the IEEE Conference on Electrical Performance of Electronic Packaging*, pp. 19–22, October 1997.

333. M. Popovich and E. G. Friedman, "Noise Coupling in Multi-Voltage Power Distribution Systems with Decoupling Capacitors," *Proceedings of the IEEE International Symposium on Circuits and Systems*, pp. 620–623, May 2005.

334. M. Popovich and E. G. Friedman, "Noise Aware Decoupling Capacitors for Multi-Voltage Power Distribution Systems," *Proceedings of the ACM/IEEE International Symposium on Quality Electronic Design*, pp. 334–339, March 2005.

335. C. R. Paul, *Analysis of Linear Circuits*, McGraw-Hill, New York, New York, 1989.

336. R. Senthinathan and J. L. Prince, "Simultaneous Switching Ground Noise Calculation for Packaged CMOS Devices," *IEEE Journal of Solid-State Circuits*, Vol. 26, No. 11, pp. 1724–1728, November 1991.

337. A. Vaidyanath, B. Thoroddsen, and J. L. Prince, "Effect of CMOS Driver Loading Conditions on Simultaneous Switching Noise," *IEEE Transactions on Components, Packaging, and Manufacturing Technology, Part B: Advanced Packaging*, Vol. 17, No. 4, pp. 480–485, November 1994.

338. S. R. Vemuru, "Accurate Simultaneous Switching Noise Estimation Including Velocity-Saturation Effects," *IEEE Transactions on Components, Packaging, and Manufacturing Technology, Part B: Advanced Packaging*, Vol. 19, No. 2, pp. 344–349, May 1996.

339. S.-J. Jou, W.-C. Cheng, and Y.-T. Lin, "Simultaneous Switching Noise Analysis and Low-Bounce Buffer Design," *IEE Proceedings on Circuits, Devices, and Systems*, Vol. 148, No. 6, pp. 303–311, December 2001.

340. H.-R. Cha and O.-K. Kwon, "A New Analytic Model of Simultaneous Switching Noise in CMOS Systems," *Proceedings of the IEEE Electronic Components and Technology Conference*, pp. 615–621, May 1998.

341. S. R. Vemuru, "Effects of Simultaneous Switching Noise on the Tapered Buffer Design," *IEEE Transactions on Very Large Scale Integration (VLSI) Systems*, Vol. 5, No. 3, pp. 290–300, September 1997.

342. P. Heydari and M. Pedram, "Ground Bounce in Digital VLSI Circuits," *IEEE Transactions on Very Large Scale Integration (VLSI) Systems*, Vol. 11, No. 2, pp. 180–193, April 2003.

343. T. J. Gabara, "Ground Bounce and Reduction Techniques," *Proceedings of the IEEE ASIC Conference*, pp. 13.–2.1–13.–2.2, September 1991.

344. A. Vittal, H. Ha, F. Brewer, and M. Marek-Sadowska, "Clock Skew Optimization for Ground Bounce Control," *Proceedings of the IEEE/ACM International Conference on Computer-Aided Design*, pp. 395–399, November 1996.

345. M. D. Pant, P. Pant, D. S. Wills, and V. Tiwari, "An Architectural Solution for the Inductive Noise Problem due to Clock-Gating," *Proceedings of the IEEE International Symposium on Low Power Electronics and Design*, pp. 255–257, August 1999.

346. J. Oh and M. Pedram, "Multi-Pad Power/Ground Network Design for Uniform Distribution of Ground Bounce," *Proceedings of the ACM/IEEE Design Automation Conference*, pp. 287–290, June 1998.

347. H. Chen, "Minimizing Chip-Level Simultaneous Switching Noise for High-Performance Microprocessor Design," *Proceedings of the IEEE International Symposium on Circuits and Systems*, pp. 544–547, May 1996.

348. A. Zenteno, V. H. Champac, M. Renovell, and F. Azais, "Analysis and Attenuation Proposal in Ground Bounce," *Proceedings of the IEEE Asian Test Symposium*, pp. 460–463, November 2004.

349. M. Badaroglu *et al.*, "Digital Ground Bounce Reduction by Supply Current Shaping and Clock Frequency Modulation," *IEEE Transactions on Computer-Aided Design of Integrated Circuits and Systems*, Vol. 24, No. 1, pp. 65–76, January 2005.

350. F. Moll and M. Roca, *Interconnect Noise in VLSI Circuits*, Kluwer Academic Publishers, Norwell, Massachusetts, 2003.

351. M. E. Van Valkenburg, *Network Analysis*, Prentice Hall, Upper Saddle River, New Jersey, 1974.

352. M. Takamiya and M. Mizuno, "A 6.7 fF/μm^2 Bias-Independent Gate Capacitor (BIGCAP) with Digital CMOS Process and Its Application to the Loop Filter of a Differential PLL," *IEEE Journal of Solid-State Circuits*, Vol. 40, No. 3, pp. 719–725, March 2005.

353. D. Lee, D. Blaauw, and D. Sylvester, "Gate Oxide Leakage Current Analysis and Reduction for VLSI Circuits," *IEEE Transactions on Very Large Scale Integration (VLSI) Circuits*, Vol. 12, No. 2, pp. 155–166, February 2004.

354. M. Anis and Y. Massoud, "Power Design Challenges in Deep-Submicron Technology," *Proceedings of the IEEE International Midwest Symposium on Circuits and Systems*, pp. 1510–1513, December 2003.

355. D. Deleganes, J. Douglas, B. Kommandur, and M. Patyra, "Designing a 3 GHz, 130 nm, Intel Pentium 4 Processor," *Proceedings of the IEEE Symposium on VLSI Circuits*, pp. 130–133, June 2002.

356. R. McGowen *et al.*, "Power and Temperature Control on a 90-nm Itanium Family Processor," *IEEE Journal of Solid-State Circuits*, Vol. 41, No. 1, pp. 229–237, January 2006.

357. S. Naffziger *et al.*, "The Implementation of a 2-Core, Multi-Threaded Itanium Family Processor," *IEEE Journal of Solid-State Circuits*, Vol. 41, No. 1, pp. 197–209, January 2006.

358. T. Hubing, "Effective Strategies for Choosing and Locating Printed Circuit Board Decoupling Capacitors," *Proceedings of the IEEE International Symposium on Electromagnetic Compatibility*, pp. 632–637, August 2005.

359. Mathematica 5.2, Wolfram Research, Inc.

360. M. P. Goetz, "Time and Frequency Domain Analysis of Integral Decoupling Capacitors," *IEEE Transactions on Components, Packaging, and Manufacturing Technology, Part B: Advanced Packaging*, Vol. 19, No. 3, pp. 518–522, August 1996.

361. T. Murayama, K. Ogawa, and H. Yamaguchi, "Estimation of Peak Current Trough CMOS VLSI Circuit Supply Lines," *Proceedings of the ACM Asia and South Pacific Design Automation Conference*, pp. 295–298, January 1999.

362. W. K. Henson *et al.*, "Analysis of Leakage Currents and Impact on Off-State Power Consumption for CMOS Technology in the 100-nm Regime," *IEEE Transactions on Electron Devices*, Vol. 47, No. 7, pp. 1393–1400, July 2000.

363. Y. Taur, "CMOS Design Near the Limit of Scaling," *IBM Journal of Research and Development*, Vol. 46, No. 2/3, pp. 213–221, March/May 2002.

Index

About the Authors

Mikhail Popovich was born in Izhevsk, Russia in 1975. He received the B.S. degree in electrical engineering from Izhevsk State Technical University, Izhevsk, Russia in 1998, and the M.S. degree in electrical and computer engineering from the University of Rochester, Rochester, NY in 2002, where he is completing the Ph.D. degree in electrical engineering.

He was an intern at Freescale Semiconductor Corporation, Tempe, AZ, in the summer 2005, where he worked on signal integrity in RF and mixed-signal ICs and developed design techniques and methodologies for placing distributed on-chip decoupling capacitors. His professional experience also includes characterization of substrate and interconnect crosstalk noise in CMOS imaging circuits for Eastman Kodak Company, Rochester, NY. He has authored several conference and journal papers in the areas of power distribution networks in CMOS VLSI circuits, placement of on-chip decoupling capacitors, and the inductive properties of on-chip interconnect. His research interests are in the areas of on-chip noise, signal integrity, and interconnect design including on-chip inductive effects, optimization of power distribution networks, and the design of on-chip decoupling capacitors.

Mr. Popovich received the Best Student Paper Award at the ACM Great Lakes Symposium on VLSI in 2005 and the GRC Inventor Recognition Award from the Semiconductor Research Corporation in 2007.

Andrey V. Mezhiba has graduated from the Moscow Institute of Physics and Technology in 1996 with Diploma in Physics. He continued his studies at the University of Rochester where he received the Ph.D. degree in electrical and computer engineering in 2004. Andrey's doctorate research interests were in the area of power distribution networks, on-chip inductance, circuit coupling, and signal integrity. Andrey is now with Intel corporation working on low-voltage analog and mixed-signal design in nanometer-scale CMOS technologies.

Eby G. Friedman received the B.S. degree from Lafayette College in 1979, and the M.S. and Ph.D. degrees from the University of California, Irvine, in 1981 and 1989, respectively, all in electrical engineering.

From 1979 to 1991, he was with Hughes Aircraft Company, rising to the position of manager of the Signal Processing Design and Test Department, responsible for the design and test of high performance digital and analog ICs. He has been with the Department of Electrical and Computer Engineering at the University of Rochester since 1991, where he is a Distinguished Professor, the Director of the High Performance VLSI/IC Design and Analysis Laboratory, and the Director of the Center for Electronic Imaging Systems. He is also a Visiting Professor at the Technion — Israel Institute of Technology. His current research and teaching interests are in high performance synchronous digital and mixed-signal microelectronic design and analysis with application to high speed portable processors and low power wireless communications.

He is the author of more than 300 papers and book chapters, numerous patents, and the author or editor of nine books in the fields of high speed and low power CMOS design techniques, high speed interconnect, and the theory and application of synchronous clock and power distribution networks. Dr. Friedman is the Regional Editor of the *Journal of Circuits, Systems and Computers*, a Member of the editorial boards of the *Analog Integrated Circuits and Signal Processing*, *Microelectronics Journal*, *Journal of Low Power Electronics*, and *Journal of VLSI Signal Processing*, Chair of the *IEEE Transactions on Very Large Scale Integration (VLSI) Systems* steering committee, and a Member of the technical program committee of a number of conferences. He previously was the Editor-in-Chief of the *IEEE Transactions on Very Large Scale Integration (VLSI) Systems*, a Member of the editorial board of the *Proceedings of the IEEE* and *IEEE Transactions on Circuits and Systems II: Analog and Digital Signal Processing*, a Member of the Circuits and Systems (CAS) Society Board of Governors, Program and Technical chair of several IEEE conferences, and a recipient of the University of Rochester Graduate Teaching Award and College of Engineering Teaching Excellence Award. Dr. Friedman is a Senior Fulbright Fellow and an IEEE Fellow.